Modelling and Simulation in Plasma Physics for Physicists and Mathematicians

Modelling and Simulation in Plasma Physics for Physicists and Mathematicians

Geoffrey J. Pert
University of York, Heslington, York
United Kingdom

WILEY

This edition first published 2024

Registered Offices
John Wiley & Sons, Inc., 111 River Street, Hoboken, NJ 07030, USA
John Wiley & Sons Ltd, The Atrium, Southern Gate, Chichester, West Sussex, PO19 8SQ, UK

For details of our global editorial offices, customer services, and more information about Wiley products visit us at www.wiley.com.

Wiley also publishes its books in a variety of electronic formats and by print-on-demand. Some content that appears in standard print versions of this book may not be available in other formats.

Library of Congress Cataloging-in-Publication Data

Names: Pert, G. J. (Geoffrey J.), author.
Title: Modelling and simulation in plasma physics for physicists and mathematicians / Geoffrey J. Pert, University of York, Heslington, York, United Kingdom.
Description: Hoboken, NJ : Wiley, 2024. | Includes bibliographical references and index.
Identifiers: LCCN 2024008761 (print) | LCCN 2024008762 (ebook) | ISBN 9781394239207 (hardback) | ISBN 9781394239221 (adobe pdf) | ISBN 9781394239214 (epub)
Subjects: LCSH: Plasma (Ionized gases) | Plasma (Ionized gases)–Mathematical models. | Plasma (Ionized gases)–Computer simulation.
Classification: LCC QC718 .P3755 2021 (print) | LCC QC718 (ebook) | DDC 530.4/4011–dc23/eng/20240324
LC record available at https://lccn.loc.gov/2024008761
LC ebook record available at https://lccn.loc.gov/2024008762

Cover Design: Wiley
Cover Image: © BokehStore/Shutterstock

Set in 9.5/12.5pt STIXTwoText by Straive, Chennai, India

Contents

Preface

This volume is the fruit of a career embracing the development of computational modelling from the 1960s to the present, mainly involving plasma physics. The author has been active as both user and developer of sophisticated computer models for plasma over the past 50 years, starting in the late 1960s. This book is intended as a continuation of our earlier study of the fundamental principles underlying the present understanding of the physics of plasmas and ionised gases in laboratory environments. The author started to study plasma using computational models over 50 years ago and has maintained an activity at the forefront of the field.

The theoretical development of the behaviour of such systems involves complex sets of equations whose calculation is beyond current analytic methods. As a consequence, the theory has increasingly relied on computational modelling methods. These generally involve complex programmes with many interacting elements. The development of these codes has relied heavily on some standard procedures, whose foundations were developed in the early days of computational techniques by pioneers such as von Neumann, Richtmyer, Metropolis and Harlow principally at Los Alamos. Building on these methods has resulted in effective methods for solving the partial differential equations involved in studying the dynamics of plasmas, such as computational fluid dynamics, Monte-Carlo, finite-difference methods. In this text, we review these methods applied to the multiple problems involved in solving the behaviour of plasma systems.

At low densities, typical of most terrestrial plasma, the space charge field between the positively charged ions and negatively charged electrons is sufficiently strong at quite small separations that the two ionic species are 'tied' together, and the plasma behaves as a single fluid. In this case, the fluid has the particularly simple equation of state of a monatomic polytropic gas, with polytropic index $\gamma = 5/3$. In the absence of external magnetic fields, and when the ion and electron densities and temperatures are equal, the plasma dynamics may be treated as a standard gas by the well-established routines of compressible gas dynamics. The early sections see Chapter 1 of this book are therefore devoted to a review of the development of computational gas dynamics. The distinctive application to magneto-hydrodynamics is considered later. We note however that energy transfer and consequent equilibration is much slower between electrons and heavy particles and that consequently the energy balance of each particle must be treated separately.

The modelling described above is known as the *continuum approximation*, which yields a method of sampling the pseudo-distribution function of the computed plasma. The output is the local thermodynamic state and fluid mean velocity.

Alternatively, we may sample the spatial and temporal velocity of a representative group of particles by calculating the dynamics of a group and the electromagnetic fields associated with them. Such an approach is known as a *particle method*, of which particle-in-cell is the most important.

The development of small, inexpensive yet powerful personal computers and laptops has allowed modelling simulations of complex interactive systems at a scale that a decade ago required the use of large multi-processor vector and parallel processing computer machines. Correspondingly, the expected accuracy, number and complexity of different elements treated in a single computer run have greatly increased. The laboratory facilities and manpower required to run such codes have increased and the costs correspondingly have grown markedly. Thus, we have the situation where we can perform the most accurate calculations which can be compared directly with experiment but at a high price and as a result, only at a limited number of facilities. Alternatively, we can relax the accuracy and perform representative calculations using desktop computers. The enormous increase in CPU power and memory has allowed the most complex calculations that were possible in the 1960s and 1970s to be performed on top-of-the-range desktop and laptop machines. As a consequence, many of the routines developed in the earliest days of computational physics can now be used as the foundation for practical desktop calculations. A very useful reference for these programmes is the set of annual texts 'Methods in Computational Physics'.[1] It is this latter requirement that this book will address. We will not consider the most sophisticated models or the construction of an accurate picture, but only the elements which tied together enable us to easily run and manipulate programmes, thereby allowing us to quickly identify the basic scaling of potential plasma configurations, i.e. as a design tool.

Like many aspects of kinetic physics, there are simplifications that make theoretical studies tractable, for example the markedly different time scales in the Bogobuliev treatment of gas kinetic theory and the Chapman–Enskog approximation in gas transport theory (§4.5 4.11). In this case, it is the different scales relative to the Debye length, where scale lengths are short compared to the Debye length or plasma frequency.

It is a requirement for the existence of the electric field cancellation associated with the Debye length that the plasma be dilute, i.e. the number of electrons contained within the Debye sphere is large. This requirement ensures that the electron distribution in the neighbourhood of any charge imbalance is Maxwellian. The Debye length is then the distance over which the field exists given by

$$\lambda_D = \sqrt{\frac{\epsilon_0 \kappa T}{n_e e^2}} \tag{1}$$

We define a number of parameters which characterise the condition of the plasma from the Debye length.

- The Debye parameter is defined by

$$N_d = \frac{4\pi}{3} n_e \lambda_D^3$$

---

1 Methods in Computational Physics, edited by B. Alder, S. Fernbach and M. Rotenberg: published by Academic Press, New York, 1963–1977.

- The plasma or Spitzer parameter by

$$\Lambda = 4\pi n_e \lambda_D^3 = 3N_D$$

The Spitzer parameter plays an important role as the cut-off for long-range weak collisions. It differentiates between the effects of single strong collisions and weak multiple ones – weak multiple collisions being $\ln \Lambda$ times more effective than strong.

- Characteristic times are defined by the plasma frequency

$$\omega_p = \sqrt{\frac{n_e e^2}{\epsilon_0 m_e}}$$

The plasma frequency and the Debye length are related by $\Lambda_D\, \omega_p = \overline{v}_e$ an average electron velocity.

Laboratory plasma physics in almost all experimental configurations involves dilute plasma with a relatively large plasma parameter ($\ln \Lambda \gtrapprox 10$) whose values greatly limit the range of practical computational complexity. Values outside this range tend to be limited to astronomical plasma.

The short Debye length in experimental situations has another important effect, namely since electric fields are screened out by the electron mobility it follows from Poisson's equation that the space charge $\sigma \approx 0$, i.e. that the plasma is quasi-neutral. This has a very important consequence on the plasma dynamics, namely the plasma behaves as a single fluid, with the electrons and ion tied together by their mutual electric field. We may therefore treat the simulation of the plasma using a single fluid model with two separate energy relations if the electron and ion temperatures are unequal. We will therefore examine in detail the computer models for fluid simulation.

The quasi-neutral behaviour of the plasma results in the density of ions and electrons being equal. However, it does not prevent the mobile electrons from carrying current and generating magnetic fields, behaviours taken into account in the magneto-hydrodynamics relationships.

It follows from the small value of the Debye length that the plasma (or Spitzer) parameter is small. Consequently, the fraction of weak multi-particle collisions compared to strong single interactions is small. This result ensures the effectiveness of multiple weak particle interactions in cell models, where the interaction occurs via the self-generated fields rather than direct collisions.

Acknowledgements

I am grateful to several people who have contributed to the work reported in this book:

Firstly, my many graduate students and research fellows who have acted as 'guinea pigs' for many of the methods and routines described;
Secondly, my editorial team at Wiley who saved the day when my health failed;
Finally, my colleague and friend Prof G. J. Tallents for enabling me to maintain a continued interaction with the real world of experimental physics.

Preamble

Plasma is defined as a conducting fluid. As such, it is an extensive state of matter within the universe, generally consisting of a fully ionised gas. It is found in many situations both cosmically and terrestrially. In this work, we will be principally concerned with only terrestrial applications, related to controlled fusion research. This encompasses magnetically and inertially confined plasma, together with magnetospheric and applied (engineering) applications.

Experimental studies of plasma in these regimes involves large and complex apparatus, such as tokamaks and high-power lasers. As a result, theoretical studies play an important role. Again, since the plasma is a material state with a large number of degrees of freedom, theoretical study generally falls into one of two methodologies:

1) **Analytic studies**: These are relatively simple models which can be described by a limited number of parameters and involve a restricted number of coupled equations. They generally describe fundamental activity within the plasma, for example Landau damping.
2) **Computer simulation**: These are more complex models attempting to give an accurate picture of the behaviour of plasma in an experimental environment. The large number of coupled equations demands the application of substantial computer power and storage. Originally this required top-of-the-range machines, but increasingly simple desktop personal computers have had sufficient power to perform substantial realistic simulations. A major advance has been the development of parallel processing systems, essentially a series of coupled PCs operating simultaneously in parallel.

The nature of the theoretical model is determined by the environment in which it is applied. This is characterised by the scale of the system ℓ with respect to the Debye length λ_D. Again the models fall into one of two classes:

1) Debye length small ($\ell \gg \lambda_D$). In this case, the inter-particle electrostatic forces dominate the motion of the charged particles, whose overall motion is, in turn, dominated by their mutual coupling. This is the region of collective motions characterised by plasma waves. In the context of plasma theory, it is a regime of individual single-particle motion generally known as *particle-in-cell*. At this level of approximation, the self-consistent inter-particle fields are treated directly through Maxwell's equations.

   This level of approximation corresponds to dilute plasma where the plasma parameter (namely the ratio of the Debye length to electron mean separation) and is the normal plasma state [see (Pert 2021; §1.3.2)].

Modelling and Simulation in Plasma Physics for Physicists and Mathematicians, First Edition. Geoffrey J. Pert.

2) Debye length large ($\ell \ll \lambda_D$). The space charge fields established by even a relatively small, spatial charge separation are so large that no space charge density can exist ($\sigma \approx 0$). The negative electrons and positive ions therefore tend to move as a single unit with mass approximately equal to the ion mass ($m \sim m_i$), i.e. the plasma dynamics are that of a single fluid. In this regime, the plasma is described by standard fluid mechanics, and the simulation follows the methods of computational physics fluid dynamics (CFD).

   Under this level of approximation, the basic equations are supplemented by two further conditions.

   a) If the characteristic time scale of the plasma development is long compared to the mean free time (ion/electron collision time), the heavy particles (ions) and electrons are in thermal equilibrium. The particle energy is equilibrated and the energy is calculated by a single first law of thermodynamics relation – *single fluid model.*
   b) On the other hand, if equilibration is slow, the energy relation for the heavy particles and ions must be calculated separately with an equalisation term between the two – *two fluid model.*

      Finally, if there are magnetic fields produced by plasma currents or by external fields or pressure gradients, transport coefficients and energy relations are modified. The magnetic fields can be calculated from the Faraday and Biot–Savart laws. When magnetic fields are important, the *magneto-hydrodynamic approximation* (MHD) may be valid.

By the early 1960s, the basic equations describing the motion of plasma had been well established and it was clear that in the fluid regime, plasma acted as a continuum and could be described by modified fluid dynamics equations. Under these conditions, computational methods based on CFD were appropriate. By the end of the 1960s to the early 1970s these were well established.

On the other hand, in the particle regime, the dynamic equations of motion are simply established based on Newton's laws of motion and Maxwell's laws of electromagnetism. The scale of modelling was however limited by the speed and storage of the basic computer.

The development of small, inexpensive yet powerful personal computers and laptops has allowed modelling simulations of complex interactive systems at a scale that a decade ago required the use of large multi-processor vector and parallel processing computer machines. Correspondingly, the accuracy, number and complexity of different elements treated in a single computer run have greatly increased. The laboratory facilities and manpower required to run such machines have increased and the costs correspondingly have grown markedly. Thus, we have the situation where we can perform the most accurate calculations which can be compared directly with experiment but at a high price and as a result only at a limited number of facilities. Alternatively, we can relax the accuracy and perform representative calculations using desktop computers. The enormous increase in CPU power and memory has allowed the most complex calculations that were possible in the 1960s and 1970s to be performed on top-of-the-range desktop and laptop machines. As a consequence, many of the routines developed in the earliest days of computational physics can now be used as the foundation for practical desktop calculations. A very useful reference for these programmes is the set of annual texts 'Methods in Computational Physics'.[1] It is this latter requirement that this book will address. We will not consider the

1 Alder, Fernbach and Rotenberg (1963–1977).

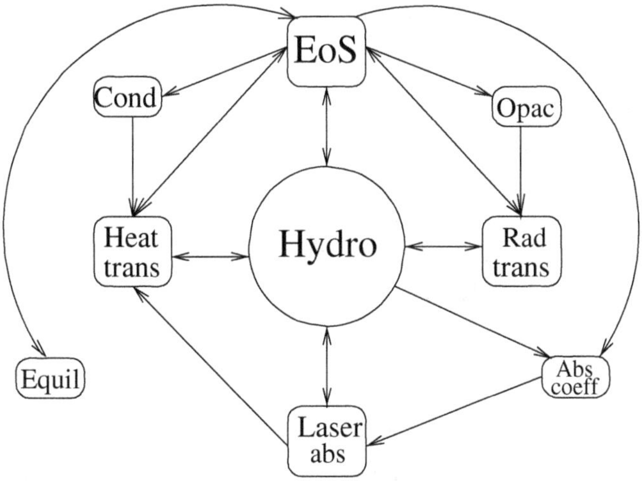

Figure I Sketch to illustrate the key interactions in the simulation of a laser–plasma interaction indicating how different physical processes interact.

most sophisticated models, which generally require a high-end parallel machine for the construction of an accurate picture, but only the elements which tied together enable us to easily run and manipulate programmes, thereby allowing us to quickly identify the basic scaling of potential plasma configurations i.e. as a design tool. As a particular case, it is convenient to imagine the plasma formed by the interaction of a powerful laser beam with a solid target (see Pert, 2013). Many of the elements and the interactions involved in such a model are identified schematically in Figure I.

References

Alder, B, S Fernbach and M Rotenberg. 1963–1977. *Methods in Computational Physics.* New York: Academic Press.

Courant, R, K Friedrichs and H Lewy. 1967. "On the partial difference equations of mathematical physics." *IBM J Res Dev* 11(2):215–234.

Pert, G J. 2013. *Introduction to Fluid Mechanics.* London: Wiley.

Pert, G J. 2021. *Introduction to Plasma Physics.* London: Wiley.

Richtmyer, R D and K W Morton. 1967. *Difference Methods for Initial Value Problems.* New York: Interscience.

Part I

Continuum Methods

1

Foundations of Computational Fluid Mechanics

1.1 Basic Concepts of Fluid Mechanics

Computational fluid dynamics attempts to solve the dynamical structures of systems within the fluid approximation, namely when the behaviour of the fluid is dominated by collisions between the constituent molecules. This implies that the mean free path is everywhere smaller than the characteristic dimensions of the problem, and that the mean collision interval short compared to the characteristic time so that the system may be treated as a continuum and individual molecular motions are treated by averaging. These averaging techniques may be applied to plasma when the Debye length is small and the averaging leads to the magneto-hydrodynamic (MHD) approximation (Pert, 2021). The derivation and validity of the dynamical MHD equations are developed in Pert (2021). Only plasma within the MHD approximation will be treated here. The collision-free model of plasma are fully treated in Section 8.1 and in more detail by Birdsall and Langdon (1991) and Hockney and Eastwood (1988).

The key underlying equation of motion in the MHD equations are those of compressible fluid mechanics: MHD requiring the introduction of additional effects. Making use of the concept of time splitting discussed earlier these additional behaviours may be treated by separate additional routines. The key fluid routines thus form a basic core to which additional physics is tied. It is therefore appropriate to examine with some care the treatment of the compressible fluid initially and subsequently the additional terms.

Compressible fluid is described by its thermodynamic state and by the velocity of flow. These values must be specified at each point within the flow field. The thermodynamic state is specified by two state variables (e.g. pressure p and density ρ), the remainder being then known from the equation of state. The velocity is defined by the three-vector components v. There are thus five independent quantities, which vary in space $\mathbf{r}$ and time t, and must be determined subject to the constraints on the system. The equations of fluid mechanics contain five equations, which are simply the basic conservation laws of physics applied to moving fluid, whose derivation is well known (Pert, 2013).

Typical problems investigated using these basic fluid techniques include aerodynamics (e.g. aeronautical design or car design), meteorology, oceanography and plasma dynamics.

Let us now examine the characteristic structure of compressible fluid mechanics which, as we shall see, are hyperbolic in form, which leads to problems in formulating numerical approximations.

1.2 The Basic Equations of Fluid Mechanics

1.2.1 Eulerian Frame

The simplest co-ordinate system, and in many cases the simplest, in which to define the flow of fluid is the stationary one which is relative to the laboratory, known as the Eulerian frame.

In the laboratory (or Eulerian) frame, the general equations of fluid mechanics can be written in several forms, of which the general conservation form is most appropriate for a general discussion (Pert, 2013). In Cartesian tensor form, these are:

$$\frac{\partial}{\partial t}[\rho] + \frac{\partial}{\partial x_j}(\rho v_j) = 0 \tag{1.1a}$$

$$\frac{\partial}{\partial t}\left[\rho v_i\right] - \frac{\partial}{\partial x_j}\pi_{ij} = 0 \tag{1.1b}$$

$$\frac{\partial}{\partial t}\left[\rho\left(\epsilon + \frac{1}{2}v^2\right)\right] + \frac{\partial}{\partial x_j}\left[\rho v_j\left(h + \frac{1}{2}v^2\right)\right] + \frac{\partial}{\partial x_j}q_j = 0 \tag{1.1c}$$

where ϵ is the specific internal energy of the fluid, $h = \epsilon + p/\rho$ the specific enthalpy and $\mathbf{q}$ the heat flux vector. $\boldsymbol{\pi}$ is the momentum flux tensor

$$\pi_{i,j} = \rho\, \mathrm{v}_i\, \mathrm{v}_j - \tau_{i,j} \tag{1.2}$$

$\tau_{i,j}$ is the total stress tensor

$$\tau_{i,j} = \sigma_{i,j} - p\,\delta_{i,j} \tag{1.3}$$

where σ_{ij} is the viscous stress tensor and $\delta_{i,j}$ the Kronecker delta (1 if $i = j$ and 0 otherwise).

The heat flow vector may be written in terms of the temperature T gradient and the thermal conductivity (κ)

$$\mathbf{q} = \kappa \nabla T \tag{1.4}$$

The temperature is determined from the equation of state in the form $T(\rho, \epsilon)$.

In the laboratory (or Eulerian) frame, the general equations of fluid mechanics can be written in several forms, of which the general conservation form is most appropriate for a general discussion (Pert, 2013).

$$\begin{aligned}
&\frac{\partial \rho}{\partial t} + \nabla \cdot (\rho \mathbf{v}) && = 0 \\
&\frac{\partial}{\partial t}(\rho \mathbf{v}) + \nabla \cdot \underline{\boldsymbol{\pi}} && = 0 \\
&\frac{\partial}{\partial t}\left[\rho\left(\epsilon + \frac{1}{2}\mathrm{v}^2\right)\right] + \nabla \cdot \left[\rho \mathbf{v}\left(h + \frac{1}{2}\mathrm{v}^2\right)\right] + \nabla \cdot \mathbf{q} && = 0
\end{aligned} \tag{1.5}$$

where ϵ is the specific internal energy of the fluid, $h = \epsilon + p/\rho$ the specific enthalpy, and q_i the heat flux vector. π_{ij} is the momentum flux tensor

$$\pi_{i,j} = \rho\, \mathrm{v}_i\, \mathrm{v}_j - \tau_{i,j} \tag{1.6}$$

$\tau_{i,j}$ is the total stress tensor

$$\tau_{i,j} = \sigma_{i,j} - p\,\delta_{i,j} \tag{1.7}$$

where $\sigma_{i,j}$ is the viscous stress tensor and $\delta_{i,j}$ the Kronecker delta

$$\delta_{ij} = \begin{Bmatrix} 1 & \text{if } i = j \\ 0 & \text{otherwise} \end{Bmatrix}.$$

These five equations contain six unknown quantities, namely $\rho, \mathbf{v}, \epsilon, p$, to which we must include an additional term, e.g. entropy to obtain closure. This is normally the equation of state $p \equiv p(\rho, \epsilon)$: for example for a polytropic gas $p = \rho\epsilon/(\gamma - 1)$, where γ is the adiabatic index of the gas. The equation of state normally includes coefficients like viscosity and thermal conductivity, which allow the viscous stress and thermal conduction flux to be calculated knowing the spatial gradients of velocity and internal energy.

Neglecting the viscous stress σ, the momentum balance equation may be cast in a more familiar form, Euler's equation, by subtracting the first equation from the second

$$\frac{\partial \underline{v}}{\partial t} + \underline{v} \cdot \nabla \underline{v} = -\frac{\nabla p}{\rho} = \frac{c^2}{\rho} \nabla \rho \tag{1.8}$$

where c is the appropriate sound speed, (namely $c^2 = \partial p/\partial \rho$) Conservation Law form.

The set of equations are of the general conservative form

$$\frac{\mathrm{d}u}{\mathrm{d}t} + \nabla \cdot \underline{f} = 0 \tag{1.9}$$

where $[\mathrm{d}/\mathrm{d}t = \partial/\partial t + \mathbf{v} \cdot \nabla]$ is the time derivative moving with fluid, u is a general conserved quantity and $\boldsymbol{f}$ the flux associated with it.

Clearly, for any volume V enclosed by a bounding surface S,

$$\frac{\mathrm{d}}{\mathrm{d}t} \int_V u \mathrm{d}V + \int_S \mathbf{f} \cdot \mathbf{ds} = 0 \tag{1.10}$$

which expresses the same result in an integral form expressing global balance within the system.

1.2.2 Lagrangian Frame

Alternatively, it is frequently convenient to consider the fluid particles stationary within their own frame, known as the Lagrangian frame. The transformation from one frame to the other is accomplished by the advection operator. Thus, if the particle velocity is $\mathbf{v}$, it can be easily seen that if the time derivative in the Lagrangian frame, namely with respect to the stationary particle, is $\mathrm{d}\mathbf{u}/\mathrm{d}t$, then the transformation from Lagrangian to Eulerian systems for the general quantity $\mathbf{u}(\mathbf{r}, t)$ is

$$\frac{\mathrm{d}\mathbf{u}}{\mathrm{d}t} = \frac{\partial \mathbf{u}}{\partial t} + \mathbf{v} \cdot \nabla \mathbf{u} \tag{1.11}$$

which we call the advection equation.

1.3 Ideal (Dissipationless) Flow – Hyperbolic Equations

In many fluids, viscosity and thermal conduction are weak (in particular in gases), i.e. the fluid behaves as an ideal dissipationless continuum. To a first approximation in plasma, viscosity can also be neglected and the medium treated as a two-component fluid (electrons and ions), but thermal conduction is an important element and dissipational, which may be treated separately within our numerical scheme. We will therefore seek to develop the numerical approximations in an ideal inviscid fluid and subsequently introduce corrections to the flow model necessary to take account of dissipative effects. In these circumstances, the equations of fluid dynamics (1.1) form a set of quasi-linear hyperbolic equations.

1.3.1 One-Dimensional Isentropic Flow

One-dimensional isentropic flow

Let us consider the simple one-dimensional case of the time-dependent Eulerian non-dissipative system, which can be expressed in the general form

$$\frac{\partial u}{\partial t} + A\frac{\partial u}{\partial x} = 0 \tag{1.12}$$

where u and f are general 'vector sets' (column matrix) of the fluid variables

$$\begin{aligned} \boldsymbol{U} &= \left\{\rho, \rho \mathrm{v}, \rho\left(\epsilon + \tfrac{1}{2}\mathrm{v}^2\right)\right\}^{\mathrm{T}} \\ \boldsymbol{F} &= \left\{\rho \mathrm{v}, (\rho \mathrm{v}^2 + p), \rho \mathrm{v}\left(h + \tfrac{1}{2}\mathrm{v}^2\right)\right\}^{\mathrm{T}} \end{aligned} \tag{1.13}$$

and A is the 3×3 matrix $\partial \boldsymbol{F}/\partial \boldsymbol{U}$. This set of three equations represents the general conservation equation in one dimension, where $\boldsymbol{F}$ is explicitly a function of the variables $\boldsymbol{U}$ alone. It therefore has the form of three first-order quasi-linear differential equations in two variables x and t.

Since the mathematics is messy if the full set of Eqs. (1.54) are used, it is convenient to reduce them to a simpler form by explicitly forming Euler's equation (1.14) from the mass (1.1a) and momentum conservation (1.1b) equations using the pressure/density relation for the adiabatic sound speed $c^2 \partial p/\partial \rho|_S = p/\rho$, and taking account of the entropy constancy. We correspondingly replace the energy conservation Eq. (1.1c) by the isentropic condition, such that the entropy S of a fluid particle remains constant, Eq. (1.14b).

$$\frac{\partial \rho}{\partial t} + \upsilon\frac{\partial \rho}{\partial x} = -\rho\frac{\partial \upsilon}{\partial x} = 0 \tag{1.14a}$$

$$\frac{\partial S}{\partial t} + \upsilon\frac{\partial S}{\partial x} = 0 \tag{1.14b}$$

$$\frac{\partial \upsilon}{\partial t} + \upsilon\frac{\partial \upsilon}{\partial x} = -\frac{1}{\rho}\nabla p = -\frac{1}{\rho}\frac{\partial p}{\partial \rho}\frac{\partial \rho}{\partial x} = -\frac{c^2}{\rho}\frac{\partial \rho}{\partial x} \tag{1.14c}$$

where S is the entropy and $c^2 = \partial p/\partial \rho|_s$ is the sound speed squared. For this set

$$u = \{\rho, S, \upsilon\} \tag{1.15}$$

$$f = [(c^2/\rho), 0, \upsilon] \tag{1.16}$$

The array $\mathbf{A}$ is given by substitution

$$\mathbf{A} = \left\{\begin{matrix} \upsilon & 0 & \rho \\ 0 & \upsilon & 0 \\ c^2/\rho & 0 & \upsilon \end{matrix}\right\} \tag{1.17}$$

The fluid equations are hyperbolic if the eigenvalues of $\boldsymbol{A}$ are real. In this case of an ideal fluid, they can be shown to be $v+c, v, v-c$, where c is the isentropic sound speed: $c^2 = \partial p/\partial \rho|_s$. To find the eigenvalues λ, solve the determinantal equation

$$\det(\boldsymbol{A} - \lambda \boldsymbol{I}) = (v-\lambda)\left[(v-\lambda)^2 - c^2\right] = 0 \tag{1.18}$$

to give $\lambda = [(v+c), v, (v-c)]$. The eigenvectors are

$$\begin{pmatrix} 1 \\ 0 \\ c/\rho \end{pmatrix} \begin{pmatrix} 0 \\ 1 \\ 0 \end{pmatrix} \begin{pmatrix} 1 \\ 0 \\ -c/\rho \end{pmatrix} \tag{1.19}$$

respectively. $\boldsymbol{I}$ is the identity matrix: $(\boldsymbol{I} \equiv \delta_{ij})$.

Since the eigenvalues are distinct we may diagonalise $\boldsymbol{A}$ by a similarity transformation $\boldsymbol{A}' = \boldsymbol{P}\boldsymbol{A}\boldsymbol{P}^{-1}$ so that

$$\boldsymbol{P}\frac{\partial \boldsymbol{U}}{\partial t} + \boldsymbol{A}'\boldsymbol{P}\frac{\partial \boldsymbol{U}}{\partial x} = 0 \tag{1.20}$$

where $\boldsymbol{A}'$ is diagonal and $\boldsymbol{P}$ is a function of the variable $\boldsymbol{u}$ only. Hence, substituting for $\boldsymbol{P}\mathrm{d}\,\boldsymbol{U}$

$$\left[\frac{\partial}{\partial t} + \lambda_i \frac{\partial}{\partial x}\right] \Gamma_i = 0 \tag{1.21}$$

for the eigenvalues λ_i. The functions Γ_i, obtained from $\int P\,\mathrm{d}u$, are constant on the lines $\frac{\mathrm{d}x}{\mathrm{d}t} = \lambda_i$, and are known as Riemann invariants. For this case, they take the form

$$\Gamma_i = [J_+ \quad J_0 \quad J_-] = \left[v + \int c\mathrm{d}\rho/\rho\right], \quad S, \quad \left[v - \int c\mathrm{d}\rho/\rho\right]$$

for the ideal fluid equations.[1]

In principle, a knowledge of the initial values of these quantities (ρ, v and ϵ) will allow us to integrate along the characteristics in space-time and determine the flow. Since given the initial values at time zero u_0, we may form an expansion in terms of the Riemann invariants $\boldsymbol{P}\boldsymbol{u}_0$ and follow them in time along the characteristics. These results form the basis for much of the study of compressible fluids and by extension of simple plasma neglecting thermal conduction.

1.4 Formal Solution

At any point (x, t), we may determine the three values of the variables $\boldsymbol{U}$ if the Riemann invariants are known, provided we know the equation of state, i.e. $p(\rho, s)$ or more succinctly $c(\rho, s)$ by tracing the development of the fluid variables in space and time along

---

1 This equation for the characteristics is more easily obtained directly from the first two equations of the set (1.5)

$$\left[\frac{\partial}{\partial t} + (v \pm c)\frac{\partial}{\partial x}\right]\left(\left[v + \int \frac{c}{\rho}\mathrm{d}\rho\right]\right) \tag{1.22}$$

$$= \underbrace{\left(\frac{\partial v}{\partial t} + v\frac{\partial v}{\partial x} \pm \frac{c^2}{\rho}\frac{\partial \rho}{\partial x}\right)}_{\text{Euler's equation}} \pm c\frac{\partial \not{c}}{\not{\partial} x} + \frac{c}{\rho}\underbrace{\left[\frac{\partial \rho}{\partial t} + v\frac{\partial \rho}{\partial x} + \rho\frac{\partial v}{\partial x}\right]}_{\text{Continuity}} \mp c\frac{\partial \not{c}}{\not{\partial} x} \tag{1.23}$$

the characteristics. Thus, in principle, we integrate along the field of characteristics to determine the flow. Clearly, the value of u at the point (x_0, t_0) is determined only by points which lie within the outermost characteristics through (x_0, t_0) – its domain of dependence. Similarly, it can only influence future events within its outermost outgoing characteristics – range of influence. The existence of such solutions has been formally demonstrated (Courant and Hilbert, 1962).

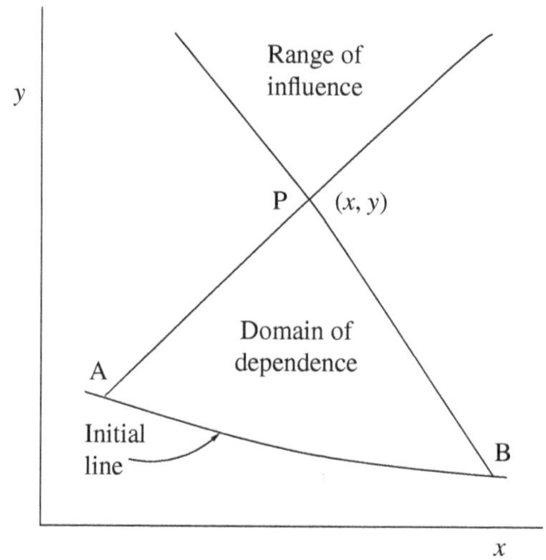

For this system, the following uniqueness theorem is valid (Courant and Friedrichs, 1948; Courant and Hilbert, 1962).

> Consider a solution with continuous second-order derivatives in the region ABP bounded by two characteristics through P, and the domain of dependence AB cut-off by them on the initial line I. Suppose another solution with continuous second derivatives is given in ABP, which assumes the same values on AB as the first. Then, the second solution is identical to the first within the domain ABP.

1.5 Discontinuities

The uniqueness theorem quoted earlier requires that the solution be continuous up to the second order everywhere within the domain of dependence. However, discontinuities arise naturally within fluid mechanics by virtue of the non-linearity of the governing equations (Courant and Friedrichs, 1948; Landau and Lifshitz, 1959). These occur in two forms: strong and weak discontinuities. Strong discontinuities involve a jump in the value of the flow variables and include shocks and contact discontinuities. Shocks involve flow through the discontinuity, whereas tangential discontinuities do not. Weak discontinuities are continuous, but with a change of gradient. Weak discontinuities, which include the head of rarefaction waves, propagate along the characteristics (Landau and Lifshitz, 1959).

If two characteristics were to intersect, the Riemann invariant would become indeterminate. Such intersections can occur in compression, where they lead to shocks. In this neighbourhood, the ideal flow condition breaks down. When the characteristics approach over distances of the order of the mean free path, large gradients in the flow are established, and dissipation becomes important. This leads to an entropy increase in accordance with the second law of thermodynamics across the shock. The thickness of the 'discontinuity' is of the order of the mean free path, which is vanishingly small in the continuum theory. In general fluid theory, the collisional behaviour at a molecular level gives rise to dissipational processes, namely viscosity and thermal conduction. Within the fluid description, the shock 'discontinuity' is controlled by dissipation, principally viscosity. The shocks therefore appear as narrow zones within which viscous effects are dominant, which in molecular terms are typically a few mean paths thick (Zel'dovich and Raizer, 1966).

Considering the case of a general fluid with dissipation, the general conservation law in integral form (1.10) requires that across a normal shock in the laboratory frame

$$\rho_r v_r = \rho_\ell v_\ell \tag{1.24}$$

$$(\rho_r v_r)\left(\rho_r v_r^2 + p_r\right) = (\rho_\ell v_\ell)\left(\rho_\ell v_\ell^2 + p_\ell\right) \tag{1.25}$$

$$(\rho_r v_r)\left(h_r + \frac{1}{2} v_r^2\right) = (\rho_\ell v_\ell)\left(h_\ell + \frac{1}{2} v_\ell^2\right) \tag{1.26}$$

where v_ℓ and v_r, ρ_ℓ and ρ_r, p_ℓ and p_r and h_ℓ and h_r are, respectively, the density, flow velocity, pressure and enthalpy upstream (ℓ) and downstream (r) of the shock in the laboratory frame. Alternatively and more succinctly in the shock frame

$$S = \frac{\left(\boldsymbol{F}(\boldsymbol{U}_r) - \boldsymbol{F}(\boldsymbol{U}_\ell)\right)}{\left(\boldsymbol{U}_r - \boldsymbol{U}_\ell\right)} \tag{1.27}$$

where S is the shock speed, the shock moving from left to right, $\boldsymbol{U}$ are the fluxes normal to the shock and $\boldsymbol{U}_\ell$ and $\boldsymbol{U}_r$ the values of the components of $\boldsymbol{U}$ on the upstream (left) and downstream (right) sides of the shock, respectively. *Note that this relation must be obeyed for all the elements of* $\boldsymbol{U}$ *with the same value of* S. In one dimension these are the familiar Rankine–Hugoniot relations for a normal shock (Eq. (1.27)).

Discontinuities arise naturally in the general solution of the hyperbolic equations. At such points, the differential equations break down, but these are expressions of more general integral relations, namely the integral forms of the conservation laws. We may construct generalised solutions which satisfy these relations called weak solutions by Lax (1954) (see Supplement Section M.3). Clearly, they satisfy the differential equations where the derivatives exist, and the Rankine–Hugoniot relations across discontinuities. However, in contrast to the actual physical situation, rarefaction as well as compression discontinuities can occur.

Two questions arise if we wish to use the dissipationless equations for calculation.

1) Is the solution unique? Clearly, our previous uniqueness theorem fails when discontinuities occur.
2) Is the solution without dissipation the limit of vanishing viscosity to the complete fluid dynamic problem?

In fact both questions are related. The discontinuities in ideal flow can be shown not to be uniquely specified by the Rankine–Hugoniot condition. In the real fluid, such uniqueness is imposed by the additional requirement for an entropy increase. This is established by the following results proved for one-dimensional systems (Oleinik, 1963a).

1) Solutions of dissipative equations converge to weak solutions of the hyperbolic system in the limit of vanishing viscosity (Oleinik, 1963a).
2) Such solutions of the viscous system satisfy

$$\boldsymbol{S}[\boldsymbol{U}, \boldsymbol{U}_\ell] \geq \boldsymbol{S} \geq \boldsymbol{S}[\boldsymbol{U}_r, \boldsymbol{U}] \tag{1.28}$$

known as the entropy condition, where S is the shock speed (1.27) and

$$\boldsymbol{S}[a, b] = \frac{\boldsymbol{F}(\boldsymbol{b}) - \boldsymbol{F}(\boldsymbol{a})}{(b - a)} \tag{1.29}$$

where $\boldsymbol{U}$ lies between $\boldsymbol{U}_\ell$ and $\boldsymbol{U}_r$ (Hopf, 1969; Quinn, 1971).
3) Weak solutions of the hyperbolic system satisfying the entropy condition are unique (Oleinik, 1963b; Quinn, 1971).

The entropy condition has a close relationship with many of the familiar properties of shocks. Thus, since the limit of a weak shock is a sound wave, which propagates at the local sound speed $\boldsymbol{S}[\boldsymbol{U}, \boldsymbol{U}] \to c(\boldsymbol{U})$, the shock speed is supersonic with respect to the upstream flow $S \geq c(\boldsymbol{U}_r)$ and subsonic with respect to the downstream flow $S \leq c(\boldsymbol{U}_\ell)$. Indeed the entropy condition reflects the facts that shock adiabat is generally convex, and that physical shocks are compressive as required by the second law of thermodynamics (Courant and Friedrichs, 1948). The entropy condition thus ensures that the shock satisfies the second law of thermodynamics, and that there is an entropy increase across the shock.

1.6 Plasma Fluid Dynamics

Since the viscosity of plasmas is very small, we may neglect it and the dynamics can be treated as those of an ideal fluid, with the proviso that the ions and electrons may have different temperatures T_i and T_e, respectively. As a consequence of ion/electron collisions, there is an energy exchange between the two sets of particles. The rate at which this takes place is governed by the equilibration time τ_{eq},

$$\frac{d\epsilon_i}{dt} = -\frac{d\epsilon_e}{dt} = -\frac{\epsilon_i/c_i - \epsilon_e/c_e}{\tau_{eq}} \tag{1.30}$$

where ϵ_i and ϵ_e are the specific internal energies and specific heats of ions and electrons, respectively, τ_{eq} is the ion/electron relaxation time given by Spitzer (1956).

As a result of the different internal energies, the pressures of the ions and electrons, p_i and p_e, respectively, are different; the total pressure follows from Dalton's law $p = p_i + p_e$. The individual species act as almost ideal gases, so that

$$p_i = n_i T_i = \frac{2}{3}\epsilon_i \qquad \text{and} \qquad p_e = n_e T_e = \frac{2}{3}\epsilon_e \tag{1.31}$$

Fortunately, provided the Debye length is much smaller than the characteristic lengths in the fluid, the ion and electron charge densities are equal. As a result,

$$n_e = Z\,n_i \tag{1.32}$$

where Z is the degree of ionisation. If q_i is the fractional ionisation of the ith stage of ionisation, the fractional ionisation $Z = \sum_i Z_i\,q_i$ where Z_i is the fractional charge on the ion i. The fractional ionisation in stage i is calculated from the balance between its loss by ionisation and recombination and gain from the ionisation of stage $i-1$ and recombination of stage $i+1$, namely

$$\frac{\mathrm{d}q_i}{\mathrm{d}t} = S_{i-1}\,q_{i-1} - R_i\,q_i - S_i\,q_i + R_{i+1}\,q_{i+1} \tag{1.33}$$

where S_i and R_i are the rates of ionisation and recombination, respectively.

1.7 Basic Principles of Finite Differencing

1.7.1 Introduction

We shall develop numerical solutions for the case of ideal dissipationless flow, which, as we shall see, presents serious difficulties in view of its non-linear hyperbolic nature. Such solutions are of great importance in both physics and engineering encompassing many of the applicable classical disciplines, namely aerodynamics and plasma physics. Since the central problem is most easily attacked in one dimension, we will firstly present the analysis considering only one spatial variable x and the temporal t. In general, the exact solution $u(x, t)$ is defined at each point (x, t). The numerical solution will take the form of an approximation function $v(x, t)$, which is only defined at a set of discrete points in (x, t). These points conveniently form a mesh in (x, t), which in a simple form have a constant separation, Δx in space and Dt in time. We may identify these points by integer co-ordinates (i, n).

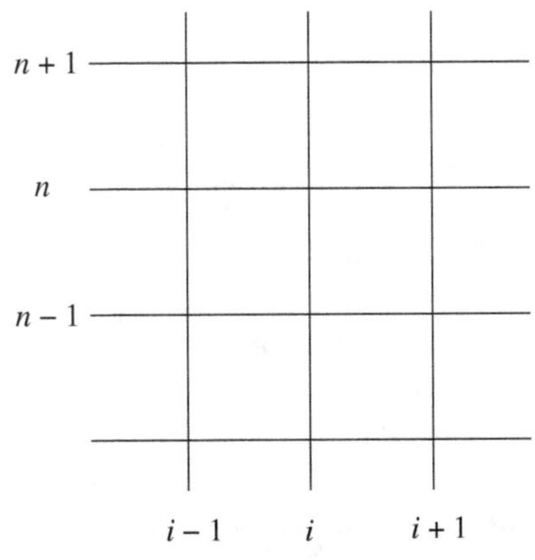

Values of $v(x, t)$ at points intermediate to the mesh are constructed by interpolation.

We require a finite-difference scheme to be constructed to allow the approximating function $\upsilon(x, t)$ to be determined from the boundary conditions. These are usually initial value conditions, i.e. we know the initial fluid state and need to calculate its subsequent history subject to boundary conditions in space. Thus, it is natural in many cases to seek to develop the solution in time. Hence, we require a relation of the type

$$\upsilon_i^{n+1} = G\left[\upsilon_{i-\ell}^{n+1}, \dots, \upsilon_{i+\ell}^{n+1}, \upsilon_{i-\ell}^{n}, \dots, \upsilon_{i+\ell}^{n}, \dots\right] \tag{1.34}$$

where υ_i^n is the value of υ at $x_i = i\Delta x$ and $t_n = nDt$. The mesh runs over I cells from the 'left' boundary at $\frac{1}{2}\Delta x$ to the right at $(I + 1/2)\Delta x$. The scheme is called implicit if the right-hand side of the equation contains terms evaluated at time $(n + 1)Dt$, and is explicit otherwise. Implicit schemes generally involve complex matrix inversion and may have other disadvantages for fluid dynamics.

In general, storage limitations usually allow only one time step to be stored, which restricts G to a one-step operator

$$\upsilon_i^{n+1} = G\left[\upsilon_{i-\ell}^{n}, \dots, \upsilon_{i+\ell}^{n}\right] \tag{1.35}$$

For an explicit two-step operator, there is a fundamental time-step constraint determined by the characteristics of the flow. Since information is carried along the characteristics with a characteristic speed υ, and since within an explicit scheme information must be carried to the neighbours only, it is seen that the time step must be limited to a maximum value $Dt < \Delta x/\upsilon$. This is known as the Courant–Friedrichs–Lewy limit following the pioneering paper by Courant, Friedrichs and Lewy (1928) in which it was first derived. The parameter $\lambda = \mathrm{v}Dt/\Delta x$ is known as the Courant number.

Since the exact solution is conserved, it is natural to expect that $\upsilon(x, t)$ obeys a similar relation, which on this simple mesh requires

$$\sum_{i=1}^{I} (\upsilon_i^{n+1} - \upsilon_i^n)/Dt = \left[-g_{I+1/2} + g_{1/2}\right]/\Delta x \tag{1.36}$$

where $g_{1/2}$ and $g_{I+1/2}$ are approximations to the boundary fluxes of $f(\frac{1}{2}\Delta x)$ and $f[(I + 1/2)\Delta x]$, respectively.

This requires

$$\frac{D\upsilon}{Dt} = \frac{\left(\upsilon_i^{(n+1)} - \upsilon_i^n\right)}{Dt} = -\frac{\Delta g}{\Delta x} = -\frac{\left(g_{(i+1/2)} - g_{(i-1/2)}\right)}{\Delta x} \tag{1.37}$$

where D denotes a time (t) difference, and Δ a space (x) difference. $g_{(i+1/2)}$ is a function of the variables $\upsilon_{i-\ell+1}^n, \dots, \upsilon_{i+\ell}^n$ and $g_{(i-1/2)}$ of $\upsilon_{i-\ell}^n, \dots, \upsilon_{i+\ell-1}^n$.

This integral (conservation) representation of the finite-difference equation allows another interpretation of υ_i^n, which may be useful, in terms of the volume average of the quantity u taken over the cell, i, at time n.

In order that the approximation be valid as an expression of the exact solution of the quantity u, we clearly require that as $\Delta x, \Delta t \to 0$, $\upsilon(x, t) \to u(x, t)$. Indeed $\upsilon(x, t)$ is a weak solution in this limit provided (Lax and Wendroff, 1960).

$$g(u, \dots, u) = f(u) \tag{1.38}$$

These solutions do not, however, necessarily satisfy the entropy condition. This is known to be rigorously satisfied only for one class of finite-difference schemes (namely monotone) Jennings (1974). Commonly used approximations can be made to converge onto the correct weak solution, with the discontinuities matching the shocks found in practice – a procedure known as shock capture.

1.7.2 Necessary Conditions in the Finite-Difference Operator *G*(*Dt*)

It is important to specify some necessary conditions for the general finite-difference operator to give satisfactory approximations to the solutions of the corresponding partial differential equations. If $G(Dt)$ is such an operator, we may identify the following necessary conditions on G (Richtmyer and Morton, 1967).

1) **Consistency:** As the spatial and temporal steps are reduced together to give a finer mesh, the solution of the difference equations must reduce to the exact solution of the corresponding differential equation. Thus, if $\Delta x = \phi(Dt)$ as $\Delta x, Dt \to 0$ and u is any genuine solution of the differential operator A to which G approximates, then G is a consistent approximation to A if

$$\left\| \left[\frac{\{G(Dt) - I\}}{Dt} - A \right] u(t) \right\| \to 0 \quad \text{as} \quad Dt \to 0 \tag{1.39}$$

For a conservative system, this is equivalent to the previous result (Eqs. (1.37) and (1.38)).
2) **Convergence:** Operating n times with the finite-difference operator $G(Dt)$ on $u(0)$ should yield an approximation to $u(nDt)$. If $u(t) = E(t)u(0)$, then the approximate solution converges to $u(t)$ for a sequence of time steps Dt_i: tending to zero, and $n_i = t/Dt_i$ such that $n_i Dt_i \to t$ as $i \to \infty$ for all possible initial states $u(0)$. Thus,

$$\| \{[G(DT_i)]^n - E(t)\}u(0) \| \to 0 \quad \text{as} \quad i \to \infty \tag{1.40}$$

for all possible initial states $u(0)$.
3) **Stability:** It is essential that during the calculation no component be amplified without limit. Thus, an approximation is said to be stable, if for some time $\tau > 0$, and the infinite set of operators

$$[G(Dt)]^n \quad \begin{cases} 0 \leq Dt \leq \tau \\ 0 \leq nDt \leq t \end{cases}$$

is uniformly bounded for specified τ and t.
Stability and convergence are related by **Lax's equivalence theorem** (Richtmyer and Morton, 1967).

> Given a properly posed initial-value problem and a finite-difference approximation to it that satisfies the consistency condition, then stability is the necessary and sufficient condition for convergence.

4) **Accuracy:** For smooth solutions $u(x, t)$, the accuracy is limited by the truncation of the Taylor's series at any point in generating the finite-difference form. In the case of

constant coefficients, we define the order of the difference scheme by the largest real number ℓ for which

$$\| u(t+Dt) - G(Dt)u(t) \| = O(Dt^{(\ell+1)}) \quad \text{as} \quad Dt \to 0 \tag{1.41}$$

for all sufficiently smooth solutions $u(x,t)$ of the differential equation. Usually, ℓ is an integer, in which case consistency requires that the equations be at least 'first-order accurate'. Clearly, ℓ gives a measure of the missing terms from the Taylor expansion of the function in the difference scheme.

5) **Positivity:** An additional physical constraint is positivity, i.e. that some quantities, for example density and energy, are always non-negative. We may require that the finite-difference scheme retain this property.
6) **Monotonicity:** Some physical behaviour maintains a monotonic profile. Finite-differences schemes with this property are known as *monotonicity preserving*.
   Closely related is a *monotone* finite-difference operator defined by

$$\frac{\partial G}{\partial u_{(i+k)}} \geq 0 \quad \text{for} \quad -\ell \leq k \leq \ell$$

   Monotone operators are always monotonicity preserving (although the converse is not true), and are necessarily first order (Jennings, 1974). They have the important and useful property that they always satisfy the entropy condition, and thus guarantee uniqueness, i.e. they guarantee correct shock capture.

1.7.3 Constant Coefficients

If the matrix A is constant, then the governing differential equations are linearised, and the characteristic equation (Eq. (1.12)) reduces to a particularly simple form, which can be solved exactly by Fourier transformation.

As an example, consider the one-dimensional single-variable case, the generalisation to many variables and/or dimensions is straightforward. The governing equation is

$$\frac{\partial u}{\partial t} + a\frac{\partial u}{\partial x} = 0 \tag{1.42}$$

subject to an initial condition $u(0,x) = u_0(x)$. The formal solution is

$$u(x,t) = u_0(x - at) \tag{1.43}$$

corresponding to a transport (*advection*) of the quantity u with velocity a.

Alternatively, let $\hat{u}(k,t)$ be the spatial Fourier transforms of $u(x,t)$. Then

$$\hat{u}(t,k) = \hat{u}_0(k) \exp[-\imath kat] \tag{1.44}$$

Since G is a linear operator, this solution also carries over into the finite-difference form

$$\hat{v}^n(k) = \hat{v}^0(k)\, G\left[\exp(-\imath k\ell\Delta x), \dots, \exp(\imath k\ell\Delta x)\right]^n \tag{1.45}$$

Since $v(x,t)$ is defined only by values at the spatial points $i\Delta x$, its Fourier transform is discrete with values of k in the range $-\pi/\Delta x$ to $\pi/\Delta x$.

In this form, the function G, with corresponding arguments, is the amplification factor of the Fourier component k. Its magnitude, $|G|$, determines the amplitude growth of the Fourier component with wave number k, and the argument, $\arg(G)$, the phase shift from step to step at a fixed point. As a result, G determines the stability of this system. Hence, we deduce that for stability

$$G^n \text{ must be bounded for } \begin{cases} 0 < Dt < \tau \\ 0 \leq nDt \leq T \end{cases}$$

for all k.

This result can be extended to give a very useful necessary condition for stability due to von Neumann (Richtmyer and Morton, 1967), namely $| G | \leq 1 + O(Dt)$.

When this result is generalised to several variables, G becomes a matrix, *amplification matrix*, with eigenvalues λ_i. In this case, von Neumann's condition is generalised to

$$| \lambda_i | \leq 1 + O(Dt) \quad \begin{cases} 0 < Dt < \tau \\ 0 \leq nDt \leq T \end{cases} \tag{1.46}$$

Although the Von Neumann condition is not sufficient, it has in practice proved extremely useful in judging the stability of finite-difference schemes. When a difference scheme is found to be conditionally stable (i.e. for a restricted range of $Dt/\Delta x$, for example) the von Neumann condition nearly always gives the correct range. Sufficient conditions for constant coefficients have been derived but are less convenient to use.

In practical problems, the differential equations do not have constant coefficients. However, practical experience shows that instabilities usually start as local phenomena. In a small region, the differential equation can be linearised to a form with constant coefficients, which can be tested for local stability. This is a useful procedure and does give necessary conditions for stability. However, it must be remembered that non-linear instabilities, which are more difficult to assess, can and do occur.

1.8 Conservative Finite-Difference Approximations

The basic fluid hydrodynamic (and more generally magneto-hydrodynamic) are in the form of a conservation law (Eq. (1.57)). For application in numerical simulation, it is natural that these should be reproduced in the finite-difference form. Thus, consider a closed system (with no boundary fluxes) in which the quantities X_i^n generated by the numerical procedure at a mesh point i at the conclusion of the nth time step are to be conserved, i.e. $\sum_i X_i^n$ is constant in time. In general, for a closed system

$$\sum_i X_i^n = \sum_i X_i^{(n-1)} = \cdots = \sum_i X_i^0 \tag{1.47}$$

In many cases, the physical quantity X is inherently non-negative, for example density, energy, etc. in which case the finite scheme should be *non-negativity maintaining*

$$\text{If} \quad X_i^n \geq 0 \quad \text{then} \quad X_i^{n+1} \quad \text{for all elements, } i \tag{1.48}$$

As a result of truncation errors introduced thereby, the resulting equations are no longer exact conservation laws. We identify different types:

1) **Strong conservation:** If the finite-difference equations obey condition (1.47) within the round-off error, then the system is *strongly conservative.*

Theorem 1.1 A strongly conservative, non-negativity maintaining system is always stable if the input data are non-negative.

Proof 1.1 The proof follows directly from the definition of stability since any element is bounded from above by $\sum_i X_i^0$. The condition in the linear data may be omitted if the scheme is linear and differential (i.e. if $X_j^0 = X_0$, a constant value, then $X_j^n = X_0$) for the system reduces to an extremal operation of the type considered in Section 10.3.3.

2) **Weak conservation:** The strong equality in Eq. (1.47) may be relaxed to allow independence of the time step *Dt* in the form

$$\sum_i X_i^n = \sum_i X_i^{(n-1)} + O(Dt^2) \tag{1.49}$$

which we call *weakly conservative*. In contrast to strongly conservative schemes, which maintain conservation exactly, weakly conservative schemes may only do so to a prescribed accuracy. Thus, consider a calculation over a total time $\tau = NDt$ for which

$$\left| \sum_i \left(X_i^n - X_i^{(n-1)} \right) \right| \leq K\,Dt^2 \tag{1.50}$$

where *K* is a positive real number, then the overall conservation error

$$\left| \sum_i (X_i^n - X_i^0) \right| \leq N\,K\,Dt^2 K\,\tau\,Dt \tag{1.51}$$

whose value may be limited by an appropriate choice of time step. One therefore introduces an accuracy condition

$$\left| \sum_i \left(X_i^{(n+1)} - X_i^n \right) \right| \leq a \left| \sum_i X_i^n \right| \tag{1.52}$$

to limit the conservation error during the calculation, *a* being preset.
Such a scheme obeys the important conditional stability theorem.

Theorem 1.2 A non-negativity maintaining, weakly conservative scheme subject to an accuracy condition is stable if the initial data are non-negative.

Proof 1.2 Suppose there exists a real, positive, finite constant *K* such that Eq. (1.50) is always obeyed, then since the overall error is given by equation (weak conservation total error), any value X_-^n is bounded from above by $\sum_i X_i^0 + K\,t\,Dt$. The existence of such a constant *K* is guaranteed by the accuracy condition (1.52); for if *K* exists for step $(n-1) \to n$, then $\sum X_i^n$ is bounded. Hence, $\sum X_i^{(n+1)}$ is bounded and *K* exists for step *n* to $N+1$. Since $\sum X_i^0$ exists *K* exists for step $0 \to 1$ and therefore all succeeding steps.

3) **Consistently conservation:** A weaker form of conservation is

$$\sum_i X_i^n = \sum_i X_i^{(n-1)} + O(Dt) \tag{1.53}$$

which we call *consistently conservative*. Such schemes cannot improve their conservation accuracy by time-step adjustment.

The above considerations apply to closed systems; however, the extension to open systems is straightforward provided the boundary fluxes are finite and does not alter the above conclusions.

Considerations of the conservative nature of any system of operations is useful as a supplement to the usual stability criterion. The von Neumann stability condition is a local, whereas a strong or weak conservation condition is global. Therefore, if both can be simultaneously obeyed, the resulting algorithm should be well behaved and reasonably accurate.

1.9 Numerical Fluid Approximations

1.9.1 Introduction

Computational fluid dynamics attempts to solve the dynamical structures of systems within the fluid approximation, namely the behaviour of the fluid is dominated by collisions between the constituent molecules. This implies that the mean free path is everywhere smaller than the characteristic dimensions of the problem, and that the mean collision interval short compared to the characteristic time (Landau and Lifshitz, 1959).

Typical problems investigated using these techniques include aerodynamics (in e.g. aeronautical design or car design), meteorology, oceanography and plasma dynamics.

The fluid is described by its thermodynamic state and the velocity of flow. These values must be specified at each point in the flow field. The thermodynamic state is specified by two state variables (e.g. pressure p and density ρ), the remainder being then known from the equation of state. The velocity is defined by the three-vector components $\underline{\mathrm{v}}$. There are thus five independent quantities, which vary in space $\underline{r}$ and time t, and must be determined subject to the constraints on the system.

The equations of fluid mechanics contain five equations, which are simply the basic conservation laws of physics applied to moving fluid.

1.9.2 The Basic Equations

In the laboratory (or Eulerian) frame, the general equations of fluid mechanics can be written in several forms, of which the general conservation form is most appropriate for a general discussion.

$$\begin{aligned} &\frac{\partial \rho}{\partial t} + \nabla \cdot (\rho \underline{\mathrm{v}}) && = 0 \\ &\frac{\partial}{\partial t}(\rho \underline{\mathrm{v}}) + \nabla \cdot \underline{\underline{\pi}} && = 0 \\ &\frac{\partial}{\partial t}\left[\rho\left(\epsilon + \frac{1}{2}\mathrm{v}^2\right)\right] + \nabla \cdot \left[\rho \underline{\mathrm{v}}\left(h + \frac{1}{2}\mathrm{v}^2\right)\right] + \nabla \cdot \underline{q} && = 0 \end{aligned} \tag{1.54}$$

where ϵ is the specific internal energy of the fluid, $h = \epsilon + p/\rho$ is the specific enthalpy, and $\underline{q}$ is the heat flux vector. $\underline{\underline{\pi}}$ is the momentum flux tensor

$$\pi_{i,j} = \rho\, \mathrm{v}_i\, \mathrm{v}_j - \tau_{i,j} \tag{1.55}$$

$\tau_{i,j}$ is the total stress tensor

$$\tau_{i,j} = \sigma_{i,j} - p\delta_{i,j} \tag{1.56}$$

where $\sigma_{i,j}$ is the viscous stress tensor and $\delta_{i,j}$ the Kronecker delta (1 if $i = j$ and 0 otherwise).

The set of equations are of the general conservative form

$$\frac{\partial u}{\partial t} + \nabla \cdot \underline{f} = 0 \tag{1.57}$$

where u is a general conserved quantity and $\underline{f}$ the flux associated with it.

Clearly, for any volume V enclosed by surface S

$$\frac{\mathrm{d}}{\mathrm{d}t}\int_V u\mathrm{d}V + \int_S \underline{f} \cdot \mathrm{d}\underline{s} = 0 \tag{1.58}$$

expresses the same result in an integral form expressing global balance within the system.

1.10 Grid Geometry

Thus far, we have not specified the nature of the co-ordinate system used, beyond noting that the co-ordinates are defined to be finite elements, often but not always, orthogonal to each other within their own co-ordinate system. The grid may be defined quite arbitrarily with respect to the fluid, but normally takes one or two forms.

1) **Eulerian frame:** The grid is fixed in the laboratory co-ordinate system. The fluid therefore flows through the grid with the fluid velocity. This is the most commonly used system.
2) **Lagrangian frame:** The grid is fixed to the fluid particles. The fluid is therefore stationary with respect to the grid, but moves through the laboratory co-ordinates with its flow velocity. This is a particularly convenient arrangement for one-dimensional laser heating problems.

More general fluid/mesh arrangements are possible. In multidimensional systems, the grid need not be orthogonal, although orthogonal grids in Eulerian systems lead to simpler finite-difference equations. Lagrangian grids are generally constructed using a transformed set of orthogonal co-ordinates. However, free (structureless) grid systems particularly in a finite element formulation are both tractable and advantageous for some problems, where the grid may able to move in an arbitrary fashion. These are frequently used in cases where the flow is not well structured, for example when the fluid is freely expanding. Arbitrary Lagrangian–Eulerian (ALE) models. Since the mesh is often totally structureless, a finite element algorithm, rather than the finite difference used here, is often applied for two-dimensional engineering application.

1.11 Control Volume Differencing

The approach we have described thus far has used simple differencing to form finite differences of the derivative form of the fluid equations. In principle, this requires that the definite-difference approximation is defined at a set of nodes (mesh points), usually on a regular Cartesian grid. Such an approach, whilst convenient to illustrate the nature of the approximation being made at a local level, is not always satisfactory and needs to be generalised in a number of ways:

1) The problem needs to be treated on a multidimensional mesh.
2) The grid requires variable spacing from element to element.
3) The mesh may not be formed from straight-line elements.

These problems may be easily circumvented using an alternative form which utilises the essentially conservative form of the fluid equations. The difference equations are based on the integral form of the fluid equations, namely the conservation equations:

$$\frac{\partial u}{\partial t} + \nabla \cdot \mathbf{f} \implies \frac{\mathrm{d}}{\mathrm{d}t}\int_V u\,\mathrm{d}V + \int_S \mathbf{f}\cdot \mathrm{dS} \tag{1.59}$$

reflecting the balance between the change in the total of quantity u per unit volume per unit time with the corresponding flux in $\mathbf{f}$. Therefore, establishing a set of finite volume cells with clearly defined volume $V_{i,j}$ and surfaces $\mathbf{S}_{i,j}$ which 'tile' the entire computational space. Defining the approximate solution u by values at the nodes (located at the cell centres) and values for the fluxes at the surfaces elements centre, the conservation equation becomes

$$\sum_{i,j} V_{i,j}\left\{u^n_{i,j} - u^{n-1}i,j\right\} + \sum_{i,j}\left\{\boldsymbol{S}_{((i+1/2),j)}\cdot \boldsymbol{f}_{((i+1/2),j)} + \boldsymbol{S}_{i,(j+1/2)}\cdot \boldsymbol{f}_{i,(j+1/2)}\right\} Dt \tag{1.60}$$

As an example, consider the cylindrical mesh with cells (i,j) having radii $R_{i,j}$ and length $Z_{i,j}$ (Figure 1.1). The surface areas are

$$SR_{j+1/2} = \frac{1}{2}\left(R^2_{(j+1/2)} - R^2_{(j-1/2)}\right) \text{ and } SZ_i = Z_{i+1/2} - Z_{i-1/2} \tag{1.61}$$

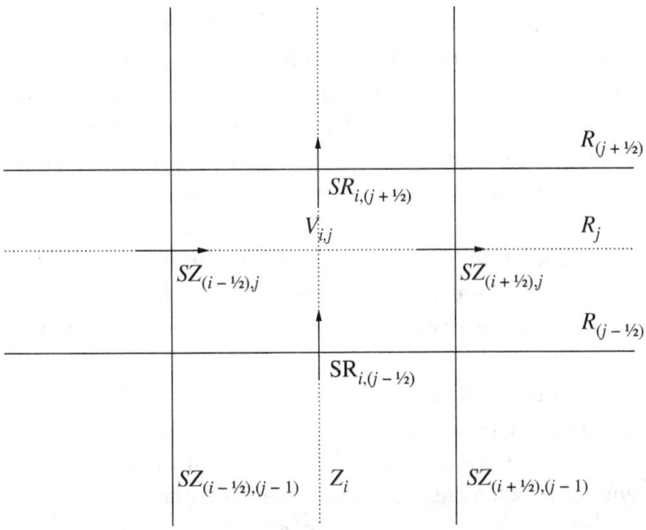

Figure 1.1 Simple mesh element in cylindrical geometry. The cell is centred at (R_j, Z) with surfaces $SR_{i,(j-1/2)}$ and $SR_{i,(j-+1/2)}$ inwards and outwards, respectively, in the transverse R direction, and $SZ_{i,(j-1/2)}$ and $SZ_{i,(j-+1/2)}$ inwards and outwards, respectively, in the axial Z direction.

and the volume

$$V_{i,j} = \left(R^2_{j+1/2} - R^2_{i,j-1/2}\right)\left(Z_{i+1/2} - z_{i-1/2}\right) \tag{1.62}$$

Although orthogonal grid lines make the final difference form simpler to use in that the surfaces and volumes are automatically regularly numbered, the control volume method is preferable for situations where the mesh is either unstructured or has an irregular or non-Cartesian form and/or is non-uniformly spaced.

1.12 Mesh Types

1.12.1 Structured and Unstructured Grids

In many cases, it is convenient to simplify the differencing of the underlying differential equations by matching the node positions evenly to mid-points on the grid. This requires that the grid is appropriately structured. Grids with a clear structure of this type are known as a *structured mesh*, as opposed the case of a set of nodal points which are distributed irregularly without systematic=d ordering. The most typical and widely used structured grid is the orthogonal Cartesian grid, often uniform in spacing.

Orthogonal Cartesian grids are well matched to Eulerian programs since the gradient operator is directly based on the node positions. The mesh is generally steady and nodes constant in time.

In a Lagrangian system, the node positions are not well defined in time, but move irregularly with the fluid, so there is no regular structure. This is known as an *unstructured grid*. In many cases, working with a multidimensional fluid system, it is convenient to order the fluid elements in a structure to enable a rapid and easy identification of neighbouring points to determine gradients, for example a quadrilateral or a triangular grouping in two dimensions. Unfortunately, this introduces severe constraints on the program at run time. For quadrilateral groups, mesh tangling and bow-tie groups terminate the run; similarly with triangular groups, the ordering is reversed (clockwise to anticlockwise). In addition, the introduction of a necessary grouping introduces a stiffness element not present in the basic physics.

Totally unstructured grids are difficult and expensive to handle, requiring significant overheads to impose some measure of structure to enable the necessary derivatives imposed by the fluid dynamic equations, for example gradients, to be identified. The best approach to avoiding these problems is to use the alternative methodology of finite elements and the Galerkin method, which has proved popular for unstructured meshes. This, however, is expensive requiring the solution of typically $6N +$ th-order simultaneous equations, where N is the number of mesh points (Zienkiewicz, Taylor and Nithiarasu, 2014). In the author's albeit limited experience, finite element codes treating the full plasma simulation work satisfactorily but are considerably more demanding of machine capabilities than fine difference. Therefore, within the philosophy underlying this work, we will not consider finite element programs further.

1.12.2 Cell-Centred and Mesh-Centred Grids

Thus far, we have not specified the relationship of the nodes (mesh points) with the grid boundaries.[2]

In performing fluid code calculations, we solve the initial value partial differential equations by means of finite-difference approximations to the exact equations using values of the variable at discrete points in space and time. These points establish a computational grid in both space and time. There are formally two approaches by which the variables may be assigned within the spatial mesh using either the derivative or the integral form of the governing equations:

a) **Cell-centred grid:** The spatial mesh is defined pointwise by the nodes. Values of the variable are assigned to each nodal mesh point alone. The finite-difference representation is established to be consistent with the governing differential Eq. (1.9).
b) **Mesh-centred grid:** The spatial mesh is defined by cells of finite volume which 'tile' (entirely fill) the space within the boundaries. Values of the variables are assigned to each cell in terms of averages over the cell volume. The finite-difference representation is established to satisfy integral conservation laws in terms of fluxes across the cell boundary equation (1.10).

It is clear that although the interpretation placed on these descriptions is different, yet as the mesh is refined the differential and integral forms must become identical.

In principle, we may define two distinct classes of spatial mesh:

i) Define the mesh points (nodes) and place the cell boundaries midway between the mesh points – *cell-centred grid.*
ii) Define the cell boundaries and place the mesh points at the cell centres – *mesh-centred grid.*

The simplest type of spatial mesh is uniform in which the spacing is constant from cell to cell. However, for many purposes, this is too simple a form, and a non-uniform grid will take either a cell-centred or a mesh-centred form. More generally, one endeavours to ensure that the finite-difference scheme is interpretative in terms of both descriptions. This is only possible for a restricted class of spatial mesh. In general, grids of type (i) are used in Eulerian codes, and those of type (ii) in irregular and Lagrangian. We show below, however, that if the scheme is to be simultaneously consistent and conservative, a cell-centred grid (type (i)) must be used. On a uniform grid, it is immaterial which approach is used, although the differential and integral approaches may lead to different forms unless care is taken. On a non-uniform grid, this is no longer true, for if the cell boundaries are midway between the mesh points, the cell centres do not lie at the mesh points, and vice versa. If the system of equations is to be both consistent and conservative, we imply that the value specified at the mesh point be regarded both as a point value and as an average over the cell volume. We consider the differencing of the simple equation

$$\frac{\partial v}{\partial t} = \frac{\partial p}{\partial z} \tag{1.63}$$

2 This section follows the analysis by the author in Pert (1983) on which this discussion is based

such that the quantity $\int v dz$ is conserved. Let the width of cell i be Δz_i, then we require $\sum v_i \Delta z_i$ to be conserved. The condition is only established if the difference

$$\partial p/\partial z \to (p_{i+1} - p_{i-1})/2\,\Delta z_i \tag{1.64}$$

for reasonable temporal integration forms. This representation is a consistent finite-difference form if the spacing between mesh points $(i-1)$ and $(i+1)$ is $2\Delta z_i$. The converse is readily established. Thus, we conclude that a consistent and conservative representation is only possible if the width Δz_i of cell i is half the spacing of neighbouring mesh points $(z_{i+1} - z_{i-1})$, i.e. a cell-centred grid.

a) **Cell-centred grid:** We define the mesh points z_i with cell boundaries $z_{i+1/2}$ such that the boundaries of each cell

$$z_{i+1/2} = 1/2\left(z_i + z_{i+1}\right) \tag{1.65}$$

by means of the cell widths

$$\Delta z_i = \left(z_{i+1/2} - z_{z-1/2}\right) = 1/2\left(z_{i+1} - .z_{i-1}\right) \tag{1.66}$$

In practice, the mesh is defined by the quantities Δz_i and by the boundaries z_b and z_t, only subject to

$$\sum_{i=1}^{N} \Delta z_i = z_t - z_b \tag{1.67}$$

when there are N mesh points, Clearly, since

$$z_2 = 2(z_b + \Delta z\_1), \qquad z_{N-1} = 2(z\_t - \Delta z_N) - z_t \tag{1.68}$$

a further boundary condition specifying the relationship between z_1 and z_2 (or z to z_N) is required to fully define the mesh. A suitable form, which satisfies an appropriate consistency relation at the boundary, is

$$z_1 = z_b + 1/2\Delta z_1, \qquad z_2 + m\Delta z_1 \tag{1.69}$$

or

$$z_{N-1} = z_t - 1/2\Delta z_N, \qquad z_{N-1} = z_t - \Delta z_N \tag{1.70}$$

Only one of these boundary conditions is independent. If by symmetry we require both be simultaneously obeyed,

$$\Delta_1 = 2\sum_{i=2}^{N-1} (-1)^i \Delta z_i \tag{1.71}$$

places an additional constraint on the values of Δz_i allowed. This condition is always satisfied if Δz_i can be expressed as the sum of two terms: one depending on cells $(i-1)$ and i alone and the other on i and $(i+1)$ alone, i.e.

$$\Delta z_i = 1/2\left(\Delta'_{i-half} + \Delta'_{i+1/2}\right) \qquad 2 \le i \le N-1 \tag{1.72}$$

provided $\Delta z'_{half} = \Delta z_1$ and $\Delta z'_{N-1} = \Delta z_N$. Hence, it follows by induction that

$$z_{i+1} = z_i + \Delta z'_{i+1/2} \tag{1.73}$$

and that $\Delta z'_{i+1/2}$ is the mesh point spacing i to $i + 1$

$$1/2\Delta z'_{1/2} + \sum_{i=1}^{N-1} \Delta z'_{i+1/2} + \Delta z'_{N-1/2} = (z_t - z_b) \tag{1.74}$$

The mesh is fully specified by the $(N - 1)$ quantities $\Delta z'_{i+1/2}$, subject to constraint (1.74). Such a form ensures the essential ordering of the mesh, $z_{i+1} > z_i$ if the quantities $\Delta z'_{i+1/2}$ are positive. In contrast, the specification of a set of positive terms Δz_i does not necessarily maintain the ordering, and furthermore may lead to tendency of the odd–even meshes to decouple.

b) **Mesh-centred grid:** We may approach the specification of this grid most generally by the set of mesh spacings

$$\Delta z''_{i+1/2} = z_{i+1} - z_i = 1/2\left(z_{i+1/2} - z_{i-1/2}\right) \tag{1.75}$$

in a similar fashion to that defined in the preceding discussion such that individual cell widths $\Delta z''_i$ are given by

$$\Delta z''_{i+1/2} = 1/2\left(\Delta z''_i + \Delta z''_{i+1}\right) \tag{1.76}$$

and

$$z_{i+half} - z_{i-1/2} + \Delta z''_{i+1/2} \tag{1.77}$$

The set $\Delta z''_i$ are subject to

$$\sum_{i=1}^{N} \Delta z''_{i+1/2} = z_t - z_b \tag{1.78}$$

For some purposes, it is convenient to specify a mesh-centred grid on a cell-centred one. The subsidiary mesh is specified by

$$\Delta z'_{i+1/2} = 1/2\left(\Delta z'_i + \Delta z'_{i+1}\right) \tag{1.79}$$

The subsidiary mesh boundaries are not unique, for operation of the recursion relation (1.79) yields different sets starting from either boundary, z_t or z_b, unless an appropriate condition is obeyed. A useful sufficient condition for this is that the set of spacings on the cell-centred grid be given by

$$\Delta z'_{i+1/2} = 1/2\left(\Delta z' i + \Delta z'_{i+1}\right) \tag{1.80}$$

with respect to a second mesh z'_i, then

$$\Delta z'_{i=1/2} \leq z_i \leq z'_{i+11/2} \tag{1.81}$$

that is if the cell centres of the mesh are correctly ordered with respect to the boundaries of the second subsidiary mesh, then the cell boundaries will lie between the cell centres of the second mesh.

References

Birdsall, C K and A B Langdon. 1991. *Plasma Physics Via Computer Simulation*. Bristol, U.K.: Adam Hilger.

Courant, R and K O Friedrichs. 1948. *Supersonic Flow and Shock Waves*. New York: Interscience.

Courant, R and D Hilbert. 1962. *Methods of Mathematical Physics: Vol II, Partial Differential Equations*. New York: Interscience.

Courant, R, K Friedrichs and H Lewy. 1928. "On the partial difference equations of mathematical physics." *Math Ann* 100:32–74. translated in Courant, Friedrichs and Lewy (1967).

Hockney, R W and J W Eastwood. 1988. *Computer Simulation Using Particles*. Abingdon, U.K.: Taylor and Francis.

Hopf, E. 1969/70. "On the right weak solution of the cauchy problem for a quasilinear equation of first order." *J Math Mech* 19:43–65.

Jennings, G. 1974. "Discrete shocks." *Pure Appl Math* 27:25–37.

Landau, L D and E F Lifshitz. 1959. *Fluid Mechanics*. London: Pergamon.

Lax, P D. 1954. "Weak solutions of nonlinear hyberbolic equations and their numerical computation." *Commun Pure Appl Math* 7:159–193.

Lax, P D and B Wendroff. 1960. "Systems of conservation laws." *Commun Pure Appl Math* 13:217–237.

Oleinik, O A. 1963a. "Construction of a generalized solution of the Cauchy problem for a quasi-linear equation of first order by the introduction of 'vanishing viscosity'." *Am Math Soc Trans Ser 2* 33:277–283.

Oleinik, O A. 1963b. "Uniqueness and stability of the generalized solution of the Cauchy problem for a quasi-linear equation." *Am Math Soc Trans Ser 2* 33:285–290.

Pert, G J. 1983. "Quasi-Lagrangain rezoning of fluid codes maintaining an orthogonal mesh." *J Comput Phys* 49:1–43.

Pert, G J. 2013. *Introduction to Fluid Mechanics*. London: Wiley.

Pert, G J. 2021. *Introduction to Plasma Physics*. London: Wiley.

Quinn, B K. 1971. "Solutions with shocks: an example of an L_1-contractive semigroup." *Commun Pure Appl Math* 24:125–132.

Richtmyer, R D and K W Morton. 1967. *Difference Methods for Initial Value Problems*. New York: Interscience.

Spitzer, Jr, L. 1956. *Physics of Fully Ionized Gases*. New York: Interscience.

Zel'dovich, Ya B and Yu P Raizer. 1966. *Physics of Shock Waves and High Temperature Hydrodynamic Phenomena*. Seventh ed. New York: Academic Press.

Zienkiewicz, O C, R L Taylor and P Nithiarasu. 2014. *Finte Element Method for Fluid Mechanics*. Seventh ed. Kidlington, Oxford, UK: Elsevier India.

2

Analytic and Quasi-Analytic Approximations

2.1 Analytic and Quasi-Analytic Methods

The general complexity of the overall laser–plasma interaction process is exemplified in Figure I. In general, the scope of analytic approximations is limited to simplified problems in which many of the interactions are neglected. Many of these were detailed by the author in Pert (2013) and Pert (2021). The results so generated give good pictures of the individual effects and scaliness of particular physical effects, but are of little value for predictive simulations.

There is one exception to this rule: namely the uniform, linear self-similar model described in detail in Pert (2013, §14.2.1). A brief account taken directly from Pert (2021, App 16.B), but based on Pert (1980) is given below to introduce the method, which has wide applicability. This simple model may be applied to the free expansion of a plasma body with well-defined profile.

2.1.1 Self-Similar Flows with Uniform Velocity Gradient

This model is based on the self-similarity condition so that we are able to separate the dynamic variables into the product of a function depending only on time and a second on the similarity variable. Imposing the second condition, namely a spatial linear velocity gradient

$$\xi_i = x_i/X(t) \tag{2.1}$$

where the set x_i are the position vectors of the Lagrangian fluid elements ζ_i. Clearly, the set of variables ξ_i are fixed for a particular fluid element, i.e. the set ξ_i is a Lagrangian variable and whose value identifies a particular fluid element.

General properties of the model follow:

1) **The fluid velocity v_i is linearly dependent on the spatial co-ordinate x_i:** The velocity of a fluid particle ξ_i is

$$v_i = \left.\frac{dx_i}{dt}\right|_{\xi_i} = \xi_i \frac{dX_i}{dt} = \frac{x_i}{X_i} V_i \tag{2.2}$$

where $V_i = dX_i/dt$.

2) **The flow is symmetric in the co-ordinates ξ_i:** The flow must have symmetry in the co-ordinates x_i, since $v_i(-x_i) = -v_1(x_i)$, etc.
3) **The model is homogeneous:** The mass of a cell dM fixed in the fluid of volume $\mathrm{d}\tau_x$ and the elementary volume $\mathrm{d}\tau_\xi$ of the Lagrangian coordinate element ξ_i. are constant in time. The Jacobian of the transformation between the configuration and Lagrangian spaces is

$$\frac{\mathrm{d}\tau_x}{\mathrm{d}\tau_\xi} = J = \frac{\partial(x_1, x_2, \dots)}{\partial(\xi_1, \xi_2, \dots)} = \prod_{i=1}^{\nu} X_i \tag{2.3}$$

Since $\mathrm{d}M = \rho\,\mathrm{d}\tau_x = \text{const}$ it follows that ρJ is an invariant of motion. Furthermore, since J is a function of time alone, the density ρ is a separable function of time and the Lagrangian space coordinates (similarity variables).

$$\rho = \rho_0(t) f(\xi_i^{\,2})$$

where $f(\xi_1^{\,2}, \xi_2^{\,2}, \dots)$ is represented as $f(\xi_i^{\,2})$ allowing for the symmetry of the function $f(\xi_i^{\,2})$ noted above. The temporal variation of density is determined by the Jacobian $\rho_0(t) = J(0)/J(t) \cdot \rho_0(0)$.
The rate of dilation $\mathrm{d}[\ln(\rho)]/\mathrm{d}t = \dot{J}(t)/J(t)$ is a function of time alone and therefore has the same spatial value everywhere. The flow is therefore homogeneous as the ratio of the volume elements in configuration space for fluid elements of identical size depends only on a factor varying in time.
4) **The thermodynamic variables are all separable:** Therefore, density, pressure, and temperature take the form $\rho(x_i, t) = \rho_0(t) f(\xi_i)$. Euler's equation determines the density and also the pressure distribution. The equation becomes

$$\frac{\partial p}{\partial \xi_i} = -\xi_i f(\xi_i^{\,2})\, \rho_0(t) X_i(t) \frac{\mathrm{d}V_i}{\mathrm{d}t} \tag{2.4}$$

and the pressure is necessarily a separable function.

$$p = p_0(t)\, \phi(\xi_i^{\,2})$$

where $\phi(\xi_1^{\,2}, \xi_2^{\,2}, \dots)$ is represented by $\phi(\xi_i^{\,2})$. It follows from the equation of state that all other thermodynamic state variables are separable as well. Therefore

$$\frac{\partial[\phi(\xi_i^{\,2})]}{\partial(\xi_i^{\,2})} = -\frac{1}{2}\lambda_i f(\xi_i^{\,2}) \qquad \text{and} \qquad p_0(t) = \lambda_i^{-1} \rho_0(t) X_i \frac{\mathrm{d}V_i}{\mathrm{d}t} \tag{2.5}$$

for all i. λ_i are separation constants, which have an arbitrary value and usually determine the relationship of the characteristic scale width X_i to the fluid boundary.
5) **The profiles of the thermodynamic variables are determined by the heating pulse:** The equation of energy conservation in Lagrangian coordinates can be written as

$$\frac{\mathrm{d}\epsilon}{\mathrm{d}t} - \frac{p}{\rho^2}\frac{\mathrm{d}\rho}{\mathrm{d}t} = Q \tag{2.6}$$

where Q is the heat release rate per unit mass and ϵ the specific internal energy. Clearly, Q must be separable

$$Q = Q_0(t)\, q(\xi_i^{\,2}) \tag{2.7}$$

For a polytropic gas where $\epsilon = p/[(\gamma - 1)\rho]$[1]

$$q(\xi_i^2) = \mu\, \phi(\xi_i^2)/f(\xi_i^2)$$

$$Q_0(t) = \mu^{-1}\left\{\frac{1}{(\gamma-1)}\, p_0(t)\dot{p}_0(t) - \frac{\gamma}{(\gamma-1)}\, p_0(t)\,\dot{\rho}(t)/[\rho(t)]^2\right\} \tag{2.8}$$

where μ is a separation constant. This equation together with Eq. (2.5) defines the pressure and density distributions from $q(\xi_i^2)$.

The condition that the heat release function be separable into the product of a function of time $Q_0(t)$ and one of the variable $q(\xi_i^2)$ is therefore clearly a necessary condition for the self-similar condition (2.1) to be upheld, and can also be shown to be sufficient. Flows of this type are therefore self-similar even though the number of dimensional variables may be larger than is normally the case.[2]

The mass M and the total energy $E(t)$ in the gas at any time, which is the sum of the kinetic E_k and thermal E_t energies, are easily found.

$$M = \int \rho\, \mathrm{d}\tau = [J\rho_0(t)] \int f(\xi_i^2)\, \mathrm{d}\tau_\xi \tag{2.9}$$

$$E_k = \int \rho\, \frac{1}{2}\sum_{i=1}^{\nu} v_i^2\, \mathrm{d}\tau = \frac{1}{2}[J\rho_0(t)] \int f(\xi_i^2) \sum_{i=1}^{\nu} \xi_i^2\, V_i^2\, \mathrm{d}\tau_\xi \tag{2.10}$$

$$E_t = \int \rho\, c_v\, T\, \mathrm{d}\tau = \frac{1}{(\gamma-1)} \int p\, \mathrm{d}\tau = \frac{1}{\nu}[J\,\rho_0(t)] \int f(\xi_i^2) \sum_{i=1}^{\nu} \xi_i^2 X_i\, \frac{\mathrm{d}V_i}{\mathrm{d}\tau}\, \mathrm{d}\tau_\xi \tag{2.11}$$

since the pressure is zero at the gas boundary.[3]

Ellipsoidal flows are a useful sub-class of flows, which are particularly easy to evaluate. Provided the initially specified variables have no preferred direction, the flow is spherical in ξ_i space but ellipsoidal in the configuration space x_i. The initial shape of the body may be ellipsoidal, rather than spherical, so that X_i and V_i take different values, but the heat distribution function and the state variables have no preferred direction in Lagrangian space. Thus, these variables are functions of the Lagrangian 'radius'

$$\zeta^2 = \sum_{i=1}^{\nu} \xi_i^2 = \sum_{i=1}^{\nu} \left(\frac{x_i}{X_i}\right)^2$$

The density distribution function f depends only on ζ^2

$$\int \xi_i^2 f(\xi_i^2)\, \mathrm{d}\tau_\xi = \frac{1}{\nu} \int \zeta^2 f(\zeta^2)\, \mathrm{d}\tau_\xi$$

---

1 Note that the separable ratio $p/\rho = P_0(t)\, p(\xi_i)/P_0(t)\, \rho(\xi_i)$ is a mean 'temperature' T_A or alternatively the squared isothermal sound speed C^2.

2 The dimensional parameters include the initial dimensions of the pellet and the heating pulse duration. The simpler problem with an infinitely small mass and heating pulse increasing as a power law in time is easily shown by dimensional analysis to yield the same self-similar form.

3 The thermal energy integral is easily performed in ellipsoidal flow by making use of the identity $\nabla(p\mathbf{X}) = \nu p + \mathbf{X}\cdot\nabla p$, and Euler's equation in the simple form: $\rho\dot{\mathbf{U}} + \nabla p = 0$. Finally, applying Gauss' theorem and setting the surface integral to zero, we obtain: Eq. (2.9)

$$E_t' = \int p\, \mathrm{d}\tau = \frac{1}{\nu}\int \mathbf{X}\cdot\dot{\mathbf{U}}\, \mathrm{d}\tau \tag{2.12}$$

Defining

$$\Psi = \frac{1}{2} \frac{\int \zeta^2 f(\zeta^2)\, \mathrm{d}\tau_\xi}{\int f(\zeta^2)\, \mathrm{d}\tau_\xi}$$

we obtain

$$\sum_{i=1}^{3} \left\{ V_i^2 + \frac{2}{\nu(\gamma-1)} X_i \frac{\mathrm{d}V_i}{\mathrm{d}t} \right\} = \nu \frac{E(t)}{\Psi M} \tag{2.13}$$

where $E(t)$ is the total energy of the gas.

This equation taken together with Eq. (2.5) in the form

$$X_1 \frac{\mathrm{d}V_1}{\mathrm{d}t} = X_2 \frac{\mathrm{d}V_2}{\mathrm{d}t} = \cdots = \lambda \frac{p_0(t)}{\rho_0(t)} \tag{2.14}$$

uniquely defines the flow, in a convenient form for numerical integration. The fluid may have a finite initial energy $E(0)$, and the temporal form of the heating pulse $Q_0(t)$ is subject to the condition that $\lim_{(t\to 0)} Q_0(t) \to A\, t^n$, $n > -1$.

The condition that the spatial heat release function $q(\xi_i^2)$ is constant is met in two simple cases. In ellipsoidal flow, defining the ellipsoidal variable $\zeta^2 = \sum \xi_i^2$, there are two simple conditions consistent with the separability of the heating pulse.

- **Adiabatic expansion:** A plasma pellet is instantaneously heated isentropically and expands with $Q(x, t) = 0$ so that

$$f(\zeta^2) = f_0(1 - \zeta^2/\zeta_0^2)^{1/(\gamma-1)} \tag{2.15}$$

- **Isothermal expansion:** If thermal conduction is strong an expanding plasma pellet maintains a uniform temperature and the density profile takes the form of a Gaussian

$$f(\zeta^2) = f_0 \exp(-\zeta^2/\zeta_0^2) \tag{2.16}$$

  and ζ_0 is the half $1/e$ width of the profile.

The linear self-similar model was introduced by several workers, and widely used, in the decades 1950/1960 (see Pert, 1980, and references therein). The model was developed as a simulation tool by Haught and Polk (1970) to predict the development of the plasma formed from small pellets by laser irradiation, which were used to fill a magnetic trap.[4]

2.1.2 Application to Expanding Plasma

The homogeneous self-similar approximation has proved very valuable for modelling the expansion of small plasmas of limited dimensions r_0 and mass m_0 heated by externally applied laser radiation of total energy. Nemchinov (1964) first drew attention to this application by showing that an initial plasma of 'zero' dimensions heated by a long pulse of

4 The governing equations are easily integrated using standard methods for ordinary differential equations, e.g. in early work we used an Adams–Bashford/Adams–Moulton fourth-order scheme (Lambert (1991, Ch 4) and Stoer and Bulirsch (1991, §7.2.6)) with self-adjusting step-length (Pert, 1974). The model is particularly useful in that several of the additional physical effects shown in Figure I can be linked by using the energy heating term to integrate the separate subroutines within the split-operator approximation (Supplement M.4) (Pert, 1974, 1976).

duration $\tau \gg R_0/\sqrt{E_0/M_0}$ where R_0 is the initial plasma heated in time long compared to the characteristic time. This application follows from the self-similar nature of a hot localised plasma body, although its profile depends on total flow history (Zel'dovich and Raizer, 1966; Pert, 1980). Numerical simulations (Pert, 1980) confirmed Nemchinov's conjecture provided $\tau \gtrsim 10$.

A useful, if unexpected, result for exploding two-dimensional foils when $\gamma = 5/3$ (appropriate to plasma) starting from rest with initial values X_0, Y_0 and $U_0 = 0, V_0 = 0$ is that for late times as $t \to \infty$ it is found that $U(\infty)/V(\infty) = \sqrt{Y_0/X_0}$ (Pert, 1987; Hunter and London, 1988) (see Appendix 2.A.2).

The method has been widely used to model the expansion and heating of small plasma spheres by Haught and Polk (1970), and Pert (1974) and several others. A particularly useful application was demonstrated by London and Rosen (1986) to examine the flow in thin foil X-ray lasers where comparison with detailed hydrodynamic code LASNEX was made and allowed estimates of the accuracy of the model to be made.

2.1.3 Matching Parameter

The most useful application of the self-similar model is for modelling the expansion of laser-heated solids, cylinders and spheres, by applying Nemchinov's conjecture using the isothermal approximation. To that end, we must define an appropriate $1/e$ radius for the model to approximate the radius. The approximation best applies to the asymptotic phase of an expansion, long after the conclusion of a heating pulse. We can compare the asymptotic profile of this prototype approach with the isothermal Gaussian model. We may do this in two ways:

1) Matching the edge of the expanding gas in the model and the prototype thus obtaining from a comparison of the plasma edge with the self-similar model to obtain a scale factor

$$\lim_{T\to\infty} \Gamma_e = R_0/r_0 = R_{edge}(T)/r_{edge}(T) \tag{2.17}$$

where r_0 is the initial $1/e$ width of the object and R_0 the initial radius of the (Gaussian) similarity model defined Lagrangian element, e.g. the $1/e$ mass point where $r_e(0)$ is the initial width of the solid and $R(0)$ the initial radius of the edge of (Gaussian) similarity profile.

2) Matching an internal point whose Lagrangian co-ordinate is the same for both prototype and model to yield a second scale factor

$$\Gamma_d = R_{1/e}/r_{1/e} = \lim_{T\to\infty} R_X(T)/r_X(T) \tag{2.18}$$

where $R_{1/e}$ is the initial value of the $1/e$ mass point X' in the similarity mass distribution and $r_{1/e}$ the equivalent point in the object X. The values obtained by comparing calculations of the similarity model with a one-dimensional von Neumann–Richtmyer (Section 4.2) code are given in Table 2.1.

The two values thus obtained are not equivalent. Γ_e reflects the equivalence at the low density edge of the plasma. On the other hand, Γ_d relates to the higher density plasma at the core. The latter is more representative of the plasma body, but the former may be appropriate for estimating the absorption width from the laser. As may be anticipated, $\Gamma_d > \Gamma_e$

Table 2.1 Matching parameters.

Spherical symmetry					
γ	Γ_e	Γ_d	Ψ	$\langle\rho\rangle/\rho_0$	$\overline{\rho}/\rho_0$
5/3	0.456	0.247	0.0457	189.2	3.60
7/5	0.444	0.283	0.0614	46.8	2.77
9/7	0.458	0.313	0.0735	28.4	2.70
Cylindrical symmetry					
γ	Γ_e	Γ_d			
5/3	0.514	0.287	0.0412	17.0	2.04
7/5	0.490	0.332	0.0550	9.90	1.96
9/7	0.496	0.362	0.0654	7.94	1.96

These results are in good agreement with Fader (1968), who found a value of 0.366 for Γ_e for $\gamma = 5/3$.

A further difficulty arises since the matching parameters vary (albeit weakly) with the heating pulse length. Simulations show that the variation in the output values is small provided the variation in pulse length is small (Pert, 1980, Table 3). Note these results also validate the Nemchinov conjecture provided the heating pulse duration is much larger than the characteristic expansion time.

2.1.3.1 The Entropy Approximation

To try to avoid these difficulties, we considered two limits for the expansion under heating: (Pert, 1987). The nature of the flow is determined by the relationship of the pulse duration τ to the characteristic expansion time $\sqrt{E/M}$, where E is the total absorbed energy and M the plasma mass (see Pert, 1987).

Long pulse, slow heating. Following Nemchinov's conjecture, the asymptotic phase of expansion is established after the pulse is switched on. The established asymptotic (self-similar) profile follows the Gaussian self-similar profile density profile with matching parameters determined directly by the initial values, namely $\Gamma_e \approx 1$ and $\Gamma_d \approx 1/e$.

Short pulse, rapid heating. Consider a centrally symmetric body of gas instantaneously heated. The subsequent expansion takes place along an adiabat characterised by the average specific entropy. An accurate approximation must therefore take place along the same adiabat for both the model and prototype. A good approximation for matching parameters is therefore given by the model radius, which generates the same entropy as the prototype. Using this approximation (Pert, 1987) generated matching parameters for a variety of input conditions.

Consider first the one-dimensional case in plane ($\nu = 1$), cylindrical ($\nu = 2$) and spherical ($\nu = 3$) geometries. The mass per sector(rdn or sterrad) M_0 and its entropy S_0 of the prototype gas at temperature T_0 are

$$M_0 = \frac{1}{\nu}\rho_0 R_0^\nu \qquad S_0 = M_0 c_v \ln(T_0/\rho_0^{\nu-1}) + M_0 K \tag{2.19}$$

Similarly, if we consider the mass and entropy of a sector of gas with self-similar (Gaussian) density profile at temperature T, the mass per sector and entropy are

$$M_s = \frac{1}{2}\Gamma\left(\frac{\nu}{2}\right)\rho' R_s^\nu \qquad S_s = M_s c_v \ln(T_s/\rho_s^{\nu-1}) + M_s K \tag{2.20}$$

where $\Gamma(x)$ is the gamma function.

Setting the mass, density and temperature equal in both prototype M_0 and model M_s, the entropy difference[5]

$$\Delta S = S_0 - S_s \approx (\gamma - 1) M_0 c_v \Gamma\{\ln(1 + \nu/2)\}\, \nu \ln\left[R_0/R_s\right] \tag{2.21}$$

Setting the entropy difference to zero yields

$$\Gamma = \frac{R_s}{R_0} = \frac{1}{\sqrt{e}\left[\Gamma(1+\nu/2)\right]^{1/\nu}} = \begin{cases} 0.6844 & \nu = 1 \\ 0.6065 & \nu = 2 \\ 0.5519 & \nu = 3 \end{cases} \tag{2.22}$$

For an ellipsoidal cylindrical mass of gas with semi-axes X^1, Y^1 and Z^1, we obtain $\Gamma_X = \Gamma_Y = 0.6844$ and $\Gamma_Z = 0.5375$. If the ellipsoid is symmetric, i.e. a cylinder radius $R^1 = \sqrt{X^{1\,2} + Y^{1\,2}}$ that $\Gamma^R = 0.6065 = \sqrt{\Gamma_x \cdot \Gamma_z}$.

In tests, these values seem to be more satisfactory for general use than those given in Table 2.1, although it should be remembered that, fortunately, the final results are not overly sensitive to the exact values of Γ_d used.

Appendix 2.A The Nemchinov Conjecture

Consider a plasma body of mass M and zero energy $E(0)$ in a ν-dimensional space with infinitely small thickness. The fluid is heated by a pulse of energy $E(t)$, whose power increases with time as a power law $\dot{E} = \alpha t^n = AW\zeta t^n$, where A is a dimensional constant and $W(\zeta)$ is distributed through the mass to cells ζ, subject to the condition that the index $n \geq 0$.

The only dimensional parameters are A and M which, together with the independent variables r and t, yield **only one**-dimensionless product $\xi = r/\{(A/M)t^n\}^{1/2}$. Therefore, dimensional analysis tells us that the motion is self-similar with characteristic radius $r = \{(A/M)t(n+2)\}g(\zeta)$, where $g(\zeta)$ is an unknown function defined for each Lagrangian element (point) ζ in the fluid.

Comparing this result with Eq. (2.2) we see that the velocity gradient is linear. Therefore it follows from our previous discussion (Section 2.1) that the motion is that of a homogeneous

5 We cannot strictly set $M_0 = M_s$ and $\rho_0 = \rho_s$, simultaneously; however, since ρ_0 and ρ_s only appear logarithmically the error incurred is small.

self-similar uniform self-similar flow. From this result, Nemchinov (1964) conjectured that the expansion of a small body of gas heated by a pulse, which is long compared to the characteristic time $R/\sqrt{E/M}$, will be self-similar, which we have shown earlier by direct calculation to be indeed correct (see Pert, 1980, fig. 5(a)).

Once the flow is established ($t > 0$), then the density is established by the nature of the plasma assumed to be either the characteristic profile for:

1) either adiabatic expansion. No thermal conduction:

$$f(\zeta^2) = f_0 \left(1 - \zeta^2 / {\zeta_0}^2\right)^{1/(\gamma-1)} \tag{2.15'}$$

2) or isothermal expansion. Strong thermal conduction:

$$f(\zeta^2) = f_0 \exp\left(-\zeta^2/{\zeta_0}^2\right) \tag{2.16'}$$

as before.

Generally, for laser-heated plasmas the isothermal condition is appropriate.

Appendix 2.B The Energy Integral

We now examine the direct development of key physical concepts as a consequence of the dynamic equations of the homogeneous similarity model using only mathematics. In two dimensions, this leads to two adiabatic invariants. The conservation of energy in a homogeneous expansion is readily established.

The temporal development of the uniform (holomorphic) self-similar model is defined by a set of dynamic equations that adhere to appropriate physical constraints e.g. number of dimensions, thermodynamic state, etc. The conservation of energy in a homogeneous expansion is readily established.

The set of simultaneous partial differential equations

$$X\frac{dU}{dt} = Y\frac{dV}{dt} = Z\frac{dW}{dt} = C(t)^2 \tag{2.14a}$$

admits three integrals in the form of a pair of invariants, provided the underlying physical constraints, e.g. mass, energy, and thermodynamic states (isothermal, isentropic, etc.) are constant.

In two dimensions, these equations represent the self-similar (homologous) expansion of a strip or disc with characteristic widths $X(t)$ and $Y(t)$ with characteristic speeds $U(t)$ and $V(t)$, respectively, in the X- and Y-directions.

In adiabatic motion with no external heat source, the isothermal sound speed $C(t)$ is determined solely by the cooling $C(0) \to C(t)$ due to adiabatic expansion from $t = 0$ to t,

$$C(t)^2 = \left[\frac{X(0)Y(0)Z(0)}{(X(t)Y(t)Z(t))}\right]^{\gamma-1} C(0)^2 = \prod_i \left[\frac{X_i(0)}{X_i(t)}\right]^{(1/(\gamma-1))} C(0)^2 \tag{2.B.1}$$

where $C(t) = \sqrt{p/\rho}$ is the isothermal sound speed at time t. The dynamical equation of the gas is instantaneously heated with energy $E(0)$, such that $E(0) = C_0^2/(\gamma - 1) = \text{const}$. There exist two exact integrals of the motion in the form of *invariants* subject to zero external heat input:

1) **The first (or energy) invariant**: Making use of the identity $\mathrm{d}\ddot{x} = \frac{1}{2}\mathrm{d}\dot{x}^2/\mathrm{d}x$ and substituting from (2.14a) and defining $\mathrm{d}E_k$ and $\mathrm{d}E_t$

$$\mathrm{d}E_k = \sum_i X_i \dot{U}_i = \frac{1}{2}\left\{\frac{1}{X}\frac{\mathrm{d}U^2}{\mathrm{d}X}\mathrm{d}X + \frac{1}{Y}\frac{\mathrm{d}V^2}{\mathrm{d}Y}\mathrm{d}Y + \frac{1}{Z}\frac{\mathrm{d}W^2}{\mathrm{d}Z}\mathrm{d}Z\right\} \tag{2.B.2}$$

$$= \frac{1}{2}\sum_i \frac{1}{X_i}\frac{\mathrm{d}\dot{X}_i^2}{\mathrm{d}X_i}\mathrm{d}X_i \tag{2.B.3}$$

$$\mathrm{d}E_t = -C(t)^2 \frac{1}{(\gamma-1)}\sum_i \frac{1}{X_i}\mathrm{d}X_i \tag{2.B.4}$$

We obtain the total differential $\mathrm{d}E = \mathrm{d}E_U + \mathrm{d}E_V + \mathrm{d}E_W = 0$, since there is no heat input. Substituting for $C_0(t)$ from Eq. (2.B.1), the total can be integrated to yield the total fluid energy per unit weighted mass, which is constant $E(t) = E(0) = \frac{1}{(\gamma-1)}C_0^2$ and give the first or energy invariant

$$\Phi = E(t) = \frac{1}{2}\left(U^2 + V^2 + W^2\right) + \frac{1}{(\gamma-1)}\prod_i\left[\frac{X(0)\,Y(0)\,Z(0)}{X(t)\,Y(t)\,Z(t)}\right]^{(\gamma-1)} C(0)^2 = E(0) \tag{I}$$

which we recognise as the conservation of energy (Eq. (2.11)). Differentiating with respect to time

$$\dot{\Phi} = (U\dot{U} + V\dot{V} + W\dot{W}) - \frac{1}{(\gamma-1)}C^2\left(\dot{X}/X + \dot{Y}/Y + \dot{Z}/Z\right) = 0 \tag{2.B.5}$$

Proof 2.1 We note that this result is a direct consequence of Eq. (2.14a) subject only to the constraint that the motion is adiabatic following an instantaneous heating subject to a homologous adiabatic expansion with no additional physics included.

2) **The second invariant**: Limiting the problem to two dimensions only (X, Y), we have only two degrees of freedom and therefore only two invariants. When the adiabatic index $\gamma = 5/3$ for plasma, the second invariant is[6]:

$$\begin{aligned}\Psi(t) &= U(t)\,V(t)\,[X(t)\,V(t) - Y(t)\,U(t)]\\ &\quad + 3\left[C(0)^2\,(X(0)\,Y(0))/(X(t)\,Y(t))\right]^{2/3}(X(t)\,U(t) - Y(t)\,V(t))\end{aligned} \tag{II}$$

Define

$$\Psi(t) = \underbrace{U(t)V(t)\,(X(t)V(t) - Y(t)U(t))}_{\Psi_1(t)} + 3\,\underbrace{C^2\,(X(t)U(t) - Y(t)V(t))}_{\Psi_2(t)} \tag{2.B.6}$$

To verify this invariant, we differentiate with respect to time.

6 This result is due to Hietarinta, see Hunter and London (1988, and references therein).

Proof 2.2 Writing

$$C^2(t) = C(0)^2 \left[(X(0)Y(0)) / (X(t)Y(t))\right]^{2/3} \tag{2.B.7a}$$

and differentiating

$$\frac{dC^2}{dt} = -\frac{2}{3}C^2\left(\frac{\dot{X}}{X} + \frac{\dot{Y}}{Y}\right) = -\frac{2}{3}C^2\left(\frac{U}{X} + \frac{V}{Y}\right) \tag{2.B.7b}$$

The derivative of $C^2(XU - YV)$ is required for the second invariant

$$\begin{aligned}
\frac{d\Psi_1}{dt} &= \frac{d}{dt}\{UV(XV - YU)\} \\
&= \left[\dot{U}V + U\dot{V}\right](XV - YU) + UY\left(X\dot{V} - Y\dot{U}\right) + \cancel{UV(UV - VU)} \\
&\qquad + C^2UV(X/Y - Y/X) + (\dot{U}V + U\dot{V})(XV - YU) \\
&= -\left[C^2\left[U^2 - UV(X/Y - Y/X)\right] - C^2\left[V^2 - UV(X/Y - Y/X)\right]\right] \\
&= -C^2\left(U^2 - V^2\right) - 2C^2UV\left(X^2 - Y^2\right)/(XY)
\end{aligned} \tag{2.B.8a}$$

and also

$$\begin{aligned}
\frac{d\Psi_2}{dt} &= \frac{d}{dt}\left[C^2(XU - YV)\right] \\
&= -\frac{2}{3}C^2(U/X + V/Y)(XU - YV) + C^2\left[\dot{X}U + X\dot{U} - \left(\dot{Y}V + Y\dot{V}\right)\right] \\
&= -\frac{2}{3}C^2\left(U^2 - V^2\right) - \frac{2}{3}C^2UV\left(\frac{Y}{X} - \frac{X}{Y}\right) + \left(U^2 - V^2\right) \\
&= \frac{1}{3}C^2\left(U^2 - V^2\right) - \frac{2}{3}C^2UV\left(\frac{Y}{X} - \frac{X}{Y}\right)
\end{aligned} \tag{2.B.8b}$$

Adding Eqs. (2.B.8a) and (2.B.8b) we see that

$$\frac{d\Psi}{dt} = \frac{d\Psi_1}{dt} + 3\frac{d\Psi_2}{dt} = 0 \tag{2.B.9}$$

and hence that the invariant $\Psi(t)$ is constant.

Note the term Ψ, Eq. (II) is only an invariant when the flow is two-dimensional and $\gamma = 5/3$, whereas the first invariant Φ, Eq. (I) is constant in two or three dimensions at arbitrary values of γ. Further invariants should exist more generally for different values of γ but have not yet been found (Hunter and London, 1988).[7]

2.B.1 Asymptotic Phase

Self-similar expansion with no heat input enters its asymptotic phase with constant asymptotic velocity. Consequently, we have $U(\infty) = \text{const} = U_\infty$ and $V(\infty) = \text{const} = V_\infty$. Hence, as $t \to \infty$ the dimensions $X \sim U(\infty)t$ and $Y \sim V(\infty)t$. Since the invariants are general integrals of motion, they must hold as the flow becomes infinitely extended. If the initial state is one of rest $U(0) = V(0) = 0$, the second invariant $\Psi = 0$. But the invariant

7 The derivation of the invariance of the second invariant in Section 2.B.8 suggests that other invariants may be found for different values of γ in similar manner, but it is easily seen to be a particular property of the value of γ when $\gamma = 5/3$.

scales as $\Psi(t) \sim t^{-7/3}$. This can only be reconciled with the invariant condition $\Psi \sim$ const if $X(0)Y(0)/U_\infty V_\infty \sim$ const, by requiring that

$$\frac{U(\infty)}{V(\infty)} \Big/ \left(\frac{Y(0)}{X(0)}\right)^{1/2} \sim \text{const} \tag{2.B.10}$$

as found earlier in numerical studies (Pert, 1987; Hunter and London, 1988).

The values of the asymptotic velocities are easily obtained from the first (energy) invariant

$$U(\infty) = \sqrt{\frac{Y(0)}{(X(0)+Y(0))}C(0)^2} \quad \text{and} \quad V(\infty) = \sqrt{\frac{X(0)}{(X(0)+Y(0))}C(0)^2} \tag{2.B.11}$$

References

Fader, W J. 1968. “Hydrodynamic model of a spherical plasma produced by Q-spoiled laser irradiation of a solid particle.” *Phys Fluids* 11:2200–2208.

Haught, A V and D Polk. 1970. “Formation and heating of laser irradiated solid particle plasmas.” *Phys Fluids* 19:2825.

Hunter, J H and R A London. 1988. “Multidimensional similarity models for exploding foils.” *Phys Fluids* 31:3102–3108.

Lambert, J D. 1991. *Numerical Methods for Ordinary Differential Systems*. Chichester, Sussex, U.K.: Wiley.

London, R A and M D Rosen. 1986. “Hydrodynamics of exploding foil X-ray lasers.” *Phys Fluids* 29:3813–3822.

Nemchinov, I V. 1964. “Dissipation of a heated mass of gas in a regular regime.” *Zh Prik Mekh Tekh Fiz* 5:18–29. Translation by Sandia Corp.

Pert, G J. 1974. “Some calculations of the heating of small pellets using the similarity model.” *Plasma Phys* 16:1051–1068.

Pert, G J. 1976. “Model calculations XUV gain+ in rapidly expanding cylindrical plasmas.” *J Phys B: Atom Mol Phys* 9:3301–3315.

Pert, G J. 1980. “Self-similar flows with uniform velocity gradient and their use in modelling the free expansion of polyropic gases.” *J Fluid Mech* 100:257–277.

Pert, G J. 1987. “The use of flows with uniform velocity gradient in modelling free expansion of polyropic gas.” *Lasers Part Beams* 5:643–658.

Pert, G J. 2013. *Introduction to Fluid Mechanics*. London: Wiley.

Pert, G J. 2021. *Introduction to Plasma Physics*. London: Wiley.

Stoer, J and R Bulirsch. 1991. *Introduction to Numerical Analysis*. Second ed. New York: Springer.

Zel’dovich, Ya B and Yu P Raizer. 1966. *Physics of Shock Waves and High Temperature Hydrodynamic Phenomena*. Seventh ed. New York: Academic Press.

3

Numerical Fluid Dynamics in the Eulerian Scheme

3.1 An Introduction to Steady-State Engineering Design Models

We will now establish workable algorithms to handle material and energy transport. The most elegant way of applying algorithms to handle inviscid compressible flow is through a general parameter and flux vectors. In two dimensions these have the form

$$\frac{\partial u}{\partial t} + \frac{\partial f}{\partial x} + \frac{\partial g}{\partial y} = 0 \tag{3.1}$$

where E

$$u = \begin{pmatrix} \rho \\ \rho \mathrm{u} \\ \rho \mathrm{v} \\ \rho E \end{pmatrix} \qquad f = \begin{pmatrix} \rho \mathrm{u} \\ \rho \mathrm{u}^2 + p \\ \rho \mathrm{uv} \\ \rho \mathrm{u} H \end{pmatrix} \qquad g = \begin{pmatrix} \rho \mathrm{v} \\ \rho \mathrm{vu} \\ \rho \mathrm{v}^2 + p \\ \rho \mathrm{v} H \end{pmatrix} \tag{3.2}$$

Here u and v are the x and y components of the flow velocity, respectively, and E the total specific energy (including both the internal energy ϵ and the kinetic energy)

$$E = \epsilon + \frac{1}{2}(\mathrm{u}^2 + \mathrm{v}^2) \tag{3.3}$$

and H the specific total enthalpy (including kinetic energy)

$$H = \mathrm{h} + \frac{1}{2}(\mathrm{u}^2 + \mathrm{v}^2). \tag{3.4}$$

For a polytropic gas, the pressure

$$p = (\gamma - 1)\, \rho\, \epsilon = (\gamma - 1)\, \rho \left[E - \frac{1}{2}(\mathrm{u}^2 + \mathrm{v}^2)\right] \tag{3.5}$$

This formulation is extremely compatible with the Lax–Wendroff/MacCormack schemes with explicit artificial dissipation. The approach is widely used for handling relatively smooth, weakly time-varying problems with complex geometrical boundaries, such as those encountered in aeronautical engineering. The artificial dissipation is introduced through additional terms into the x and y components of the fluxes, f and g.

Modelling and Simulation in Plasma Physics for Physicists and Mathematicians, First Edition. Geoffrey J. Pert.

Implicit artificial dissipation methods, on the other hand, are generally used where the problem is complex, possibly broken down into a number of interacting sub-systems, and often strongly time varying, but which are usually taking place in a simple geometrical environment, such as those occuring in plasma physics. These methods invariably require some type of donor cell transport in terms of the cell boundary velocity to generate the necessary numerical diffusion, which constitutes the dissipation; and often also include an artificial viscosity of the von Neumann/Richtmyer type to provide additional dissipation. In consequence, these methods normally cannot be easily formulated in terms of the general set of conservation laws described earlier. In these cases, it is usually convenient to treat only the convective terms directly, i.e.

$$u = \begin{pmatrix} \rho \\ \rho \mathrm{u} \\ \rho \mathrm{v} \\ \rho E \end{pmatrix} \quad f = \begin{pmatrix} \rho \mathrm{u} \\ \rho \mathrm{u}^2 \\ \rho \mathrm{uv} \\ \rho u \end{pmatrix} \quad g = \begin{pmatrix} \rho \mathrm{v} \\ \rho \mathrm{vu} \\ \rho \mathrm{v}^2 \\ \rho \mathrm{v} E \end{pmatrix} \tag{3.6}$$

i.e. $f = \mathrm{u}u$ and $\mathrm{g} = \mathrm{v}u$. The remaining terms are associated with the acceleration caused by the pressure gradients, and the corresponding mechanical work are treated separately, essentially within a Lagrangian formulation. Thus, the acceleration and convection parts of the flow are separated in a split time-step structure (Richtmyer and Morton, 1967). This separation is particularly convenient when the hydrodynamic flow is only one component of a much more complex dynamical system. In this case, the hydrodynamics is used as the basic structure to which other physical sub-systems are attached. A simple example is demonstrated by flow which incorporates thermal conduction and is easily treated within the 'Lagrangian' phase.

It is not sensible to attempt to draw any firm conclusions as to the merit of explicit versus implicit artificial dissipation schemes. Clearly, different approaches are better suited to different classes of problems. It appears that for relatively smooth flows, as in aeronautical design, Jameson's method has proved very satisfactory and is widely used. For more discontinuous problems with strong time-varying shocks, implicit methods seem better, but the best form is still controversial. Much present work indicates van Leer's form of the limiter, for example using Young's prescription, is simple and easy to programme, and seems to be only weakly damaging due to clipping. On the other hand, the most popular form at present is flux-corrected transport. Neither gives an entirely satisfying solution, but both are satisfactory.

3.1.1 Steady-State Problems

In many calculations, for example in aeronautics – it is required to find the final steady flow configuration satisfying a set of boundary conditions. It is immaterial how the intermediate steps, which are used to find this solution, are developed. However, it is clear from the non-linear nature of the problem that some form of iteration must be used.

The most direct approach is to time-march an Eulerian code, when we effectively iterating the solution. Such an approach is equivalent to using a Jacobi-type iteration process with the time-step as an iteration parameter. This is a satisfactory and popular procedure. Since we iterate until the time variation is reduced to zero, the accuracy depends solely on the spatial terms. However, the speed of convergence will vary with the method used. Ideally, therefore, we would wish to use a method that allows as large a step as possible.

Typically, a modified Lax–Wendroff scheme, such as MacCormack's, can be used, based on the full set of conservation laws from purely hydro-dynamic equations with artificial dissipation (Jameson, 1984). Such an approach is widely used, but MacCormack's scheme suffers from a number of minor inconveniences. These are associated with the order of the method (second), the ability to handle cells of widely varying size on the mesh and the symmetry with respect to upwind and downstream differences. The stepping is also limited by stability constraints to Courant numbers $\lambda \leq 1$.

Jameson, Schmidt and Turkel (1981) introduced an improved method based on fourth-order Runge–Kutta time-stepping and control volume spatial differencing. Thus, we consider a cell i of volume ΔV_i with a set of faces j of area $\Delta S_{i,j}$ and directly apply the conservation law in the differential form

$$\frac{\mathrm{d}}{\mathrm{d}t}\{u_i\,\Delta V_i\} + \sum_{j=1}^{6} f_{i,j}\,\Delta S_{i,j} = 0 \tag{3.7}$$

where $f_{i,j}$ is an average of the flux component of u over the face i, j. The artificial dissipation is included in the flux f. The control volume approach has several advantages on a non-Cartesian grid, offering direct differencing and facilitating easy application of the boundary conditions at a non-planar surface, for example an aerofoil.

Equation (3.7) forms a set of ordinary differential equations for the variables $u_i(t)$ in time, which can be integrated by one of the standard methods for ordinary differential equations. Since it is advantageous to have a scheme, which is self-starting and easy to adjust the time-step, multi-stage two-level Runge–Kutta-type schemes have clear advantages over predictor–corrector ones, and have therefore been adopted. Jameson's fourth-order scheme can expressed as the solution to the equation set at each mesh point:

$$\frac{\mathrm{d}v_i}{\mathrm{d}t} + Q_i = D_i \tag{3.8}$$

where F_i is the sum of outgoing convective and D_i the incoming dissipation fluxes. The fourth-order Runge–Kutta scheme is then

$$\begin{aligned}
v^{(0)} &= v^n \\
v^{(1)} &= v^{(1)} - \frac{1}{4}Dt\{Q^{(0)} - D^{(0)}\} \\
v^{(2)} &= v^{(2)} - \frac{1}{3}Dt\{Q^{(1)} - D^{(1)}\} \\
v^{(3)} &= v^{(3)} - \frac{1}{2}Dt\{Q^{(2)} - D^{(2)}\} \\
v^{(4)} &= v^{(4)} - \frac{1}{4}Dt\{Q^{(3)} - D^{(3)}\} \\
v^{(n+1)} &= v^{(4)}
\end{aligned} \tag{3.9}$$

We can reduce the computational work by setting $D^{(4)} = D^{(3)} = D^{(2)} = D^{(1)} = D^{(0)}$. This method has the further advantage of relaxing the Courant–Friedrichs–Lewy time-step constraint for stability to a Courant number $\lambda \leq 2\sqrt{2}$.

We can also use implicit methods where there is no Courant limit for stability, although it may be advisable to introduce a limiting time-step to maintain the solution smooth. In this form, we write the fluid equations in the form of Eq. (1.12) as

$$\frac{\mathrm{d}v}{\mathrm{d}t} = F(v) \tag{3.10}$$

Thus, defining the change $\Delta v = v^{(n+1)} v^n$ and linearising the flux term, we obtain an implicit equation for the set Δv

$$\Delta v = v^n + F(v^n)Dt + A\Delta v^{(n+1)} \tag{3.11}$$

where $A = \frac{\partial F}{\partial v} Dt$ is an $N \times N$ five-diagonal matrix. The solution of this equation can be made reasonably efficient by using preconditioned matrix inversion techniques, e.g. Incomplete Cholesky Conjugate Gradient (ICCG).

3.1.2 Multi-Grid Techniques

A general problem with the methods described above is that with fine mesh resolution, the Courant–Friedrichs–Lewy condition requires a small 'time-step' for stability, i.e. we must use a small iteration parameter if the iteration process is to be stable. This is computationally expensive. On the other hand, if we are prepared to sacrifice accuracy, we can use a large grid and iterate more rapidly. Clearly, we need to try to compromise these conditions to improve the iteration performance, but at the same time, retaining the accuracy of the final result.

A solution to this problem is known as multi-gridding. We use a large coarse grid with large 'time-steps' to rapidly expel large-scale disturbances from the system. Interpolating onto a smaller grid introduces errors, which cannot be rapidly dispersed and should be locally damped, thereby dissipating the high-frequency modes.

In steady flow calculations, we structure the fluid equations in the form of a set of residuals $R(u_i)$, which we require to vanish as the iteration proceeds:

$$R(u) = \frac{1}{\Delta V_i}[Q(u_i) - D(u_i)] \longrightarrow 0 \tag{3.12}$$

where $Q(u_i)$ is the hydrodynamic flux and $D(u_i)$ the artificial dissipation flux integrated over all cell faces, and ΔV_i the cell volume (area in two dimensions). In this form, the fluid equation (3.7) becomes

$$\frac{\mathrm{d}u_i}{\mathrm{d}t} + R(u_i) = 0 \tag{3.13}$$

A typical multi-grid calculation works on a series of meshes, each of which is twice the spacing of the previous one. The flow variables are integrated over a 'time-step' and transferred directly to the next coarser grid taking care to maintain the conservation relations, i.e. we use initial values on the coarse grid mesh $u_{2h}^{(0)}$ determined by those from the previous 'time-step' on the finer mesh u_h, where

$$u_{2h}^{(0)} = \sum \Delta V_h\, u_h / \Delta V_{2h} \tag{3.14}$$

where the sum is taken over the 4 cells (in two dimensions) or 8 cells (in three dimensions) on the finer grid, which fill that on the coarse grid.

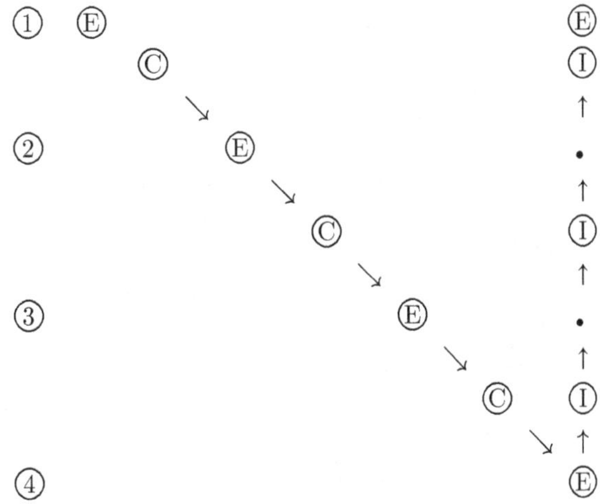

Figure 3.1 Arrangement of a four-step multi-grid stage with a sequence of Euler hydrodynamic calculations (E) on a coarsened grids (C) followed by interpolation (I) back onto the original fine mesh.

A successful strategy is to use a 'saw-tooth' formulation (Figure 3.1) in which a succession of Euler calculations Ⓔ are truncated onto progressively coarser grids, Ⓒ until the coarsest grid is reached. A series of interpolations Ⓘ then take the results back onto the finest grid without any intermediate integration. The same Courant number is used on each grid to generate the 'time-step', so that progressively larger time-steps are used on each coarser grid. In two dimensions, the number of cells scales inversely as the square of the size factor; as a result, the CPU time required for a complete two-dimensional multi-grid cycle is only

$$1 + \frac{1}{4} + \frac{1}{16} + \cdots \leq \frac{4}{3} \tag{3.15}$$

compared to that required for a single step on the fine grid, yielding a substantial increase in the effective iteration time (15 times in the particular four-step scheme illustrated). Multi-grid methods thus offer a major saving in computer time for steady-state problems.

3.2 Spatial Differencing

Thus far, we have only considered problems within simple geometrical configurations. However, application to practical systems generally involves flow around bodies with complex shapes, for example aeroplanes. In this case, it is desirable to attempt to make the mesh reflect the expected behaviour of the flow. In particular, we would like to have a fine resolution near the surface of the body, where the flow gradients are large. However, it would require a very large number of cells and therefore be computationally very expensive to treat the entire flow with a small mesh limited in this way. Far from the body, the flow variation is slow and can tolerate larger cells. There is therefore a requirement for non-uniform spatial meshes. The control volume approach for establishing the spatial finite differences, described earlier, is well suited to this condition. Some further practical constraints which influence the choice of mesh are as follows:

- The mesh should be parallel to the body in contact with the surface. This simplifies the treatment of the boundary condition of zero velocity normal to the surface.

- An orthogonal mesh reduces the number of spatial terms required for accurate differencing and improves the accuracy of the spatial representation.
- The mesh should attempt to follow the expected streamlines. This allows the variation of mesh dimension required for good sampling, discussed earlier, to be taken into account.

In aeronautical modelling, this has led to three characteristic grid forms illustrated in Figure 3.2. The C-grid is particularly suited to representation of the flow around aerofoils, matching the incoming undisturbed flow and the outgoing wake. For other components of the complete aeroplane, other grids may be appropriate.

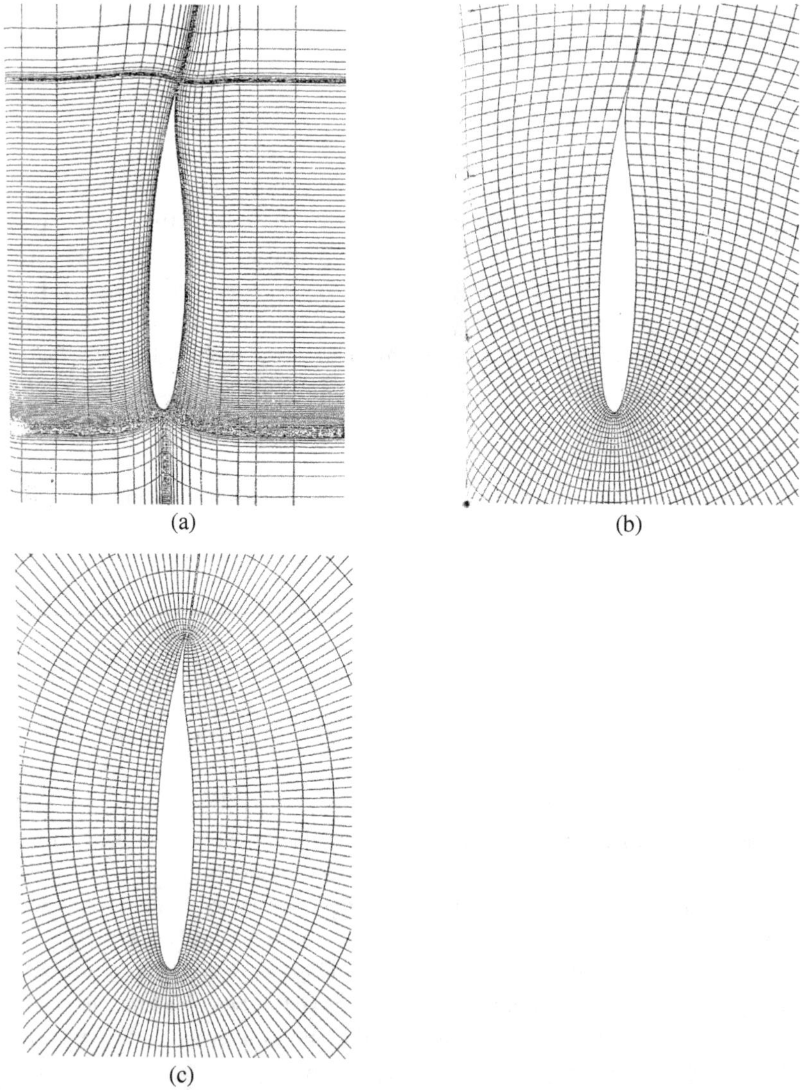

Figure 3.2 Representative forms of the standard grid configurations used for aeronautical modelling. (a) H-grid, (b) C-grid, and (c) O-grid.

Grid generation is a complex topic, more of an art than science. Since ideally we require the mesh to be approximately orthogonal, direct methods can be used. These can be numerical where we solve a set of partial differential equations with boundary conditions matching those of our problem. For example in two dimensions, the set of lines $\phi(x,y) = const.$ and $\psi(x,y) = const.$ are each established by functions $\phi(x,y)$ and $\psi(x,y)$, each of which satisfies Laplace's equation $\nabla^2\phi = \nabla^2\psi = 0$. Alternatively, we may use algebraic methods to interpolate internal points from the boundary data. As an example, in two dimensions, it may be convenient to use the fact that the if complex function $\phi + \imath\psi$ is analytic, ϕ and ψ form an orthogonal co-ordinate set.

3.2.1 Equipotential Zoning

An effective method of zoning is based on assigning the grid lines by means of potential functions using a set of ordered boundary values (Crowley, 1962). Thus, the grid points are defined by the integral values of the potential functions (ϕ, ψ) within the space (x, y) with values $k = 0, 1, 2, 3, \ldots$ and $l = 0, 1, 2, 3, \ldots$ on the boundary. Thus, we need to solve Laplace's equations

$$\nabla^2\phi = \nabla^2\psi = 0 \tag{3.16}$$

and take the mesh lines as the contours $\phi = k = 0, 1, 2, 3, \ldots$ and $\psi = l = 0, 1, 2, 3, \ldots$. Winslow (1963) showed how this could be achieved effectively and efficiently. Following this method, we transform from the (x, y) plane to the (hodograph) plane (ϕ, ψ) using the usual transformation based on Jacobians to get

$$\begin{aligned} \alpha\frac{\partial^2 x}{\partial\phi^2} - 2\beta\frac{\partial^2 x}{\partial\phi\partial\psi} + \gamma\frac{\partial^2 x}{\partial\psi^2} = 0 \\ \alpha\frac{\partial^2 y}{\partial\phi^2} - 2\beta\frac{\partial^2 y}{\partial\phi\partial\psi} + \gamma\frac{\partial^2 y}{\partial\psi^2} = 0 \end{aligned} \tag{3.17}$$

where

$$\alpha = \left(\frac{\partial x}{\partial\psi}\right)^2 + \left(\frac{\partial y}{\partial\psi}\right)^2 \tag{3.18}$$

$$\beta = \frac{\partial x}{\partial\psi}\frac{\partial x}{\partial\phi} + \frac{\partial y}{\partial\psi}\frac{\partial y}{\partial\phi} \tag{3.19}$$

$$\gamma = \left(\frac{\partial x}{\partial\phi}\right)^2 + \left(\frac{\partial y}{\partial\phi}\right)^2 \tag{3.20}$$

provided the Jacobian $J = \frac{\partial x}{\partial\psi}\frac{\partial y}{\partial\phi} - \frac{\partial y}{\partial\psi}\frac{\partial x}{\partial\phi}$ is non-zero.

To solve these equations on a mesh with quadrilateral elements, we use the finite-difference approximations

$$\frac{\partial^2 x}{\partial\phi^2} = x_{k-1,l} - 2x_{k,l} + x_{k+1,l} \tag{3.21}$$

$$\frac{\partial^2 y}{\partial\phi^2} = y_{k-1,l} - 2y_{k,l} + y_{k+1,l} \tag{3.22}$$

$$\frac{\partial^2 x}{\partial\psi^2} = x_{k,l-1} - 2x_{k,l} + x_{k,l+1} \tag{3.23}$$

$$\frac{\partial^2 y}{\partial \psi^2} = y_{k,l-1} - 2y_{k,l} + y_{k,l+1} \tag{3.24}$$

$$\frac{\partial^2 x}{\partial \phi \partial \psi} = \frac{1}{4}\left(x_{k-1,l-1} - x_{k+1,l-1} - x_{k-1,l+1} + x_{k+1,l+1}\right) \tag{3.25}$$

$$\frac{\partial^2 y}{\partial \phi \partial \psi} = \frac{1}{4}\left(y_{k-1,l-1} - y_{k+1,l-1} - y_{k-1,l+1} + y_{k+1,l+1}\right) \tag{3.26}$$

Hence, we obtain

$$\alpha_{k,l} = \left(x_{k,l+1} - x_{k,l-1}\right)^2 + \left(y_{k,l+1} - y_{k,l-1}\right)^2 \tag{3.27}$$

$$\beta_{k,l} = \frac{1}{2}\left[\left(x_{k,l+1} - x_{k,l-1}\right)\left(x_{k+1,l} - x_{k-1,l}\right) + \left(y_{k,l+1} - y_{k,l-1}\right)\left(y_{k+1,l} - y_{k-1,l}\right)\right] \tag{3.28}$$

$$\gamma_{k,l} = \left(x_{k+1,l} - x_{k-1,l}\right)^2 + \left(y_{k+1,l} - y_{k-1,l}\right)^2 \tag{3.29}$$

Finally, substituting these terms into the governing equations (3.17) and solving we obtain

$$\begin{aligned} x_{k,l} &= \frac{1}{2(\alpha_{k,l} + \gamma_{k,l})}\{\alpha_{k,l}(x_{k+1,l} + x_{k-1,l}) + \gamma_{k,l}(x_{k,l+1} + x_{k,l-1}) \\ &\quad + \beta_{k,l}(x_{k+1,l-1} - x_{k+1,l+1} + x_{k-1,l+1} - x_{k-1,l-1})\} \\ y_{k,l} &= \frac{1}{2(\alpha_{k,l} + \gamma_{k,l})}\{\alpha_{k,l}(y_{k+1,l} + y_{k-1,l}) + \gamma_{k,l}(y_{k,l+1} + y_{k,l-1}) \\ &\quad + \beta_{k,l}(y_{k+1,l-1} - y_{k+1,l+1} + y_{k-1,l+1} - y_{k-1,l-1})\} \end{aligned} \tag{3.30}$$

The calculation proceeds by applying these formulae iteratively to solve progressively for $(x_{k,l}, y_{k,l})$ using the set from the previous iteration to calculate the next. This is usually done as a Gauss–Siedel iteration, updating the values of α, β and γ as the new values are obtained.

The method is remarkably effective. Although the values of $(x_{k,l}, y_{k,l})$ finally obtained are weighted averages of their eight neighbours, the nature of the weighting tends to generate approximately square cells of similar size. If the user requires cells of differing size in different parts of the mesh, this can be achieved by imposing specified additional weights into the averages (Winslow, 1981).

The mesh is only orthogonal if $\beta_{k,l}$ is zero everywhere. In general, this is unlikely to occur; however, the nature of the iteration is to reduce $\beta_{k,l}$.

3.2.2 Multi-Block Techniques

Single grids can be chosen for most two-dimensional applications. However, when we extend the calculations to consider the flow in three dimensions around, say a complete aircraft, a single grid with fine resolution around the aircraft profile becomes prohibitively expensive to use. To overcome this problem, multi-block techniques have been devised, whereby different sections of the aircraft are modelled with separate grids, which are then joined satisfying necessary conditions of continuity of the values and fluxes of the flow parameters. The methodology is relatively straightforward, although the programming is complex. The flow region is divided into a set of sub-domains, each of which is described by its own grid of appropriate form. The boundaries are identified with the meshes involving smooth, continuous grid lines; non-smooth, discontinuities of gradient of the grid lines or overlapping, discontinuous grid lines. Each form requires an appropriate boundary

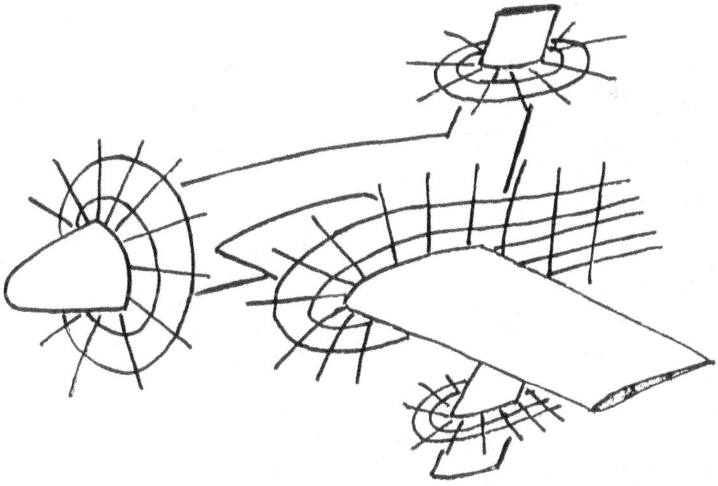

Figure 3.3 Schematic representation of a block structured grid around an aircraft.

treatment. The block structure is defined by the topology of the aircraft configuration (e.g. Figure 3.3), generally via a hands-on interaction using a complex graphics representation.

Multi-block techniques are best suited to certain types of configurations. In particular, close coupled configurations, which involve highly stretched or highly skewed grids, grid cross-over or non-smoothness, cause severe difficulty with the basic multi-grid strategy. The flow solvers must therefore be robust and forgiving of grid irregularity, as detailed local grid control may be impossible. Nonetheless, with these limitations, the technique has proved workable and useful for those problems to which it can be applied.

3.3 Generalised Euler Schemes

3.3.1 Introduction

In this section, we will examine the simplest problem in fluid mechanics, which nonetheless encompasses a major difficulty in designing schemes to perform accurate calculations under varying conditions. We consider the case in which the non-linearity inherent in fluid mechanics is eliminated by examining the basic advection problem with constant coefficients in one dimension. From a practical point of view, this represents the simple transfer of material with constant velocity. There is therefore no compression nor acceleration. Despite the simplicity of this problem, it is a taxing one to treat numerically with arbitrary coefficients, and moreover forms the basis for more general treatments of convection.

The basic partial differential equation, which we seek to model, is the general advection equation

$$\frac{\partial u}{\partial t} + a\frac{\partial u}{\partial x} = 0 \tag{3.31}$$

It is the simplest one-dimensional hyperbolic equation. It has a characteristic speed a. If a is a constant, it has the general solution representing a travelling pulse with speed a, namely $u(x, t) = u_0(x - at)$. We may test our numerical solution against this analytic solution.

3.3.2 First Order Centred Difference Scheme

The simplest numerical scheme for the advection equation is the centred difference in space, forward difference in time

$$\frac{v_i^{n+1} - v_i^n}{Dt} + a\frac{v_{i+1}^n - v_{i-1}^n}{2\Delta x} = 0 \tag{3.32}$$

or

$$v_i^{n+1} = v_i^n - \frac{\lambda}{2}(v_{i+1}^n - v_{i-1}^n) \tag{3.33}$$

where $\lambda = a\,Dt/\Delta x$ is the Courant number.

This scheme is conservative with

$$g_{i+1/2} = \frac{\lambda}{2}(v_{i+1}^n - v_i^n) \tag{3.34}$$

However, it is also unstable for

$$\begin{aligned} G &= 1 - \frac{\lambda}{2}\left[\exp(\imath k\Delta x) - \exp(-\imath k\Delta x)\right] \\ &= 1 - \imath\lambda\sin(k\Delta x) \end{aligned} \tag{3.35}$$

and therefore

$$|G|^2 = 1 + \lambda^2\sin^2(k\Delta x) \tag{3.36}$$

Indeed, it is absolutely unstable, since $|G| > 1$ for all values of λ and $k\Delta x$. It is therefore unsuitable for practical use.

A simple test considering the propagation of a square pulse demonstrates this behaviour, Figure 3.4. We note the growing oscillation behind the pulse characteristic of instability, despite the relatively small value of the Courant number, λ. This instability is due to propagation along different characteristics from elements spaced $\sim \Delta x$ in front and behind the measurement point.

3.3.2.1 Upwind or Donor Cell Differencing

The numerical errors described above arise from a failure to reflect the propagation along the characteristics. Flux is largely determined by upstream values. The spatial difference of velocity can be calculated upstream using:

$$v_i^{n+1} = v_i^n - \lambda \begin{cases} (v_i^n - v_{i-1}^n) & \text{if } \lambda > 0 \\ (v_{i+1}^n - v_i^n) & \text{otherwise} \end{cases} \tag{3.37}$$

in which case

$$\begin{aligned} G &= 1 - \lambda \begin{cases} (1 - \exp[-\imath k\Delta x]) & \text{if } \lambda > 0 \\ (\exp[\imath k\Delta x] - 1) & \text{otherwise} \end{cases} \\ &= 1 \mp 2\imath|\lambda|\exp[\mp\imath k\Delta x/2]\sin(k\Delta x/2) \end{aligned} \tag{3.38}$$

$$|G|^2 = 1 - 2|\lambda|(1 - |\lambda|)[1 - \cos(k\Delta x)] \tag{3.39}$$

This condition is clearly stable if $|\lambda| \leq 1$.

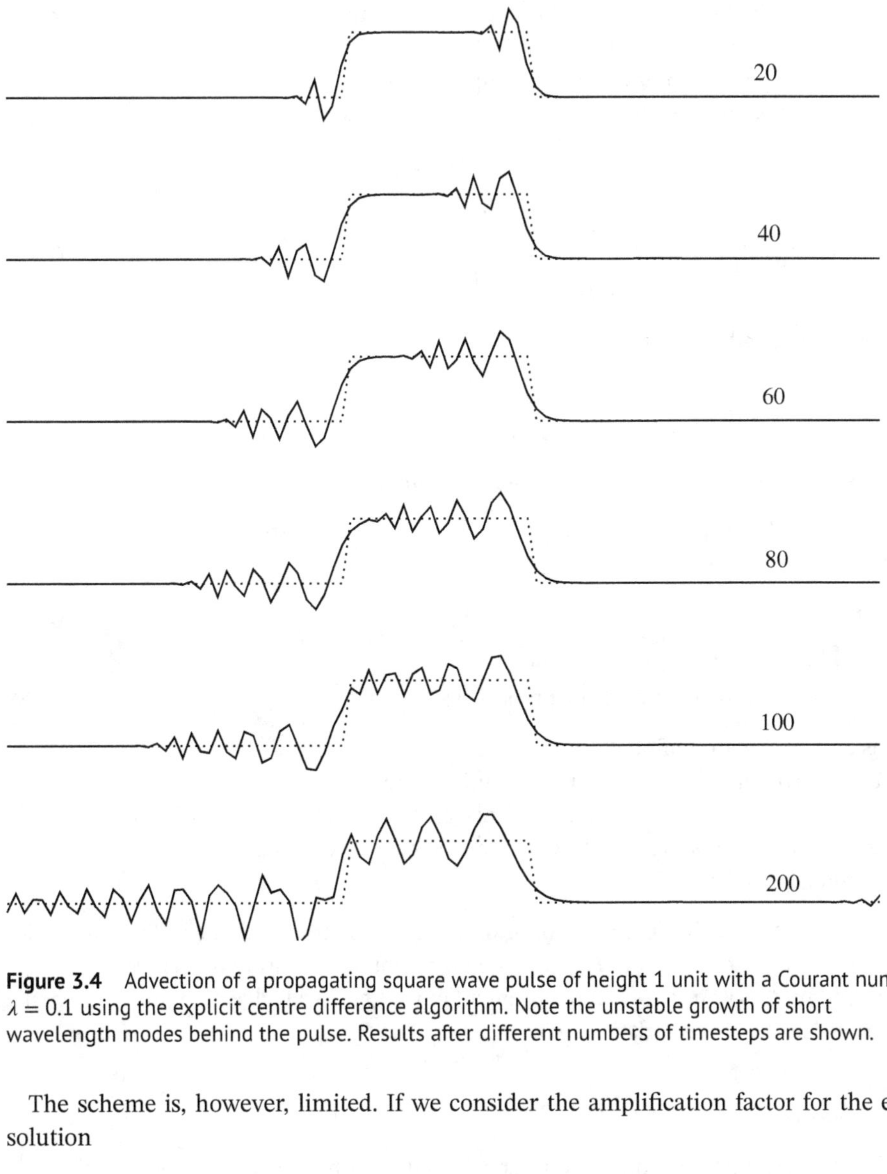

Figure 3.4 Advection of a propagating square wave pulse of height 1 unit with a Courant number $\lambda = 0.1$ using the explicit centre difference algorithm. Note the unstable growth of short wavelength modes behind the pulse. Results after different numbers of timesteps are shown.

The scheme is, however, limited. If we consider the amplification factor for the exact solution

$$G = \exp(-\imath\lambda k\Delta x) \tag{3.40}$$

which represents a phase shift $\lambda k\Delta x$ per step, with no amplitude change. We conclude that

1) The scheme is dissipative (or diffusive) since $|G| < 1$.
2) The phase shift $\arg(G) = \arctan\{-|\lambda|\sin(k\Delta x)/[1 - |\lambda| + |\lambda|\cos(k\Delta x)]\}$ differs from the exact value. This represents waves of differing k moving with different velocities, i.e. dispersion. Note, however, if $|\lambda| = 1/2$, $\arg(G) = \mp(k\Delta x)/2$, which is the exact value.

Thus, a general wave pulse is distorted in two ways: by diffusion, i.e. its amplitude is reduced, and the pulse is broadened; and by dispersion, i.e. individual wave elements move separately and distort the profile of the pulse. Usually, the former dominates in first-order schemes, for example donor cell (Figure 3.5).

The practical implementation of simple upstream differencing in a fluid code where the advection coefficient is not constant may be accomplished in two slightly different ways.

1) One-sided differencing based on the flux vector

$$g_{i+1/2} = \begin{cases} f(v_{i+1}) \text{ if } \dfrac{f(v_{i+1}) - f(v_i)}{v_{i+1} - v_i} < 0 \\ f(v_i) \quad \text{otherwise} \end{cases} \tag{3.41}$$

2) Donor cell based on the flow velocity

$$g_{i+1/2} = \mathrm{v}_{i+1/2} \begin{cases} v_{i+1} \text{ if } \mathrm{v}_{i+1/2} < 0 \\ v_i \quad \text{otherwise} \end{cases} \tag{3.42}$$

where $\mathrm{v}_{i+1/2}$ is the 'flow velocity' across the boundary.

Referring back to the set of characteristic equations, we can expect that the full algorithm will be stable if

$$\max\left(|\mathrm{v}| + c\right)(Dt/\Delta x) \leq 1 \tag{3.43}$$

i.e. Courant–Friedrichs–Lewy condition.

The scheme has some noteworthy properties:

1) It is positivity maintaining if $\lambda \leq 1$.
2) One-sided differencing is monotonicity preserving.
3) One-sided differencing is monotone if the velocity has everywhere the same sign.
4) Donor cell is monotonicity preserving in the case of constant coefficients, when the differential equation is linear.

Despite some attractive features, donor differencing is little used due to the large diffusion of the wave. The Courant–Friedrichs–Lewy stability condition, which we meet here for the first time, is a very general requirement in well-behaved advection routines when a is identified as the characteristic speed.

3.3.3 Second-Order Schemes

The first-order solutions examined thus far exhibit two sources of error: numerical diffusion and dispersion.

If we expand the one-dimensional form in a Taylor's series

$$\begin{aligned} u(x, t + Dt) &= u(x, t) + \left.\frac{\partial u}{\partial t}\right|_{x,t} Dt + \frac{1}{2}\left.\frac{\partial^2 u}{\partial t^2}\right|_{x,t} Dt^2 \\ &= u(x, t) - A\left.\frac{\partial u}{\partial x}\right|_{x,t} Dt + \frac{1}{2}\frac{\partial}{\partial x}\left.\left(A^2\frac{\partial u}{\partial x}\right)\right|_{x,t} Dt^2 \end{aligned} \tag{3.44}$$

since for the conservative form of the equations

$$\frac{\partial}{\partial t}\frac{\partial u}{\partial t} = -\frac{\partial}{\partial t}\frac{\partial f}{\partial x} = -\frac{\partial}{\partial x}\left(A\frac{\partial u}{\partial t}\right) = \frac{\partial}{\partial x}\left(A^2\frac{\partial u}{\partial x}\right) \tag{3.45}$$

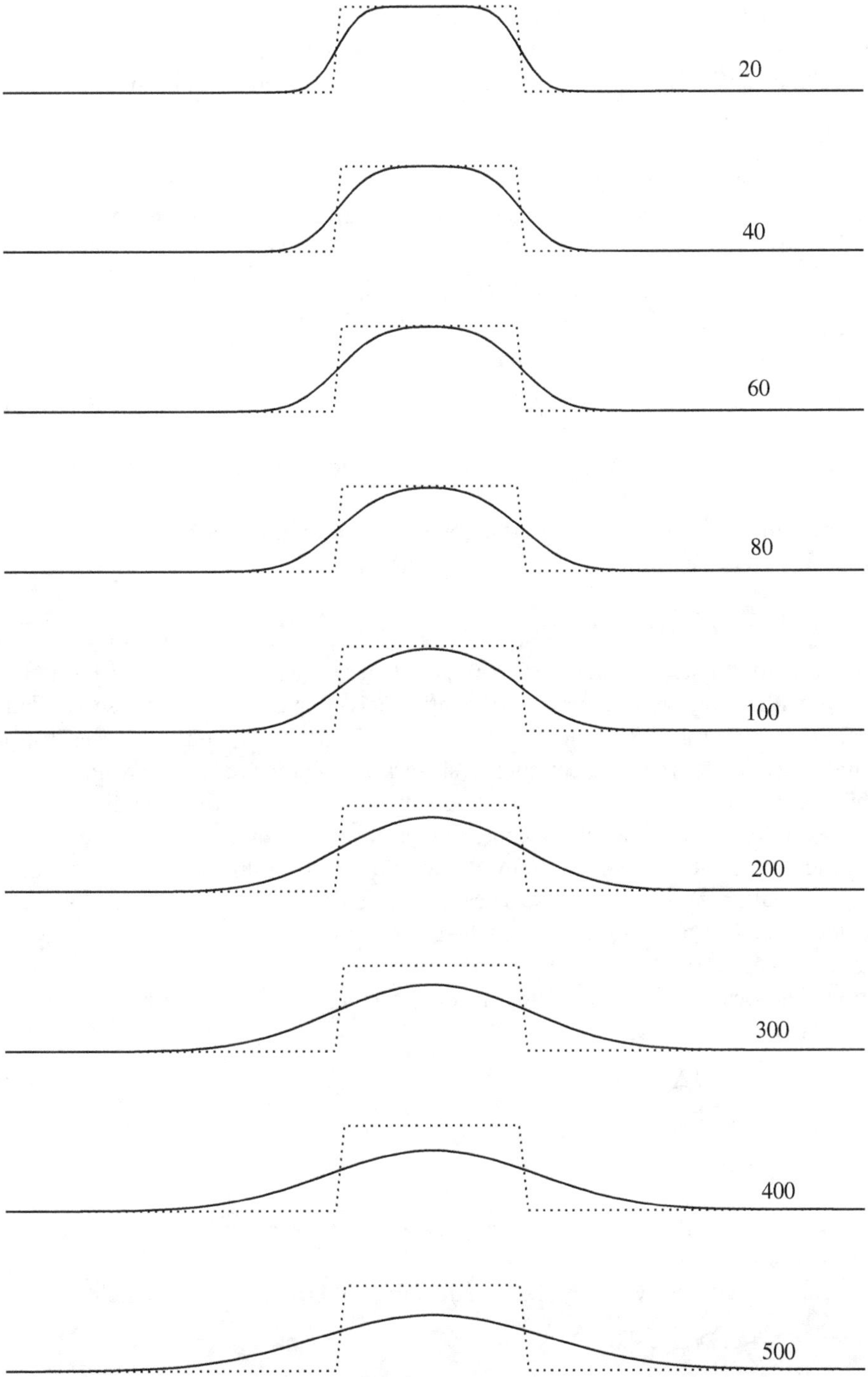

Figure 3.5 Advection of a propagating square wave pulse of height 1 unit with a Courant number $\lambda = 0.5$ using first-order upwind differencing. Note the strong numerical diffusion of the solution characteristic of first-order schemes.

In first-order schemes, we only represent the first two terms, and the truncation error's lowest-order term $\frac{1}{2}\frac{\partial}{\partial x}(A^2\frac{\partial u}{\partial x})\,Dt^2$ has the form of a diffusion term, which is the origin of the numerical diffusion seen in the first-order schemes. To eliminate this diffusion, we must include second-order terms.

3.3.3.1 Lax–Wendroff Scheme

We improve the accuracy by explicitly adding a finite-difference form of the lowest-order truncation term from the Taylor's series (Lax and Wendroff, 1960), i.e.

$$v_i^{n+1} = v_i^n - \frac{Dt}{2\Delta x}A_j^n(v_{i+1}^n - v_{i-1}^n) + \frac{Dt^2}{2\Delta x^2}[{A_{i+1/2}^n}^2(v_{i+1}^n - v_i^n) - {A_{i-1/2}^n}^2(v_i^n - v_{i-1}^n)] \tag{3.46}$$

or more simply as an addition to the flux,

$$g_{i+1/2} = \frac{1}{2}(f_{i+1}^n + f_i^n) - \frac{1}{2}\,\frac{Dt}{\Delta x}\,A_{i+1/2}^n(f_{i+1}^n - f_i^n) \tag{3.47}$$

where $A_{i+1/2}^n = A(1/2[v_{i+1}^n + v_i^n])$ and $f_i^n = f(v_i^n)$. This form is clearly conservative and consistent.

We can modify g without violating these conditions by adding an arbitrary term which vanishes if $v_i = v_{i+1}$, i.e. with flux term

$$g'_{i+1/2} = g_{i+1/2} + \frac{1}{2}\,Q(v_i, v_{i+1})\,(v_{i+1} - v_i) \tag{3.48}$$

where Q is matrix of appropriate form. It is clear that Q is an additional diffusion term; by analogy with its role for the momentum equation, it is often called artificial viscosity, but it is more general. Its role is to provide an additional dissipation term, similar to viscosity in a real fluid, in regions of large gradient, such as shocks or stagnation points where non-linear instability can arise, and to ensure shock capture by satisfying the entropy condition.

The Lax–Wendroff scheme offers considerable advantage over first-order methods. The strong numerical diffusion is removed, but with artificial dissipation, the essential entropy generation necessary for shock capture can be readily introduced. However, dispersion is now a problem, and the scheme is neither positivity nor monotonicity preserving (Figure 3.6).

With no artificial viscosity the amplification factor for the one variable constant coefficient case is

$$G = 1 - \imath\lambda\sin(k\Delta x) - \lambda^2\,[1 - \cos(k\Delta x)] \tag{3.49}$$

so that

$$|G|^2 = 1 - \lambda^2\,(1 - \lambda^2)\,[1 - \cos(k\Delta x)]^2 \tag{3.50}$$

The scheme is stable if $|\lambda| \leq 1$. For the general case, we recover the Courant–Friedrichs–Lewy condition.

The addition of artificial viscosity slightly strengthens this result to, for example

$$\max\,(|\mathrm{v}| + c)\,Dt/\Delta x \leq 0.78 \tag{3.51}$$

The implementation of the Lax–Wendroff scheme to the full set of equations is cumbersome due to the matrix multiplication. For this reason, the two-step form described earlier is usually used. The Lax–Wendroff technique has proved very successful but has problems

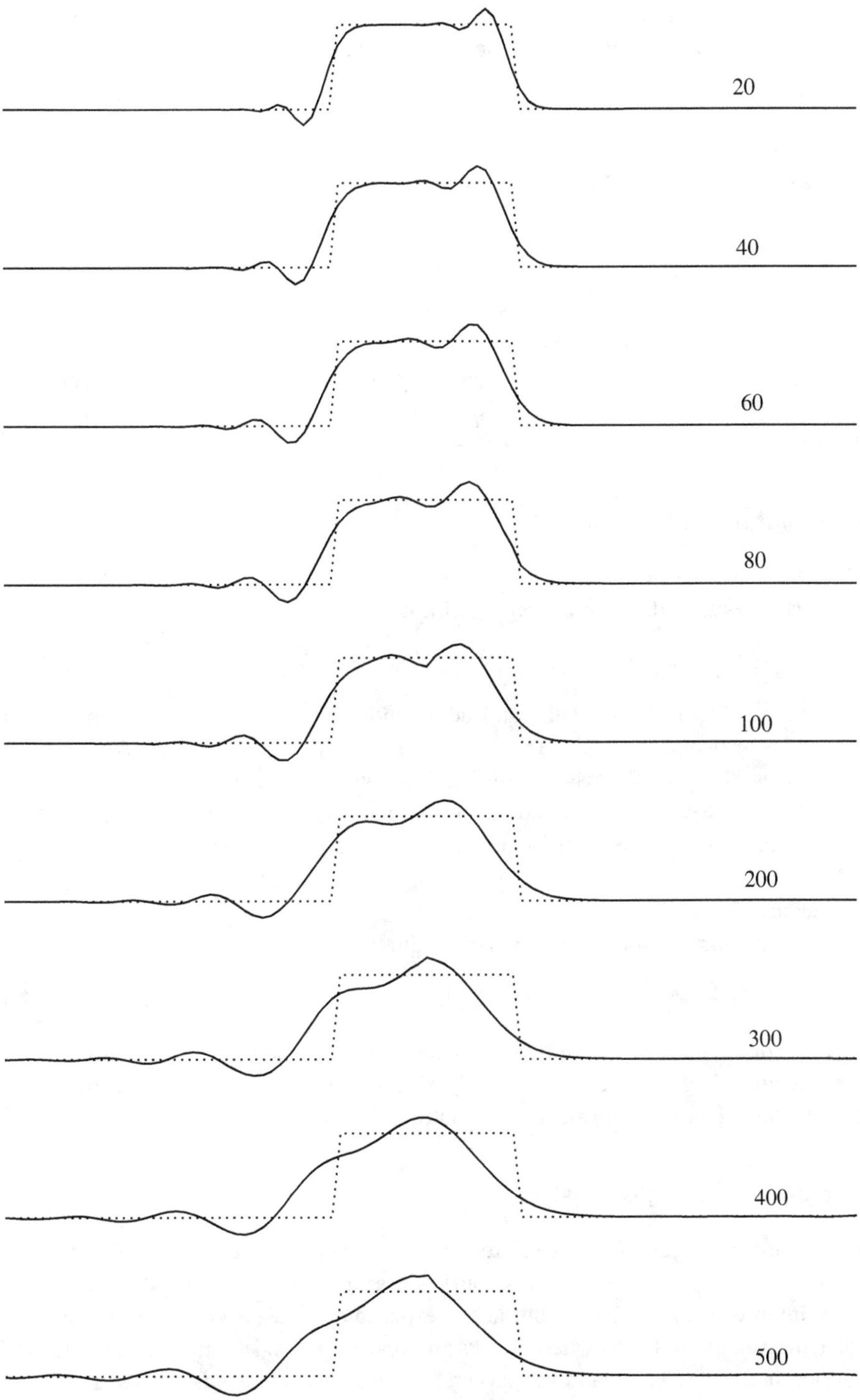

Figure 3.6 Advection of a square pulse of height 1 unit with a Courant number $\lambda = 0.5$ using the Lax–Wendroff algorithm. Note the distortion of the pulse associated with phase errors.

with spurious oscillations at shock fronts – and relatively large dispersion, which are characteristic of second-order centred difference schemes and can be controlled with artificial dissipation.

3.3.3.2 MacCormack's Method

Another similar two-step method due to MacCormack uses alternate backward and forward differences

$$\begin{aligned} \overline{v}_i^{n+1} &= v_i^n - \frac{Dt}{\Delta x}(f_{i+1}^n - f_i^n) \\ v_i^{n+1} &= \frac{1}{2}\left(v_i^n + \overline{v}_i^{n+1}\right) - \frac{1}{2}\frac{Dt}{\Delta x}\left(\overline{f}_i^{n+1} - \overline{f}_{i-1}^{n+1}\right) \end{aligned} \tag{3.52}$$

A variant uses forward and backward differencing on alternate steps. Although not obviously Lax–Wendroff in type or even consistent, the method is extremely successful and is widely used.

3.3.4 Time-Centred Schemes

3.3.4.1 Leap-Frog Scheme

The obvious time-centred, second-order algorithm:

$$v_i^{(n+1)} = v_i^{(n-1)} - \lambda\left(v_{(i+1/2)}^n - v_{(i-1/2)}^n\right) \tag{3.53}$$

involves the use of two meshes. This can lead to a non-linear oscillation between the two meshes $[n, (i+1/2)]$ and $[(n+1), i]$. This instability is easily suppressed by weakly coupling the two meshes. However, there is an additional storage requirement for the second mesh, which may make the scheme unattractive. The method also has poor dispersion properties. It is not recommended in this form for general fluid dynamic problems.

3.3.4.2 Implicit Scheme

A simple implicit formulation of the difference equations

$$v_i^{n+1} = v_i^n - \frac{\lambda}{4}\left[v_{i+1}^n + v_{i+1}^{n+1} - v_{i-1}^n - v_{i-1}^{n+1}\right] \tag{3.54}$$

involves the solution of a set of simultaneous equations. Although in one dimension, this is easily accomplished using the Thomas algorithm, these complications, together with significantly large phase errors, make it an unattractive scheme.

3.3.5 Zero Phase Average Methods

Phase errors due to dispersion can be extremely damaging in problems where the solution is relatively smooth. For these problems, fourth-order methods have been devised, which markedly improve these problems, but at the expense of increased computing time and storage. However, for problems where shocks are weak or non-existent, e.g. meteorological simulation, such methods have proved successful and have been extensively used.

An alternative approach for reducing phase errors is due to Fromm, who noticed that the phase error of a backward step is frequently opposite in sign to the phase error of a forward one. Thus, if in a constant coefficient system, a shift of $\lambda = 1$ simply moves v_i to v_{i+1} with no phase shift the operation

$$v_i^{n+1} = \frac{1}{2}[G(\lambda) + G(\lambda - 1)G(1)]\, v_i^n \tag{3.55}$$

should give a correspondingly reduced phase shift. Fromm suggested using a Lax–Wendroff operator for G, although the method is more general. Thus, for example first-order upstream differencing is a zero-phase average method of first order.

If equal weight is given to each term, Fromm's method defines the upstream difference (i.e. centred at the upstream cell boundary) whose phase properties are in general always superior to those of centred differences (i.e. centred on the mesh point), as demonstrated by the square wave test (Figure 3.7).

However, in this simple form, the method has serious limitations in that it is neither monotonicity nor positivity preserving. Shock overshoot will therefore occur, and shock capture cannot be guaranteed to occur correctly.

A more severe limitation is that upstream differencing cannot in general be put into conservative forms due to the asymmetric form of difference.

3.3.6 'Clipped' Schemes

Fromm's scheme shows a distinct improvement over the basic Lax–Wendroff in the handling of dispersion. However, it still suffers from serious defects associated with non-positivity and programme construction, which mitigate against its application in a full hydrodynamic simulation. To overcome these difficulties, methods have been developed, involving a measure of 'clipping', which changes the order of the calculation when necessary.

3.3.6.1 Monotonicity Preserving Schemes

These result from the observation that for constant coefficients the fluid equations preserve monotonicity. However, the normal high-order schemes do not preserve monotonicity due to the short wavelength oscillations, which are generated. Van Leer suggested clipping the flux to prevent overshoot in regions of strong gradient. The underlying philosophy of this approach is based on the fact that a high-order scheme essential passes a piece-wise continuous polynomial of second or third order through the mesh values. Since the polynomial may contain local extrema which are not present in the mesh values, these will be transported by convection and give a spuriously high value. Alternatively, we may argue that the value at the cell boundary calculated from the polynomial interpolation is too large, and the corresponding flux across the boundary too high. Van Leer's remedy is to limit the interpolated value at the cell boundary to that of the neighbouring cell, thereby forcing the transport to first-order form (van Leer, 1973, 1974, 1977a,b, 1979).

Van Leer's algorithm for handling these situations was not particularly easy to implement. However, Youngs (1982) has shown that an easier form of the condition is obtained from the fluxes $g_{i+1/2}$, which may be expressed in donor cell form by considering the velocity $\mathrm{v}_{i+1/2}$ at the cell interface.

$$g_{i+1/2} = \mathrm{v}_{i+1/2}\, v_{i+1/2}^{*} \tag{3.56}$$

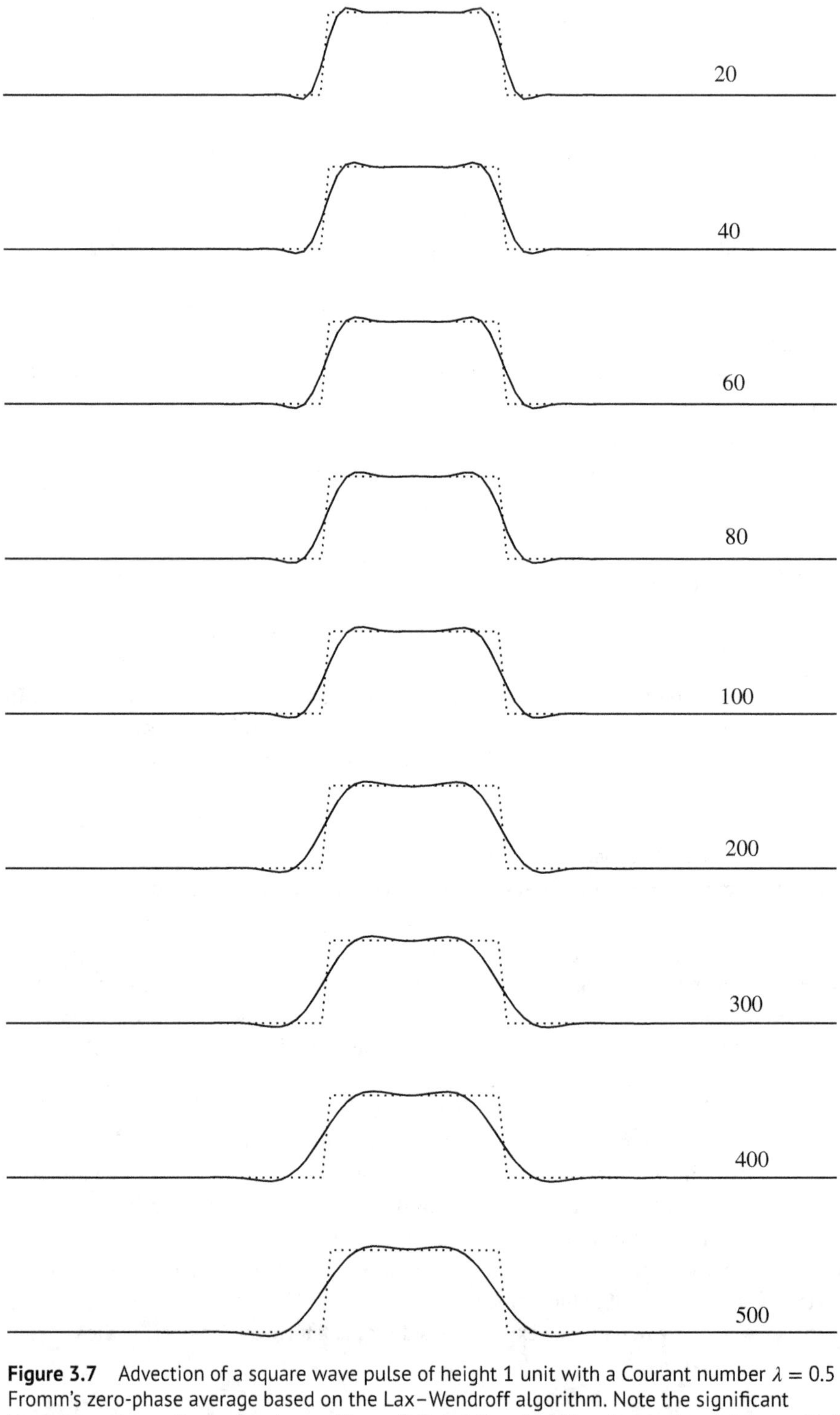

Figure 3.7 Advection of a square wave pulse of height 1 unit with a Courant number $\lambda = 0.5$ using Fromm's zero-phase average based on the Lax–Wendroff algorithm. Note the significant improvement over the basic scheme (Figure 3.6), but the algorithm does not maintain positivity or monotonicity.

We may quite generally write

$$
\begin{aligned}
v^*_{i+1/2} &= v_i \quad + \frac{1}{2}(1 - \lambda_{i+1/2})D_{i+1/2} \\
&= v_{i+1} + \frac{1}{2}(1 + \lambda_{i+1/2})D'_{i+1/2}
\end{aligned}
\tag{3.57}
$$

For first-order upstream difference

$$
\begin{cases}
D_{i+1/2} = 0 \ \text{if}\ \lambda_{i+1/2} > 0 \\
D'_{i+1/2} = 0 \ \text{otherwise}
\end{cases}
\tag{3.58}
$$

For second-order Lax–Wendroff

$$
D_{i+1/2} = v_{i+1} - v_i = -D'_{i+1/2}
\tag{3.59}
$$

For third-order upstream difference

$$
\begin{cases}
D_{i+1/2} = (v_{i+1} - v_i) - \dfrac{1 + \lambda_{i+1/2}}{3}(v_{i+1} - 2v_i + v_{i-1}) \ \text{if}\ \lambda_{i+1/2} > 0 \\
D'_{i+1/2} = (v_i - v_{i+1}) - \dfrac{1 - \lambda_{i+1/2}}{3}(v_i - 2v_{i+1} + v_{i+2}) \ \text{otherwise}
\end{cases}
\tag{3.60}
$$

To construct a monotonicity preserving scheme, we limit $g_{i+1/2}$ through $D_{i+1/2}$ to prevent too large a flux introducing an overshoot by

$$
\begin{aligned}
&\text{For } \lambda_{i+1/2} > 0 \textit{ then } D_{i+1/2} = S \min\left\{ |\overline{D}_{i+1/2}|,\ 2|v_{i+1} - v_i|,\ 2|v_i - v_{i-1}| \right\} \\
&\qquad S = \begin{cases} 0 & \text{if sign}\,(v_{i+1} - v_i) \neq \text{sign}\,(v_i - v_{i-1}) \\ \text{sign}(v_i - v_{i-1}) & \text{otherwise} \end{cases} \\
&\text{and for } \lambda_{i+1/2} < 0 \textit{ then } D'_{i+1/2} = S \min\left\{ |\overrightarrow{D}'_{i+1/2}|,\ 2|v_{i+2} - v_{i+1}|,\ 2|v_{i+1} - v_i| \right\} \\
&\qquad S = \begin{cases} 0 & \text{if sign}\,(v_{i+2} - v_{i+1}) \neq \text{sign}\,(v_{i+1} - v_i) \\ \text{sign}(v_{i+2} - v_{i+1}) & \text{otherwise} \end{cases}
\end{aligned}
\tag{3.61}
$$

where $\overrightarrow{D}'$ and $\overrightarrow{D}'$ are the second (Eq. 3.59)- or third (Eq. 3.60)-order approximations in the two cases. We note that at an extremum the scheme reverts to donor cell $S = D_{i+1/2}$ or $\overline{D}_{i+1/2} = 0$.

Figure 3.8 shows the result of the square pulse test for the third-order scheme. We note that diffusion is greatly reduced by the high-order finite-difference approximation, and dispersion minimised by the use of upstream differencing. However, these schemes are difficult to express in a general conservation law form, which may limit their application for some purposes.

3.3.6.2 Flux-Corrected Transport (FCT)

The underlying philosophy of these schemes is to perform a two-step calculation. The first step uses a first-order diffusive scheme, and the second step removes that diffusion, subject

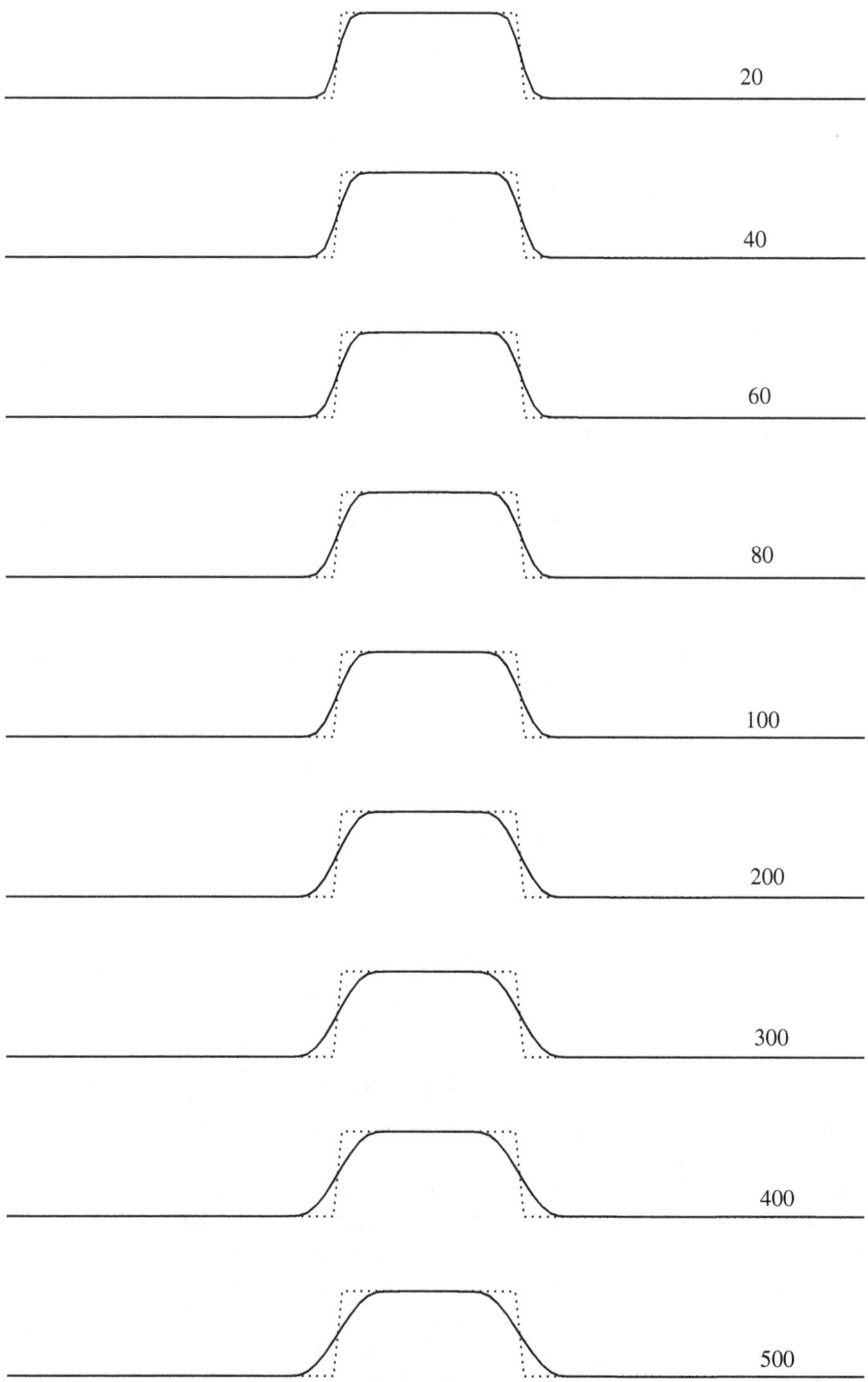

Figure 3.8 Advection of a square pulse of height 1 unit with a Courant number $\lambda = 0.5$ using the Young's third-order algorithm. Note the weak diffusion and absence of phase errors.

to an extremum condition. The first-order scheme removes the ripples by excess diffusion, which is then removed without recreating the high frequencies.

In principle, the scheme can be used with any pair of transport operators. In particular, one can therefore take advantage of some of the useful properties of first-order upwind differencing, positivity, monotonicity preserving or monotone. In particular, the scheme should have good shock capture properties.

Since the overall amplification factor of the scheme is

$$G = G_A \, G_T \tag{3.62}$$

where G_A and G_T are the amplification factors of the anti-diffusion and transport stages respectively, and since $\arg(G_A) = 0$, one can use upstream differencing to get good phase properties.

As an example, consider the donor cell form of upstream differencing as transport step.

$$\tilde{v}_i^{n+1} = v_i^n - \{\lambda_{i+1/2}^- \, v_{i+1}^n + (\lambda_{i+1/2}^+ - \lambda_{i-1/2}^-)\, v_i^n - \lambda_{i-1/2}^+ \, v_{i-1}^n\} \tag{3.63}$$

where $\lambda_{i+1/2} = \mathrm{v}_{i+1/2}\, Dt/\Delta x$ and

$$\lambda_{i+1/2}^+ = \max(0,\, \lambda_{i+1/2}) \qquad \lambda_{i+1/2}^- = \min(0,\, \lambda_{i+1/2})$$

The flux correction yields

$$v_i^{n+1} = \tilde{v}_i^{n+1} - \phi_{i+1/2} + \phi_{i-1/2} \tag{3.64}$$

where ϕ is chosen to reduce v_i^{n+1} to Lax–Wendroff form for constant velocity, i.e.

$$\tilde{\phi}_{i+1/2} = \frac{1}{2}\, |\lambda_{i+1/2}|\, (1 - |\lambda_{i+1/2}|)\, (\tilde{v}_{i+1} - \tilde{v}_i) \tag{3.65}$$

To eliminate spurious oscillation during anti-diffusion the flux ϕ is limited:

$$\phi_{i+1/2} = S \max[\,0,\, \min\{\, S\,\Delta_{i-1/2},\, |\tilde{\phi}_{i+1/2}|,\, S\,\Delta_{i+3/2}\}] \tag{3.66}$$

where $\Delta_{i+1/2} = \tilde{v}_{i+1} - \tilde{v}_i$ and $S = \mathrm{sign}(\Delta_{i+1/2})$.

The stability of flux limiters cannot be evaluated analytically due to the non-linear nature of the cut-off of the limiter. As a rough estimate one may expect this will limit the short wavelengths alone ($k\Delta x \sim \pi$). The amplification matrix for the anti-diffusion stage is:

$$G_A = 1 + 2\eta\,[1 - \cos(k\Delta x)] \tag{3.67}$$

where $\eta = \frac{1}{2}|\lambda|(1 - |\lambda|)$. Thus for donor cell FCT the complete amplification factor is

$$|G|^2 = 1 - 12\eta^2[1 - \cos(k\Delta x)]^2 - 16\eta^3[1 - \cos(k\Delta x)]^3 \tag{3.68}$$

and is stable for $0 \le \eta \le 3/8$, i.e. $0 \le |\lambda| \le 1$ as for the uncorrected scheme. Note also the scheme maintains positivity for $0 \le |\lambda| \le 1/2$. For the complete fluid equations these reduce to a Courant–Friedrichs–Lewy condition.

Figure 3.9 shows the result of the square pulse test applied to donor cell flux-corrected transport. As expected diffusion is weak to the inherent correction in the method, and dispersion as a result of upwind differencing. However the donor cell algorithm has the problem that as $|\lambda| \to 0$, $|G| \to 1$, i.e. the damping is lost in regions of zero flow velocity,

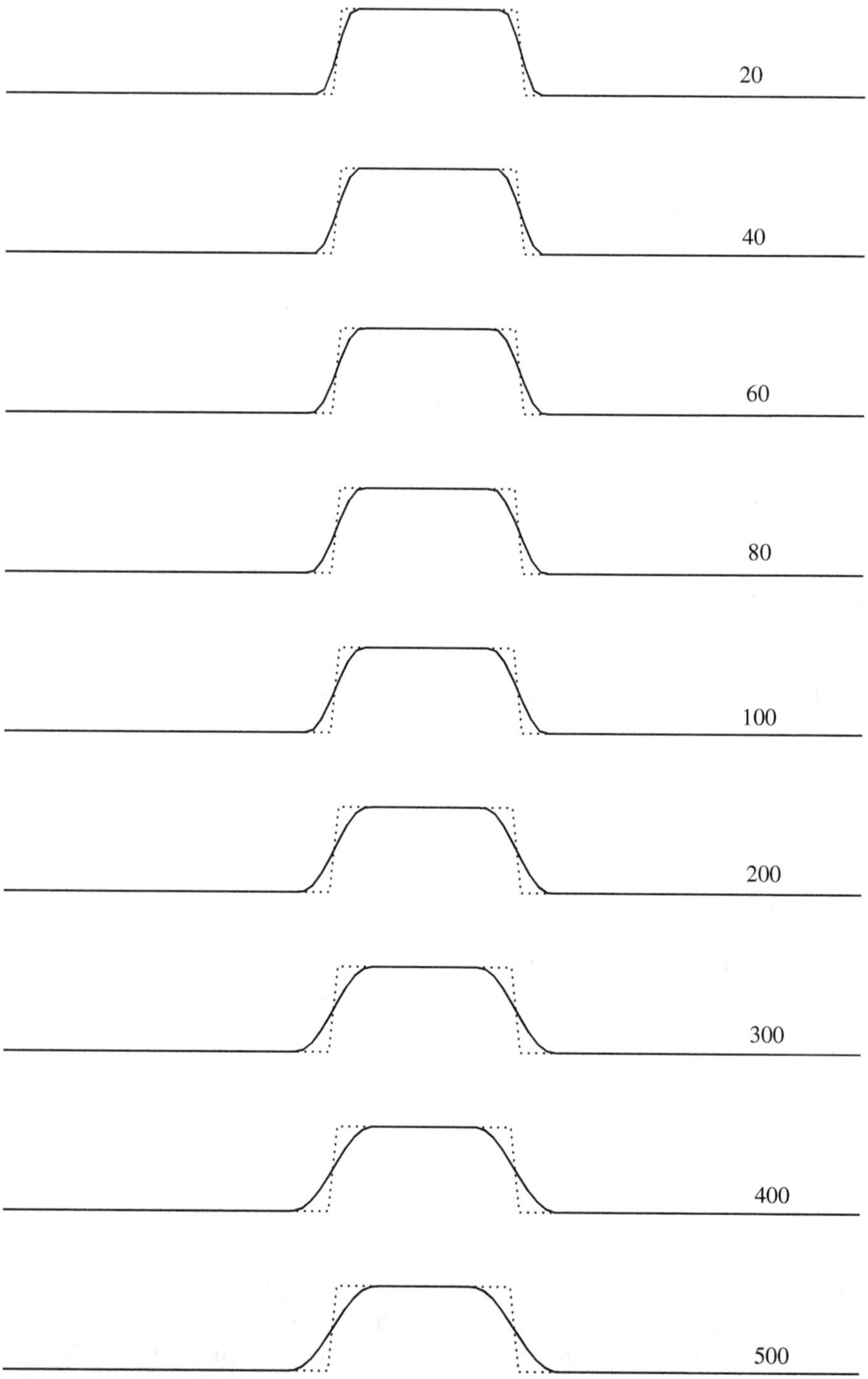

Figure 3.9 Advection of a square pulse of height 1 unit with a Courant number $\lambda = 0.5$ using the donor cell FCT algorithm. Note the weak diffusion and absence of phase errors.

which can lead to undesirable growth. An alternative more diffusive first-order scheme is sharp and smooth transport algorithm (SHASTA), which in conservative form is

$$g_{i+1/2} = \frac{1}{2}\left[\left(\frac{1}{2} + Q_{i+1/2}\right)^2 v_i - \left(\frac{1}{2} - Q_{i+1/2}\right)^2 v_{i+1}\right] \tag{3.69}$$

where

$$Q_{i+1/2} = \frac{1}{2}\left(\mathrm{v}_i + \mathrm{v}_{i+1}\right) Dt/\left[\Delta x + (\mathrm{v}_{i+1} - \mathrm{v}_i) Dt\right] \tag{3.70}$$

and the flux corrector

$$\phi_{i+1/2} = \frac{1}{8}\left(\tilde{v}_{i+1} - \tilde{v}_i\right) \tag{3.71}$$

In this case, the stability condition $|\lambda| \leq \sqrt{7/12}$ but positivity maintenance also requires $|\lambda| \leq 1/2$.

Figure 3.10 again displays the good properties of the FCT algorithm in the application to the square pulse test – indeed the results of the donor cell and SHASTA tests are almost identical, and very slightly better than Young's monotonicity preserving form.

3.3.6.3 TVD Schemes

A third method of identifying the condition for limiting the flux based on the fact that the total variation

$$TV = \int_{-\infty}^{\infty} \left|\frac{\partial u}{\partial x}\right| \mathrm{d}x \tag{3.72}$$

of a solution of the conservation law

$$\frac{\partial u}{\partial t} + \frac{\partial}{\partial x}[f(u)] = 0 \tag{3.73}$$

cannot increase. Thus, the finite-difference form

$$TV = \sum |v_{i+1} - v_i| \tag{3.74}$$

should not increase. Since the total variation will increase if an initially monotonic profile ceases to be monotone, this property ensures monotonicity preservation.

Schemes with this property, namely total variation diminishing (TVD) have a sufficient condition

$$\frac{\mathrm{d}u_i}{\mathrm{d}t} = c^+_{i+1/2}(u_{i+1} - u_i) - c^-_{i+1/2}(u_i - u_{i-1}) \tag{3.75}$$

if $c^+_{i+1/2}$ and $c^-_{i+1/2}$ are non-negative. TVD schemes are based on limitation to maintain this condition.

3.3.7 Artificial Viscosity

At this stage, we have concentrated on the development of satisfactory algorithms for treating simple hyperbolic equations. In fact, as we have seen, the equations of fluid mechanics are non-linear and possess solutions which involve the union of the simple characteristics of inviscid flow. In reality, in compressive flow, this leads to the formation of shock

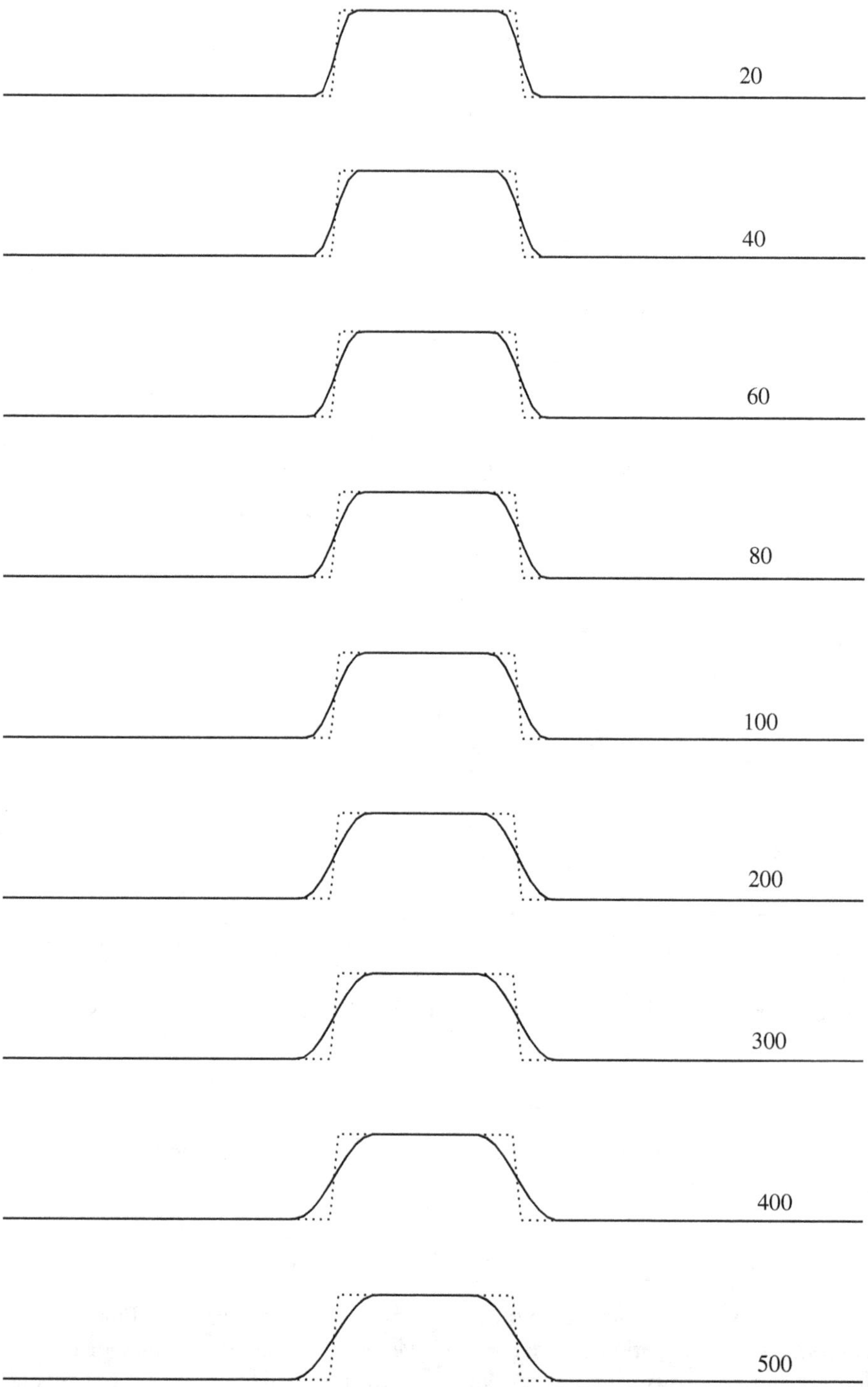

Figure 3.10 Advection of a square pulse of height 1 unit with a Courant number $\lambda = 0.5$ using the SHASTA FCT algorithm. Again note the weak diffusion and absence of phase errors.

waves propagating through the undisturbed medium. We have seen that, at least in one dimension, such shocks involve entropy generation associated with the omitted dissipative viscosity and thermal conduction terms in hyperbolic flow. To overcome this omission, which will always take place in compressive flow, we need to supply some numerical alternative, known as *artificial dissipation* to supply this deficiency and guarantee an accurate shock capture. Thus, artificial dissipation of some sort is essential in any fluid code. Specifically in more detail, the reasons for this are threefold:

1) To control shock capture and damp out high-frequency oscillations across the shock. These occur as a consequence of Gibb's phenomenon, namely the aliasing of the spatial high frequencies associated with the shock discontinuity into lower order modes represented on the mesh.
2) To prevent any odd/even point solution decoupling and any other unstable modes which may be transparent in the differencing scheme.
3) To cut-off the high-frequency energy transfer by non-linearity which may result in aliasing. This transfer of energy to high frequencies occurs as a result of the non-linear nature of the fundamental fluid equations in the same way as the turbulent energy cascade from long to short waves. In the latter case, this gives rise to the Kolmogorov spectrum.

Aliasing is the generation of wave energy at high frequency and its transfer to low frequency due to the discrete nature of the finite-difference mesh (Figure 3.11). Since the mesh can only support waves with spatial wave number $k < \pi/\Delta x$, we cannot distinguish between a wave k and $k + \pi/\Delta x$. Any attempted energy transfer into modes with spatial frequency $> \pi/\Delta x$ is reflected into waves with $k - \pi/\Delta x$ on the mesh. Artificial dissipation is required to damp waves $k \sim \pi/\Delta x$ to prevent this behaviour.

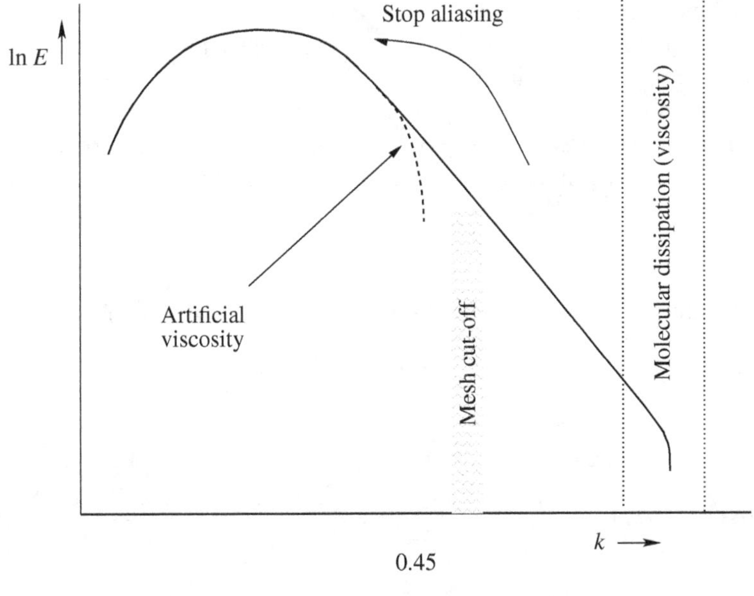

Figure 3.11 Energy spectrum due to non-linear hydrodynamic coupling.

Artificial dissipation can be included in two ways:

1) Explicitly by including an additional flux term

$$\frac{\partial u}{\partial t} + \frac{1}{\Delta V} \sum_{\substack{cell\\faces}} \text{fluxes} = D \tag{3.76}$$

where D is the added dissipation and ΔV the cell volume.

2) Implicitly by a term associated with the truncation error in the algorithm. Upwind schemes are the prototype of such schemes.

3.3.7.1 Explicit Artificial Dissipation

Artificial dissipation was included in the earliest numerical fluid dynamics algorithms, principally to handle the necessary entropy generation and introduce smoothing to overcome the Gibb's effect associated with shocks (von Neumann and Richtmyer, 1950). This had a relatively simple form as an artificial viscosity term which took the role of the physical term in compression. By adding a viscous stress proportional to the square of the velocity gradient ($\tau \propto \rho(\frac{\partial v}{\partial x})^2$), the shock was smeared over several cell widths and thereby smoothed. We shall return to this form of the term in more detail in treating Lagrangian systems in Chapter 6. In Eulerian systems, it was found necessary to introduce additional smoothing near stagnation points, where the flow velocity is small compared to the sound speed. To this end, an artificial viscosity proportional to the sound speed and velocity gradient was used ($\tau \propto \rho c \frac{\partial v}{\partial x}$) (Gentry, Martin and Daly, 1966). Such empirical methods have been fairly generally satisfactory and widely used, but are not easily generalised and have now been superseded by approaches based on the deficiencies of the numerical approximations.

We noted earlier that an additional dissipative flux is easily incorporated into the general conservative set of differential equations. This may be constructed to quite generally satisfy specific constraints, which provide the necessary damping available in the actual medium, but lacking otherwise in an inviscid calculation.

Lax and Wendroff (Richtmyer and Morton, 1967) initially attempted to resolve these problems by including artificial dissipation fluxes to increase the order of the difference equation. Considering $u(x, t)$ to be an exact solution of the differential equation to be fourth order

$$\frac{\partial u}{\partial t} + A\frac{\partial u}{\partial x} = Qu \tag{3.77}$$

then we require Q such that the truncation error in the Lax–Wendroff integration

$$v_i^{n+1} - v_i^n + \frac{1}{2}\frac{ADt}{\Delta x}(v_{i+1}^n - v_{i-1}^n) - \frac{1}{2}\left(\frac{ADt}{\Delta x}\right)^2 (v_{i+1}^n - 2v_i^n + v_{i-1}^n) = O[(Dt^4)] \tag{3.78}$$

instead of third order as in the usual case. For the constant coefficient case, we find by Taylor's expansion that we require

$$Qu = -\frac{1}{6}A[\Delta x^2 - A^2 Dt^2]\frac{\partial^3 u}{\partial x^3} \tag{3.79}$$

Unfortunately, when this term is included, the resultant is dispersive but not dissipative. To include further dissipation, we must replace the truncation error term $O(Dt^4)$ by one $O(Dt^5)$, which is obtained by an additional term Q' with Q, namely

$$Q'u = -\frac{1}{8}A^2 Dt[\Delta x^2 - (ADt)^2]\frac{\partial^4 u}{\partial x^4} \tag{3.80}$$

which introduces the required dissipation.

Lax and Wendroff (1960) showed that a suitable form of Q when u is a n dimensional form is the polynomial

$$Q = \alpha_0 I + \alpha_1 A + \cdots + \alpha_{n-1}A^{n-1} \tag{3.81}$$

such that the eigenvalues of $Q(a, b)$ are equal to the absolute values of the differences of the eigenvalues of $A(a)$ and $A(b)$ multiplied by arbitrary constants of order 1. The actual values of $\alpha_0 \dots \alpha_{n-1}$ are relatively easy to determine, but complicated in form – the interpolation polynomial being conveniently expressed in the Lagrange form. For a one-dimensional calculation, the coefficients α_0, α_1 and α_2 are given by Emery (1968).

Artificial dissipation in this standard form proposed by Lax and Wendroff is not easy to implement. A simpler algorithm is provided by the two-step form described by Richtmyer and Morton (1967), and has been widely used. In the first step, a set of intermediate values $v_{i+1/2}^{n+1/2}$ are calculated at the centred space $(i + 1/2)\Delta x$ and time $(n + 1/2)Dt$ steps

$$\begin{aligned} v_{i+1/2}^{n+1/2} = &\frac{1}{2}(v_i^n + v_{i+1}^n) - \frac{1}{2}\left(\frac{Dt}{\Delta x} + \alpha_2^+\right)(f_{i+1}^n - f_i^n) \\ &- \frac{1}{2}\alpha_1^+(v_{i+1}^n - v_i^n) \end{aligned} \tag{3.82}$$

and from these

$$\begin{aligned} v_i^{n+1} = &v_i^n - \frac{Dt}{\Delta x}(f_{i+1/2}^{n+1/2} - f_{i-1/2}^{n+1/2}) \\ &- \frac{1}{2}\frac{Dt}{\Delta x}[\alpha_0^+(v_{i+1}^n - v_i^n) - \alpha_0^-(v_i^n - v_{i-1}^n)] \end{aligned} \tag{3.83}$$

where $\alpha_0^+, \alpha_1^+, \alpha_2^+$ are the coefficients of the polynomial that yields $Q_{i+1/2}$ in terms of A_i and A_{i+1}, and $\alpha_0^-, \alpha_1^-, \alpha_2^-$ those for $Q_{i-1/2}$.

Figure 3.12 shows the behaviour of the Lax–Wendroff algorithm in a shock wave test. It can be seen that artificial dissipation removes the overshoot and stabilises the shock, but at the expense of an increased shock width.

However, even in this modified form, the Lax–Wendroff prescription was difficult to use, particularly in more than one spatial dimension due to the required matrix multiplication. Lapidus (1967) introduced a simplified smoothing algorithm for multiple dimensions using a third-order term proportional to the velocity gradients, thereby incorporating the artificial viscosity concept within the generalised flux.

Modern methods add a simpler vector term. The most successful scheme is due to Jameson. The idea is to add a high-order dissipation throughout the domain to give a base level of damping to prevent non-linear instability, but not sufficient to prevent oscillations

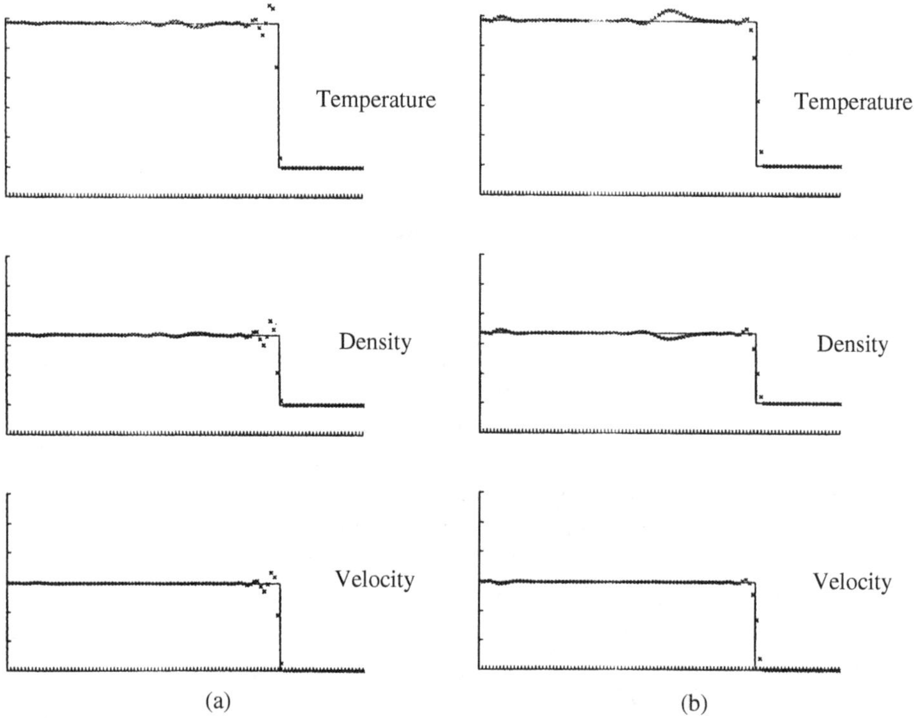

Figure 3.12 The development of a Mach 4 shock wave in a polytropic gas of index 5/3 is shown after 200 time-steps, calculated with a Courant number $(|v| + c)Dt/\Delta x$ behind the shock of 0.5 – the standard shock wave test calculated with the Lax–Wendroff scheme without artificial viscosity (a) and with an artificial viscosity constant α set equal to 1 (b). Note the ripples behind the shock in (a), where the shock thickness is about two cells. In (b) the ripples are eliminated but the shock width is about four cells. (a) No viscosity and (b) with viscosity.

in the neighbourhood of shock waves. In order to capture shocks, additional low-order dissipation is added locally by a sensor designed to detect discontinuities.

To identify the form of the low-order dissipation, let us examine the first-order upwind difference which is equivalent to a second-order centred difference plus diffusion.

$$\begin{aligned} a\,Dt\,\Delta^u v &= \frac{a\,Dt}{\Delta x}(v_i - v_{i-1}) \\ &= \frac{a\,Dt}{2\Delta x}(v_{i+1} - v_{i-1}) - \frac{a\,Dt}{2\Delta x}(v_{i+1} - 2v_i + v_{i-1}) \\ &= a\,Dt\,\Delta^c v - \frac{a\,Dt\,\Delta x}{2}\,\Delta^2 v \end{aligned} \tag{3.84}$$

This suggests that the low-order artificial dissipation should be of the form

$$\frac{1}{2}\,\epsilon^{(2)}\,|a|\,Dt\,\Delta x\,\Delta^2 v \tag{3.85}$$

and added to a basic second-order scheme.

For higher accuracy, we could try to make our basic second-order scheme into a second-order upwind scheme.

$$a\,Dt\,\Delta^u v = \frac{a\,Dt}{2\,\Delta x}(3v_i - 4v_{i-1} + v_{i-2}) \tag{3.86}$$

which is equivalent to a centred difference plus a scaled centred fourth-order difference

$$a\,Dt\,\Delta^c v + \frac{a\,Dt\,\Delta x^3}{4}(v_{i+2} - 4v_{i+1} + 6v_i - 4v_{i-1} + v_{i-2}) \tag{3.87}$$

suggests a dissipation

$$-\frac{1}{4}\epsilon^{(4)}\,|a|\,Dt\,\Delta x^3\,\Delta^4 v \tag{3.88}$$

to be added to the basic second-order scheme.

The coefficients $\epsilon^{(2)}$ and $\epsilon^{(4)}$ cannot be identified from this analysis but must be determined empirically on the basis of experimentation.

Firstly, we must select the appropriate value of a to use in the general fluid case. Jameson sets the value equal to the spectral radius of the Jacobean matrix A, i.e. in the size dimensional case $|v| + c$.

To switch on the low-order dissipation requires a shock wave sensor, for which the second difference of pressure is convenient

$$\mathrm{v}_i = \left|\frac{p_{i+1} - 2p_i + p_{i-1}}{p_{i+1} + 2p_i + p_{i-1}}\right| \tag{3.89}$$

Let

$$\overline{\mathrm{v}}_{i+1/2} = \max\left[\mathrm{v}_{i+2}, \mathrm{v}_{i+1}, \mathrm{v}_i, \mathrm{v}_{i-1}\right] \tag{3.90}$$

then

$$\epsilon^{(2)}_{i+1/2} = \min\left[\frac{1}{2}, k^{(2)}\vec{\overline{\mathrm{v}}}\{i + {}^1\!/\!_2\}\right] \tag{3.91}$$

$$\epsilon^{(4)}_{i+1/2} = \max\left[0, (k^{(4)} - \alpha\overline{\mathrm{v}}_{i+1/2})\right] \tag{3.92}$$

where $k^{(2)}$, $k^{(4)}$ and α are constants. Typically, $k^{(2)} \approx 1, k^{(4)} = 1/32$ and $\alpha = 2$. This prescription turns *on* the second order and turns *off* the fourth-order terms at a shock.

The dissipation is included by adding additional terms to the flux in the conservative scheme.

$$g'_{i+1/2} = g_{i+1/2} - d_{i+1/2} \tag{3.93}$$

where

$$\begin{aligned} d_{i+1/2} &= \epsilon^{(2)}_{i+1/2}\,R_{i+1/2}\,(v_{i+1} - v_i) \\ &\quad - \epsilon^{(4)}_{i+1/2}\,R_{i+1/2}\,(v_{i+2} - 3v_{i+1} + 3v_i - v_{i-1}) \end{aligned} \tag{3.94}$$

where

$$R_{i+1/2} = \frac{1}{2}\left[\frac{\mathrm{V}_{i+1}}{Dt^*_{i+1}} - \frac{\mathrm{V}_i}{Dt^*_{i+1}}\right] Dt \tag{3.95}$$

where Dt^*_i and Dt^*_{i+1} are the limiting time-steps in cells i and $i+1$, respectively.

3.3.7.2 Implicit Artificial Dissipation

We have already noted the existence of an alternative form of artificial dissipation, namely that directly associated with the truncation terms in the finite-difference form. For example the upwind differencing scheme

$$v_i^{n+1} = v_i^n - \lambda \begin{cases} v_i^n - v_{i-1}^n & \lambda > 0 \\ v_{i+1}^n - v_i^n & \text{otherwise} \end{cases} \tag{3.96}$$

which may be compared to the stable Lax–Wendroff form

$$\begin{aligned} v_i^{n+1} = v_i^n & - \frac{\lambda}{2}(v_{i+1}^n - v_{i-1}^n) \\ & + \frac{\lambda^2}{2}(v_{i+1}^n - 2v_i^n + v_{i-1}^n) \end{aligned} \tag{3.97}$$

which is only weakly damped. This upstream differencing contains the additional term

$$|\lambda|(1 - |\lambda|)(v_{i+1}^n - 2v_i^n + v_{i-1}^n) \tag{3.98}$$

which has the form of diffusion and is the source of the numerical difference in this algorithm.

Clearly, upstream finite differencing has this implicit diffusion provided $\lambda \neq 0$ or 1. It therefore avoids the problems discussed, earlier, but with a penalty, namely strong diffusion elsewhere.

Implicit artificial dissipation is designed to take advantage of the strong dissipation of first-order schemes where required, but to revert to higher order elsewhere. The problem is thus to identify the regions where diffusion is required to smooth out non-physical effects. Three such methods were introduced in this chapter. All are closely related and have been used with varying success.

- **Monotonicity preserving scheme:** We have earlier discussed the general philosophy underlying the monotonicity preserving schemes introduced by van Leer (1973, 1974, 1977a,b, 1979). In these, a high-order calculation is 'clipped' to a first-order scheme when necessary to maintain a monotone profile. As noted earlier, this approach can be used with any high-order convective operator, the low-order correction is automatically donor cell.
- **Flux-corrected transport (FCT):** The underlying philosophy of these schemes is to perform a two-step calculation. The first step uses a first-order diffusive scheme, and the second step removes that diffusion, subject to an extremum condition. The first-order scheme removes the ripples by excess diffusion, which is then removed without recreating the high frequencies. In principle, this scheme can be used with any pair of transport operators. Since the first-order scheme can be arbitrarily chosen to be strongly or weakly chosen, the level of artificial dissipation can be varied in different regions of flow to enhance shock capture.
- **TVD schemes:** A third method of identifying the condition for limiting the flux based on the fact that the total variation of a solution of the conservation law cannot increase. This property is used to introduce an artificial compression into the characteristics of the flow to prevent their spreading which leads to the development of ripples. This approach has been little used.

The first two methods have been extensively investigated. The early FCT algorithms of Boris and Book (1973) were not easily incorporated in multiple dimensions, but improvements by Zalesak (1979, 1981) overcame this deficiency. Similarly, in its original form van Leer's algorithm was not easy to programme, however Young's form is convenient to use. Comparison between the two approaches stems around the order in which the first-order and second-order terms are handled. In van Leer's form, the higher order is the primary calculation with the low order as a correction. In FCT, the converse applies. The latter therefore smooths over a wider range, in that the correction is applied after diffusion. Comparisons between the two methods (Wellnitz, 1984; Woodward and Colella, 1984) suggest a balance in favour of monotonicity preserving schemes. However, it worth noting that with a suitable choice FCT can be operated without any additional dissipation.

Appendix 3.A Energy Conservation in Fluid Codes

Most of the transport algorithms which we have considered conserve the totality of the quantity being transported. This is a very useful property ensuring that if the scheme is also positivity maintaining, it is stable against gross error growth. However, the conservation of energy between kinetic and particle internal energy consequent on pressure forces is not guaranteed. It is possible in a well-designed Euler code to fully ensure complete energy conservation, this is not possible in other geometries where a measure of energy deficit must be accepted.

It is not difficult to show that on a uniform grid, any well-defined difference scheme is conservative. However, if the mesh is non-uniform, it can be shown that only a cell-centred scheme can be made conservative. For example consider the simple consistent difference of the acceleration term due to pressure

$$\frac{\partial v}{\partial t} = \frac{\partial p}{\partial z} \quad \rightarrow \quad \frac{(p_{i+1} - p_{i-1})}{2\Delta z_i} \tag{3.A.1}$$

then momentum conservation

$$\sum_{i=1}^{I} v_i \, \Delta z_i = -(p_I - p_1)Dt \tag{3.A.2}$$

only if the cell width Δz_i is half the spacing between mesh points $z_{(i+1)}$ and $z_{(i-1)}$, i.e. a cell centred grid (Eq. 1.65).

The major problem lies with energy conservation, namely the two terms in the fluid equations due to acceleration and the gain in kinetic energy per unit time must correspond

$$\rho\mathbf{v} \cdot \frac{\partial \mathbf{v}}{\partial t} = -\rho\mathbf{v} \cdot \nabla p \tag{3.A.3}$$

and the corresponding work done by the pressure force per unit volume per unit time

$$-p\nabla \cdot \{\rho\mathbf{v}\} \tag{3.A.4}$$

Thus, the total energy change due to pressure (and other internal stress) on the volume V per unit time

$$E = \int_V \left[-\rho\mathbf{v} \cdot \nabla p - \mathbf{v}\nabla \cdot (\rho\mathbf{v})\right] \mathrm{d}V = -\int_V \nabla \cdot (p\rho\mathbf{v}) \, \mathrm{d}V = \int_S p\mathbf{v} \cdot \mathrm{d}S \tag{3.A.5}$$

where S is the bounding surface of V. These terms represent the introduction of pressure into enthalpy to represent the total internal energy in Eq. (1.1c), expressing the first law of thermodynamics.

We apply finite differencing to these terms on a one-dimensional mesh centred non-uniform orthogonal grid of I cells

$$\frac{(\tilde{u}_i^{(n+1)} - u_i^n)}{Dt} = -\frac{p_{(i+1)}^n - p_{(i-1)}^n}{2\rho_i^n \, \Delta z_i^n} \tag{3.A.6a}$$

$$\frac{(\tilde{\epsilon}_i^{(n+1)} - \epsilon_i^n)}{Dt} = -p_i^n \frac{u_{(i+1)}^w - u_{(i-1)}^w}{2\rho_i^n \, \Delta z_i^n} \tag{3.A.6b}$$

Forming the total kinetic energy change in time Dt

$$DE_k = \sum_{i=1}^{I} u_i^w (p_{(i+1)}^n - p_i^n) Dt \tag{3.A.7a}$$

and the work done

$$DW = \sum_{i=1}^{I} p_i^n (\tilde{u}_{(i+1)}^w - u_{(i-1)}^w + \tilde{u}_{(i+1)}^n - u_{(i-1)}^n) Dt/2 \tag{3.A.7b}$$

where $\tilde{u}_i^w$ is a velocity at the intermediate time t^w. By adding these two terms to the total energy change on the mesh in time Dt and taking into account the intermediate time $t^w = (t^n + t^{(n+1)})$, the centred difference is given by

$$DE_k + DW = \left\{ \left(p_1 \, u_0 + p_0 \, u_1 \right) - \left(p_{(I-1)} \, u_I + p_I \, u_{(I-1)} \right) \right\} Dt \tag{3.A.8}$$

which is a boundary term expressing work done by pressure forces at the boundaries $z_{1/2}$ and $z_{(I+1/2)}$. The equations are therefore strongly conservative in energy (see Section) when the velocity is centre differenced. The flux between cells i and $i+1$ is the ZIP differenced form

$$\frac{1}{2} \left\{ p_{(i+1)}^n \tilde{u}_i^n + p_i^n \tilde{u}_{i+1}^n \right\} \tag{3.A.9}$$

Unfortunately, Eqs. (3.A.6) are unstable, being essentially a centred difference advection equation (Section 3.3.2). The solution (Pert, 1983) is to follows a two-step procedure as in the Lax–Wendroff difference (Section 3.3.3.1) and difference the equations forming an intermediate density $\rho_i^{(n+1/2)}$, which for an ideal gas with polytropic constant γ is determined using the equation of state

$$\tilde{\rho}_i^{n+1/2} - \rho_i^n \Big/ \left\{ 1 + \frac{1}{4} (u_{i+1}^n - u_i^n) Dt / \Delta z_i^n \right\} \tag{3.A.10}$$

from which we obtain the intermediate pressure $\tilde{p}_i^{(n+1/2)}$

$$\tilde{p}_i^{n+1/2} - p_i^n \Big/ \left\{ 1 + \frac{1}{4} \gamma (\tilde{u}_{i+1}^n - u_i^n) Dt / \Delta z_i^n \right\} \tag{3.A.11}$$

We can then obtain

$$\tilde{u}_i^{n+1} = u_i^n - \left(\tilde{p}_{i+1}^{n+1/2} - \tilde{p}_{i-1}^{n+1/2} \right) Dt / 2\rho_i^n \Delta z_i^n \tag{3.A.12a}$$

$$\tilde{\epsilon}_i^{n+1} = \epsilon_i^n - \tilde{p}_i^{n+1/2} \left(u_{i+1}^w - u_{i-1}^w \right) Dt / 2\rho_i^n \Delta z_i^n \tag{3.A.12b}$$

The flux from cell i to $i + 1$ is easily shown to be the ZIP differenced form

$$\frac{1}{2}\left\{\tilde{p}_{(i+1)}^{(n+1/2)}\tilde{u}_i^n + \tilde{p}_i^{(n+1/2)}\tilde{u}_{i+1}^n\right\} \tag{3.A.13}$$

Thus, an advection scheme based on the simple Lax–Wendroff form is strongly conservative. Fortunately, all the well-behaved methods that we have investigated use the Lax–Wendroff form as a low-order basis. As a result, we conclude that to a reasonable order of approximation the forms are energy conserving, i.e. weakly conservative.

The generalisation to two or more dimensions on an orthogonal grid is straightforward.

This calculation involves the separation of a Lagrangian phase, which defines the energy balance of the fluid from the basic convection phase. In general, if re-zoning is included in the latter step, this separation is necessary. If not, the term may be included by the transport of enthalpy as in Eq. (1.1c).

References

Boris, J P and D L Book. 1973. "Flux-corrected transport I. SHASTA, A fluid transport algorithm that works." *J Comput Phys* 11:38–69.

Crowley, W P. 1962. An exponential zoner on a quadrilateral mesh. Technical report. Lawrence Livermore National Laboratory.

Emery, A F. 1968. "An evaluation of several differencing methods for inviscid fluid flow problems." *J Comput Phys* 2:306–331.

Gentry, R A, R E Martin and B J Daly. 1966. "An Eulerian differencing method for unsteady compressible flow." *J Comput Phys* 1:87–118.

Jameson, A. 1984. Transonic flow calculations for aircraft. In *Numerical Methods in Fluid Mechanics (Como 1983)*. Number 127 *in* "Lecture Notes in Mathematics" Berlin: Springer-Verlag pp. 156–244.

Jameson, A, W Schmidt and E Turkel. 1981. Numerical solutions of the Euler equations by finite volume methods using Runge-Kutta time-stepping schemes. In *AIAA 14th Fluid and Plasma Dynamics Conference (Palo Alto 1981)*. Number AIAA-81-1259.

Lapidus, A. 1967. "A detached shock calculation by second-order finite difference." *J Comput Phys* 2:154–177.

Lax, P D and B Wendroff. 1960. "Systems of conservation laws." *Commun Pure Appl Math* 13:217–237.

van Leer, B. 1973. *Towards the Ultimate Conservative Difference Scheme I. The Quest for Monotonicity*. In *Lecture Notes in Physics*. Vol. 18 Berlin: Springer-Verlag pp. 163–168.

van Leer, B. 1974. "Towards the ultimate conservative difference scheme II. Monotonicity and conservation combined in a second-order scheme." *J Comput Phys* 14:361–370.

van Leer, B. 1977a. "Towards the ultimate conservative difference scheme III. Upstream-centred finite difference-difference schemes for ideal compressible flow." *J Comput Phys* 23:263–275.

van Leer, B. 1977b. "Towards the ultimate conservative difference scheme IV. A new approach to numerical convection." *J Comput Phys* 23:276–299.

van Leer, B. 1979. "Towards the ultimate conservative difference scheme V. A second-order sequel to Godunov's method." *J Comput Phys* 32:101–136.

von Neumann, J and R D Richtmyer. 1950. "A method for the numerical calculation of shocks." *J Appl Phys* 21:232–237.

Pert, G J. 1983. "Quasi-Lagrangain rezoning of fluid codes maintaining an orthogonal mesh." *J Comput Phys* 49:1–43.

Richtmyer, R D and K W Morton. 1967. *Difference Methods for Initial Value Problems.* New York: Interscience.

Wellnitz, B A. 1984. Rezoning workshop - 1983. Technical Report LA-10112-C. Los Alamos National Laboratory.

Winslow, A M. 1963. Equipotential zoning of two-dimensional meshes. Technical Report UCRL-7312. Lawrence Livermore National Laboratory.

Winslow, A M. 1981. Adaptive-mesh rezoning by the equipotential method. Technical Report UCID-19062. Lawrence Livermore National Laboratory.

Woodward, P and P Colella. 1984. "The numerical simulation of two-dimensional fluid flow with strong shocks." *J Comput Phys* 54:115–173.

Youngs, D L. 1982. Time-dependent multi-material flow with large fluid distortion. In *Numerical Methods in Fluid Mechanics*, ed. K W Morton and M J Baines. Academic Press p. 273.

Zalesak, S T. 1979. "Fully dimensional flux-corrected transport algorithms for fluids." *J Comput Phys* 31:335–362.

Zalesak, S T. 1981. "High order 'ZIP' differencing of convective terms." *J Comput Phys* 40:497–508.

4

Lagrangian Systems

4.1 Lagrangian Fluid Dynamics

We have seen that the principal difficulty with Eulerian systems involves handling the advective or transport derivative in the fluid equations. The question arises, can we avoid this term by working in the fluid rather than the laboratory frame? This approach to fluid mechanics is called the Lagrangian picture.

We consider a fluid particle identified by some parameter which is fixed to it for all time. For example in one dimension this might be its initial position at time zero, but the choice is quite arbitrary. If we consider time changes as seen by the particle, the derivative is

$$\frac{D}{Dt} = \frac{\partial}{\partial t} + \underline{\mathrm{v}} \cdot \nabla \tag{4.1}$$

i.e. the advection term is included in the time derivative.

The Lagrangian set of fluid equations are therefore

$$\frac{D\boldsymbol{R}}{Dt} = \mathrm{v} \tag{4.2}$$

$$\rho \frac{D\mathrm{v}}{Dt} = \nabla \cdot \boldsymbol{\tau} \tag{4.3}$$

and

$$\rho \frac{D\epsilon}{Dt} = \boldsymbol{\tau} \cdot \cdot \dot{\boldsymbol{e}} - \nabla \cdot \boldsymbol{q} \tag{4.4}$$

where $\boldsymbol{I}$ is the identity tensor, δ_{ij}, $\boldsymbol{R}$ the position vector of the element, v its velocity, $\underline{\underline{\tau}}$ and $\underline{\underline{\dot{e}}}$ the total stress and strain rate tensors, respectively, and $\underline{q}$ the heat flux. Since $\underline{\underline{\tau}} = \underline{\underline{\sigma}} - p\underline{\underline{I}}$ and $\underline{\underline{I}} \cdot \cdot \underline{\underline{\dot{e}}} = \nabla \cdot \underline{\mathrm{v}} = \frac{1}{V}\frac{DV}{Dt}$, the dilation rate, where $V = 1/\rho$ is the specific volume, is

$$\underline{\underline{\tau}} \cdot \cdot \dot{\boldsymbol{e}} = -p\frac{DV}{Dt} \tag{4.5}$$

when the viscosity, $\boldsymbol{\sigma}$, is zero.

Neglecting the dissipational terms $\boldsymbol{\sigma}$ and $\underline{q}$, these equations represent sound waves for small amplitude. This suggests that as a starting point we look for one-dimensional difference forms similar to the successful wave solution, i.e. in a leap-frog form.

In one dimension, written in terms of a general spatial variable R and for an ideal dissipation-less fluid, the above equations reduce to

$$\begin{aligned} \frac{DR}{Dt} &= \mathrm{v} \\ \rho\frac{D\mathrm{v}}{Dt} &= -\frac{\partial p}{\partial R} \\ \frac{D\epsilon}{Dt} &= -p\frac{DV}{Dt} \end{aligned} \tag{4.6}$$

and

$$V = V_0\left(\frac{R}{R_0}\right)^{\nu} \tag{4.7}$$

where V_0 and R_0 are the initial values of the volume and radius, respectively, and $\nu = 1/2/3$ in Cartesian/cylindrical/spherical geometry, respectively.

These dynamic equations are supplemented by the equation of state $p = f(\rho, \epsilon)$. For the particular case of a polytropic gas of index γ, this takes the form

$$p = (\gamma - 1)\rho\epsilon \qquad \text{or} \qquad pV = (\gamma - 1)\epsilon \tag{4.8}$$

Introducing the cumulative mass as a Lagrangian co-ordinate,

$$M = \int_0^R \rho R^{\nu-1} \mathrm{d}R \tag{4.9}$$

we have

$$\begin{aligned} \frac{DR}{Dt} &= \mathrm{v} \\ \frac{D\mathrm{v}}{Dt} &= -R^{(\nu-1)}\frac{\partial p}{\partial M} \\ \frac{D\epsilon}{Dt} &= -p\frac{DV}{Dt} \end{aligned} \tag{4.10}$$

4.2 One-Dimensional von Neumann–Richtmyer Algorithm

Let us express these equations in a leap-frog finite difference form (von Neumann and Richtmyer, 1950). We initialise the system by identifying a set of fluid particles with integer $K \in (1, KT)$ at R_K. Imagine that these particles lie at the centre of a cell lying between $R_{K-1/2}$ and $R_{K+1/2}$, the latter being the grid points or nodes.

We follow the motion of the fluid points R_K and $R_{K+1/2}$

$$R^{n+1}_{K+1/2} = R^n_{K+1/2} + \mathrm{v}^{n+1/2}_{K+1/2}\, Dt \tag{4.11}$$

The volume of the cell is given by

$$(R_{K+1/2})^{\nu} - (R_{K-1/2})^{\nu} \tag{4.12}$$

Hence, if DM_K is the (constant) mass of the Kth cell $(R_{K-1/2} - R_{K+1/2})$ per unit area, (length × radian), or steradian in planar, cylindrical or spherical geometry, then

$$V^{n+1}_K = [(R^{n+1}_{K+1/2})^{\nu} - (R^{n+1}_{K+1/2})^{\nu}] \,/\, DM_K \tag{4.13}$$

and

$$\epsilon_K^{n+1} = \epsilon_K^n - \frac{1}{2}\left(p_K^{n+1} + p_K^n\right)\left(V_K^{n+1} - V_k^n\right) \tag{4.14}$$

We note that since $p = p(\epsilon, V)$ this is an implicit equation, linear only if p varies linearly with ϵ, i.e. the medium is a polytropic gas and

$$\mathrm{v}_{K+1/2}^{n+1/2} = \mathrm{v}_{K+1/2}^{n-1/2} - (R_{K+1/2}^n)^{\nu}(p_{K+1}^n - p_K^n)\, Dt \Big/ \left[\frac{1}{2}(DM_K + DM_{K+1})\right] \tag{4.15}$$

4.2.1 Stability

We linearise these equations for small perturbations

$$V_K^{n+1} - V_K^n = \frac{(R_K)^{(\nu-1)}}{DM_K}\left(\mathrm{v}_{K+1/2}^{n+1/2} - \mathrm{v}_{K-1/2}^{n+1/2}\right) Dt \tag{4.16}$$

Writing

$$\Delta\epsilon = \left.\frac{\partial\epsilon}{\partial p}\right|_V \Delta p + \left.\frac{\partial\epsilon}{\partial V}\right|_p \Delta V \tag{4.17}$$

gives

$$\left.\frac{\partial\epsilon}{\partial p}\right|_V (p_K^{n+1} - p_K^n) + \left(p + \left.\frac{\partial\epsilon}{\partial V}\right|_p\right)(V_K^{n+1} - V_K^n) = 0 \tag{4.18}$$

The sound speed can be written as

$$c^2 = V^2\left(p + \left.\frac{\partial\epsilon}{\partial V}\right|_p\right) \Big/ \left.\frac{\partial\epsilon}{\partial p}\right|_V \tag{4.19}$$

Hence,

$$\mathrm{v}_{K+1/2}^{n+1/2} - \mathrm{v}_{K+1/2}^{n-1/2} = \frac{(R_{K+1/2})^{(\nu-1)}\,(V_{K+1}^n - V_K^n)\, c^2\, Dt}{\frac{1}{2}(V_{K+1/2}^{n+1/2})^2\,(DM_K + DM_{K+1})} \tag{4.20}$$

This pair of equations are the standard leap-frog solutions of the wave equations, which are stable if

$$\left[\frac{c\, R^{(\nu-1)}\, Dt}{V\, DM}\right]^2 \le 1 \tag{4.21}$$

i.e.

$$\frac{c\, Dt}{\Delta R} \le 1 \tag{4.22}$$

where $\Delta R = R_{K+1/2} - R_{K-1/2}$. This equation will be immediately recognised as just the familiar Courant–Friedrichs–Lewy condition for this particular system.

4.2.2 Artificial Viscosity

The von Neumann–Richtmyer algorithm is essentially third order in accuracy. It, thus, has no implicit artificial dissipation within the algorithm. Nonetheless, it is not susceptible to non-linear instability and is generally well behaved, except in shocks. The system does not however contain the essential dissipation necessary to generate the entropy increase required in a shock.

Classical fluid mechanics shows that such dissipation can be provided by either viscosity or heat conduction. When viscosity is included all transitions are smooth, not discontinuous. The thickness of the transition is proportional to the coefficient of the dissipation. In practice, including a real viscosity is generally so weak that the shock would be extremely narrow, i.e. much less than the resolution of the mesh. As a result, it would not be adequate to prevent the undesirable overshoot and oscillation at the shock, which occurs in the dissipationless system.

The solution is to add an artificial viscous term, whose role is to spread the shock over about 3 or 4 mesh points. With a linear (constant coefficient) viscosity, where the stress is proportional to the strain rate, the shock width is always proportional to shock strength. However, if the viscosity is quadratic the shock width is constant, independent of the shock strength. This is an attractive feature, which has led to one particularly popular form of artificial viscosity (von Neumann and Richtmyer, 1950):

$$q = \begin{cases} \ell^2 \rho (\partial u/\partial x)^2 & \text{if} \quad \partial u/\partial x < 0 \\ 0 & \text{otherwise} \end{cases} \tag{4.23}$$

With this term, the Lagrangian equations become in Cartesian geometry

$$\begin{aligned} \frac{DX}{Dt} &= u \\ \frac{Du}{Dt} &= -\frac{1}{\rho_0}\frac{\partial(p+q)}{\partial x} \\ \frac{D\epsilon}{Dt} &= -(p+q)\frac{\partial V}{\partial t} \\ V &= \frac{1}{\rho_0}\frac{\partial X}{\partial x} \end{aligned} \tag{4.24}$$

where x is the initial co-ordinate of the fluid element and X its current value, and ρ_0 the initial density. Hence,

$$\frac{DV}{Dt} = \frac{1}{\rho_0}\frac{\partial u}{\partial x} \tag{4.25}$$

We assume an equation of state for a polytropic gas

$$\epsilon = pV/(\gamma - 1) \tag{4.26}$$

Let us look for a travelling wave solution moving in the x-direction with speed s in the Lagrangian system (von Neumann and Richtmyer, 1950). The variables are functions of $w = x - st$

$$\begin{aligned} \rho_0 s\frac{dV}{dw} + \frac{du}{dw} &= 0 \\ \rho_0 s\frac{du}{dw} - \frac{d(p+q)}{dw} &= 0 \\ \frac{d\epsilon}{dw} + (p+q)\frac{dV}{dw} &= 0 \end{aligned} \tag{4.27}$$

If $M = \rho_0 s$ – the mass swept up by the wave per unit time per unit area

$$\begin{aligned} MV + u &= C_1 \\ M^2 V + p + q &= C_2 \\ \frac{1}{2}M^2V^2 + \epsilon + (p+q)V &= C_3 \end{aligned} \tag{4.28}$$

If the variables approach asymptotic values as $x \to -\infty$ and $+\infty$ with zero derivatives

$$\begin{aligned} M(V_1 - V_2) + (u_1 - u_2) &= 0 \\ M^2(V_1 - V_2) + (p_1 - p_2) &= 0 \\ \frac{1}{2}M^2(V_1^2 - V_2^2) + (\epsilon_1 - \epsilon_2) + p_1V_1 - p_2V_2 &= 0 \end{aligned} \tag{4.29}$$

which are the Rankine–Hugoniot equations,

Since

$$pV/(\gamma - 1) = \epsilon = \frac{1}{2}M^2V^2 - C_2V + C_3$$

we obtain

$$\begin{aligned} qV &= -\frac{1}{2}(\gamma + 1)M^2V^2 + C_4V + C_5 \\ &= \frac{1}{2}(\gamma + 1)M^2(V_2 - V)\,(V - V_1) \end{aligned} \tag{4.30}$$

For $\frac{du}{dw} \geq 0$, $q = 0$ and $V = V_1$ or V_2. In the wave $\frac{du}{dw} \leq 0$, i.e. only compression waves are allowed, as happens in reality. The artificial viscosity

$$q = \frac{\ell^2}{V}\left(\frac{\mathrm{d}u}{\mathrm{d}w}\right)^2 = \frac{\ell^2M^2}{V}\left(\frac{\mathrm{d}V}{\mathrm{d}w}\right)^2 \tag{4.31}$$

gives

$$\left(\ell\frac{\mathrm{d}V}{\mathrm{d}w}\right)^2 = \frac{1}{2}(\gamma + 1)(V_2 - V)(V - V_1) \tag{4.32}$$

and

$$V = \frac{1}{2}(V_1 + V_2) + \frac{1}{2}(V_2) - (V_1)\sin\left\{\sqrt{\left[\frac{\gamma + 1}{2}\right]}\frac{(w - w_0)}{\ell}\right\} \tag{4.33}$$

This solution is only valid for a half sine wave since $\mathrm{d}V/\mathrm{d}w \geq 0$. Hence, the full solution is

$$V = \begin{cases} V_1 & (x - st) \leq \frac{-\pi\ell}{\sqrt{\{(\gamma+1)/2\}}} \\ \frac{1}{2}(V_1 + V_2) + \frac{1}{2}(V_2 - V_1)\sin\left\{\sqrt{\left[\frac{\gamma+1}{2}\right]}\frac{(x-st)}{\ell}\right\} & |x - st| \leq \frac{\pi\ell}{\sqrt{\{(\gamma+1)/2\}}} \\ V_2 & (x - st) \geq \frac{\pi\ell}{\sqrt{\{(\gamma+1)/2\}}} \end{cases} \tag{4.34}$$

A typical solution is shown in Figure 4.1. The effective shock width is slightly smaller than $\sqrt{\{2/(\gamma + 1)\}} \cdot \pi\ell$, typically ℓ.

4.2.3 Finite-Difference Form

A standard method of forming the numerical artificial viscosity is to replace ℓ by $a\Delta x$ where a is a constant (typically in the range 1.5–2.0), which defines the thickness of the shock (typically 3–5 cells). This gives

$$q_K^{n+1/2} = \begin{cases} a^2\rho_K^{n+1/2}\left(\mathrm{v}_{K+1/2}^{n+1/2} - \mathrm{v}_{K-1/2}^{n+1/2}\right)^2 & \text{if } \left(\mathrm{v}_{K+1/2}^{n+1/2} - \mathrm{v}_{K-1/2}^{n+1/2}\right) < 0 \\ 0 & \text{otherwise} \end{cases} \tag{4.35}$$

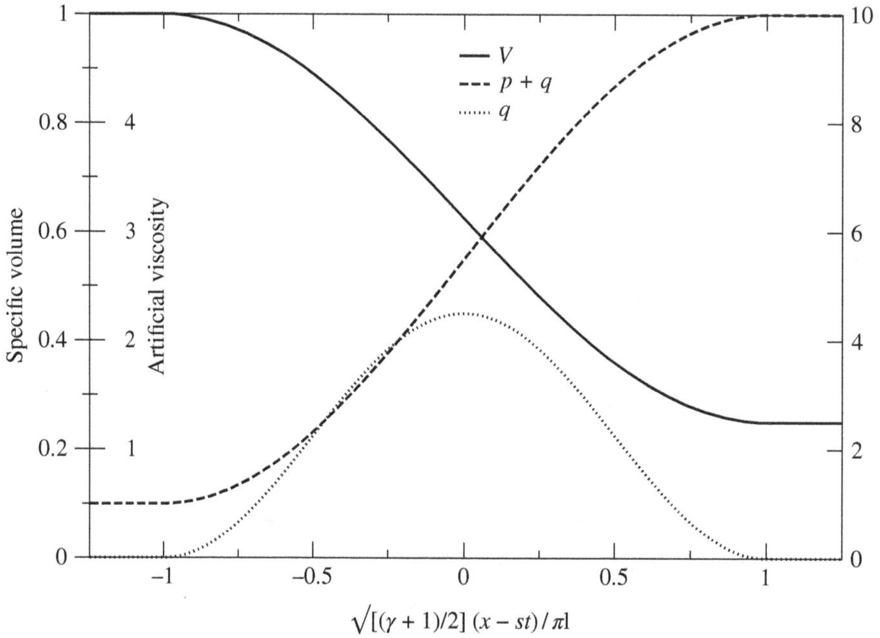

Figure 4.1 Artificial viscosity stabilised shock profile for gas of polytropic index, $\gamma = 5/3$ with a pressure jump ratio 10.

Hence, the acceleration equation becomes

$$\mathrm{v}_{K+1/2}^{n+1/2} = \mathrm{v}_{K+1/2}^{n-1/2} - \frac{\left(p_{K+1}^{n} + q_{K+1}^{n-1/2} - p_{K}^{n} - q_{K}^{n-1/2}\right)}{\left[\frac{1}{2}(DM_{K+1} + DM_{K})\right]} Dt \tag{4.36}$$

and the energy equation

$$\epsilon_{K}^{n+1} = \epsilon_{K}^{n} - \left[\frac{1}{2}(p_{K}^{n+1} + p_{K}^{n}) + q_{K}^{n+1/2}\right](V_{K}^{n+1} - V_{K}^{n}) \tag{4.37}$$

Figure 4.2 compares the result obtained from the standard shock tube test with values of a set to 1.0 and 2.0. It can be seen that the oscillations in the shock are progressively reduced, but at the expense of a broader shock.

An extension to the stability condition is required to include artificial viscosity. A suitable, and extensively used, condition is

$$\frac{cDt}{\Delta R} + \frac{4a^2 \left|V_{K}^{n+1} - V_{K}^{n}\right|}{V_{K}^{n}} \leq 1 \tag{4.38}$$

4.3 Multidimensional Lagrangian Schemes

The generalisation of the one-dimensional Lagrangian scheme to more than one dimension introduces serious difficulties. These arise from the nature of the Lagrangian description in which the mesh is tied to the fluid. In consequence, as the fluid deforms the mesh must follow it. As a result, the mesh may become severely distorted by shearing or vorticity.

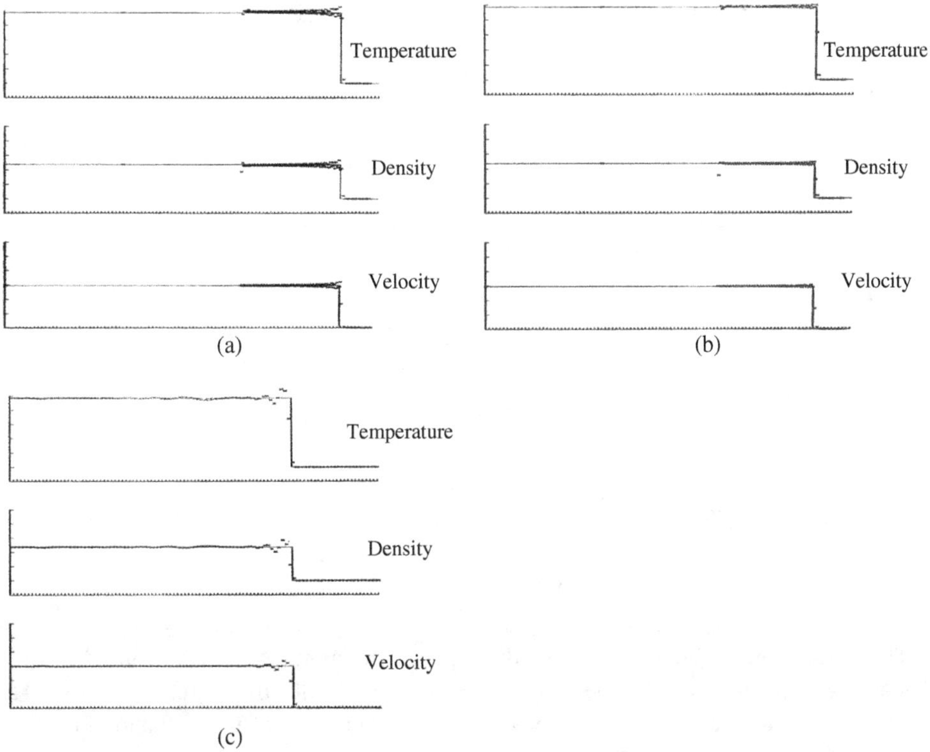

Figure 4.2 The shock wave test calculated with the von Neumann–Richtmyer Lagrangian scheme. The results are shown on the Eulerian mesh after 800 time steps, calculated with a Courant number 0.5; the artificial viscosity factor a is (a) 1.0, (b) 2.0 and (c) 4.0. Note the appreciable reduction in oscillation at the shock front as the artificial viscosity is increased. The oscillation in the boundary is associated with deviations in the initial conditions from the steady shock profile, and is spurious.

Since the mesh is no longer orthogonal, the mathematical, and consequent numerical, form is considerably more complex than in one dimension. In particular, the overall structure of the original mesh may be lost.

4.3.1 Mesh Tangling

The origin of these pathological problems lie in what are known as null space modes or zero energy modes. Null space modes refer to patterns of velocity that distort the shape and size of a cell but produce no restoring forces since there is no volume change or therefore internal energy change. Typical representations are combinations of orthogonal hourglass modes. The origin of these can be seen by comparing the number of degrees of freedom available to the fluid motion with those of the cell (Wilkins, 1969; Flanagan and Belytschko, 1981; Margolin and Pyun, 1987). In two dimensions, the fluid has a total of six degrees of freedom: two from the translational motion, one from the rotation, and three from the strain rate tensor [one volume change ($\dot{\epsilon}_{11} + \dot{\epsilon}_{22}$), one shear ($\dot{\epsilon}_{12} = \dot{\epsilon}_{21}$) and one changing the aspect ratio ($\dot{\epsilon}_{11} - \dot{\epsilon}_{22}$)]. The cell, on the other hand, has eight degrees of freedom: four boundary points each with two degrees of freedom. There are therefore two additional modes which lead to non-physical behaviour, such as that shown in Figure 4.3.

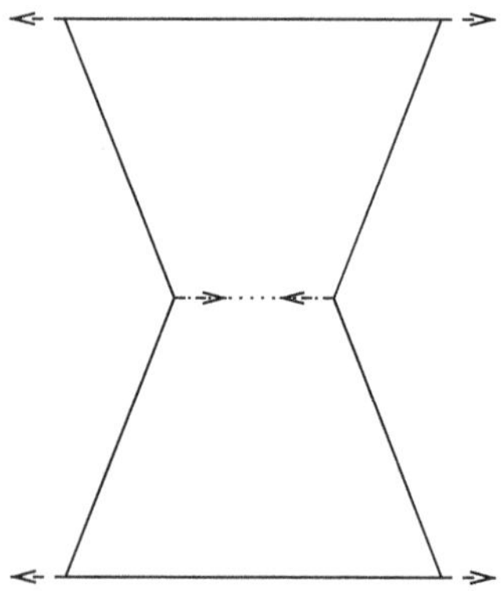

Figure 4.3 Basic hourglass mode.

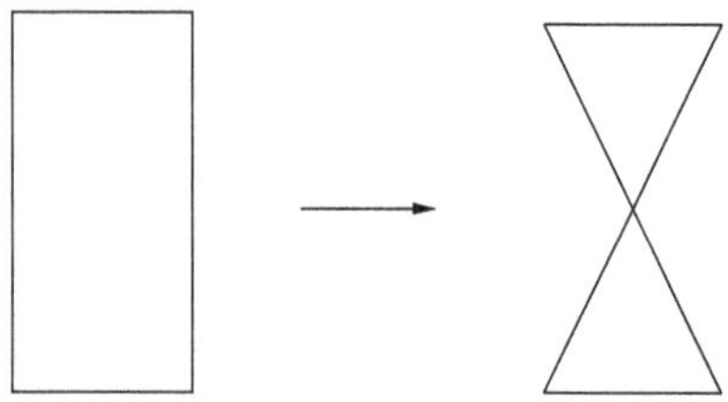

Figure 4.4 Creation of a 'bowtie' cell formed by mesh point overtaking.

The most severe problems arise from the essential differential motions within the fluid, and in the worst form lead to a general tangling of the cells illustrated in Figures 4.4–4.11. In its least severe form, this gives rise to distorted cells due to the differential motion associated with hourglass modes and other problems. In this case, the finite difference representation over the mesh is poor, but if the code is reasonably robust, will not give rise to pathological errors. The execution of the code will however fail, if pathological 'bowtie' zones are formed by one mesh overtaking its neighbour.

A number of remedies have been proposed to overcome these problems. None are completely satisfactory. Two approaches are used:

- Conventional Lagrangian methods use cells with mesh values defined at the corners. The advantage of these approaches is that the mesh structure and ordering is retained, although deformed. Mesh point neighbours are therefore unchanged, enabling finite differences to be readily calculated. However, this computational advantage is obtained at a penalty in the hydrodynamics. The simplest zones are triangular, which eliminate the problems associated with overtaking. Unfortunately, if the mesh point order is retained, these meshes impose an undesirable stiffness on the flow, in order to satisfy the constraints on the mesh. Quadrilateral zones are less stiff but must be provided with a safeguard against mesh points overlying each other. A relatively simple method (Flanagan and Belytschko, 1981; Margolin and Pyun, 1987) to alleviate the hourglass problem uses a mesh filter on the velocity to identify and then correct the velocity values to remove the modes. An alternative method (Wilkins, 1969; Browne and Wallick, 1973) uses local artificial viscosity, which establishes a repulsive stress as mesh points approach, when null space modes are identified using sub-zones of the full cell. However, this can also introduce weak stiffness into the calculation.

  Mesh deformation effects may be mitigated by re-zoning when severe distortion is established. If done automatically during the run, it introduces the problem of identifying both a limiting level of distortion and a satisfactory new mesh on to which the old can be interpolated. For obvious reasons, interpolation should be carried out as infrequently as possible, and should be conservative, so that the fluid dynamic mass, momentum and energy relations are upheld as accurately as possible.

- Alternatively, we may allow the mesh complete freedom to move as dictated by the hydrodynamic equations *Free Lagrangian Methods*. In this case, the structure of the mesh is lost at each time-step, and must be recovered before the next time-step integration can be attempted. In its most general form, this would require a full sort of all the mesh points each step, and could be prohibitively expensive in computer time. Fortunately, most of the ordering from the previous step is retained at the completion of the advance, and can be updated to generate the new structure with reasonable efficiency. Since the resultant grid may be highly deformed, finite element methods are very appropriate for calculating the spatial differences, giving rise to the time derivatives. Finite differences are appropriate for handling the temporal integration. In this case, the mesh is structure-less, and consequently there are no fluid cells, as in earlier schemes, with well-defined fluid mass, momentum and energy. The conservation laws thus have no significance in a local sense and can only be interpreted globally through integration using the finite element interpolations.

The choice of approach is dictated by the application planned. For a purely hydrodynamic problem, the free Lagrangian, finite element scheme is very attractive, despite the complex nature of the coding. However, if the fluid dynamics is only a component of a more general systems involving a number of additional physical effects, the more conventional cell-based schemes, with routine re-zoning, may be more appropriate.

4.3.2 Jacobian Methods

As an example, we consider the standard two-dimensional Lagrangian calculation based on the use of two co-ordinate systems (x, y) in the laboratory, and (k, ℓ) in the mesh (Schultz, 1964). The two systems are related by the Jacobian j of the transformation from the arbitrary (x, y) co-ordinates to the orthogonal (k, ℓ).

$$j = \frac{\partial(x,y)}{\partial(k,\ell)} = \begin{vmatrix} \dfrac{\partial x}{\partial k} & \dfrac{\partial x}{\partial \ell} \\ \dfrac{\partial y}{\partial k} & \dfrac{\partial y}{\partial \ell} \end{vmatrix} \tag{4.39a}$$

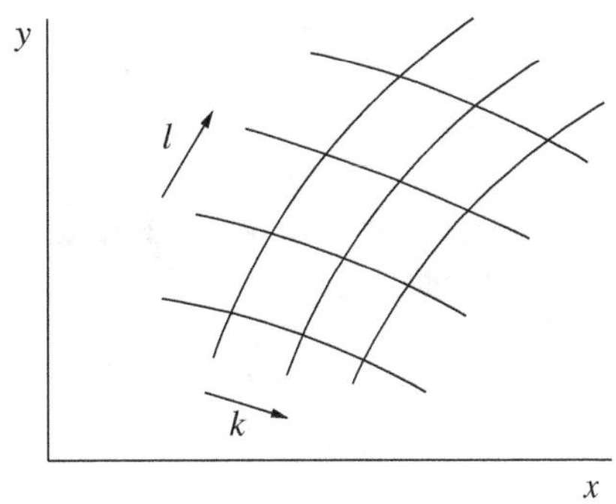

Figure 4.5 Relationship between the laboratory (x, y) and the Lagrangian (k, ℓ) co-ordinate systems.

Hence, for an arbitrary function $f(k, \ell)$

$$\frac{\partial f}{\partial x} = \frac{1}{j}\left(\frac{\partial y}{\partial \ell}\cdot\frac{\partial f}{\partial k} - \frac{\partial y}{\partial k}\cdot\frac{\partial f}{\partial \ell}\right) \tag{4.39b}$$

$$\frac{\partial f}{\partial y} = \frac{1}{j}\left(-\frac{\partial x}{\partial \ell}\cdot\frac{\partial f}{\partial k} + \frac{\partial x}{\partial k}\cdot\frac{\partial f}{\partial \ell}\right) \tag{4.39c}$$

Introducing the position vector $\underline{R} = (x, y)$ and its normal $\overline{\underline{R}} = (y, -x)$, we may write

$$\nabla f = \frac{1}{j}\left\{\frac{\partial \overline{\underline{R}}}{\partial \ell}\cdot\frac{\partial f}{\partial k} - \frac{\partial \overline{\underline{R}}}{\partial k}\cdot\frac{\partial f}{\partial \ell}\right\} \tag{4.39d}$$

and

$$j = \frac{\partial \underline{R}}{\partial k}\cdot\frac{\partial \overline{\underline{R}}}{\partial \ell} \tag{4.39e}$$

and therefore

$$\frac{\partial j}{\partial t} = j\,\nabla\cdot\underline{\mathrm{v}} \tag{4.39f}$$

In this co-ordinate system, the equations of fluid mechanics take the form:

$$\begin{aligned} \rho j &= const. \\ \rho\frac{\partial \underline{\mathrm{v}}}{\partial t} + \nabla p &= 0 \\ \frac{\partial \epsilon}{\partial t} + p\frac{\partial V}{\partial t} &= W \end{aligned} \tag{4.39g}$$

where W is the heat/energy source per unit volume due to external heating, thermal conduction and additional factors.

The equations are solved by a leap-frog algorithm in an identical manner to the von Neumann–Richtmyer LED scheme described earlier in terms of the parameters $\underline{R}^n_{k+1/2,\ell+1/2}$ and $\underline{\mathrm{v}}^{n+1/2}_{k+1/2,\ell+1/2}$ at the cell corners and $p^n_{k+1/2,\ell+1/2}, \rho^n_{k+1/2,\ell+1/2}, \epsilon^n_{k+1/2,\ell+1/2}$ at the centres, supplemented by an equation of state $p = p(\rho, \epsilon)$. In this classical form, the energy is only conserved to terms $O(Dt)$.

4.3.3 Integral Methods

We may alternatively approach this problem of formulating the Lagrangian fluid equations in numerical approximation using integral conservation law forms. Thus, we calculate the accelerations at a mesh corner $\underline{R}_{k+1/2,\ell+1/2}$ by integrating the pressure force over some surface, which encloses the corner, and by dividing by the mass of fluid enclosed by the surface. These surfaces may be defined in a number of ways, either as square or diamond volumes surrounding the mesh point (Browne and Wallick, 1973). Thus, the force acting on the mesh corner point $\underline{R}_{k+1/2,\ell+1/2}$

$$\underline{f}_{k+1/2,\ell+1/2} = -\int_s p\,\mathrm{d}\underline{s} \tag{4.40}$$

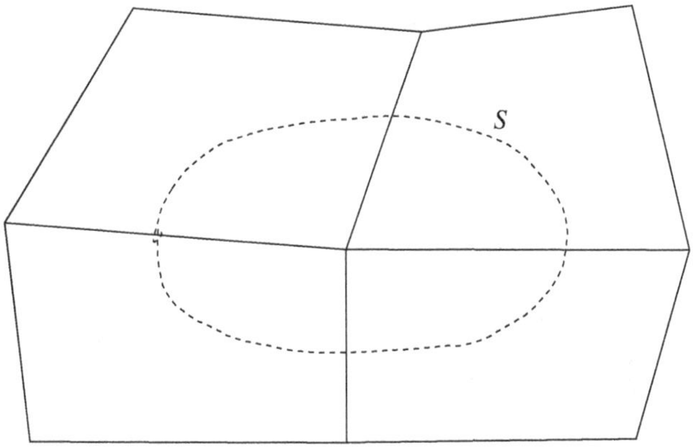

Figure 4.6 General integration surface S within the neighbouring cells surrounding the mesh point.

Figure 4.7 Sub-cell mesh with rectangular sub-zones in (k, ℓ) space.

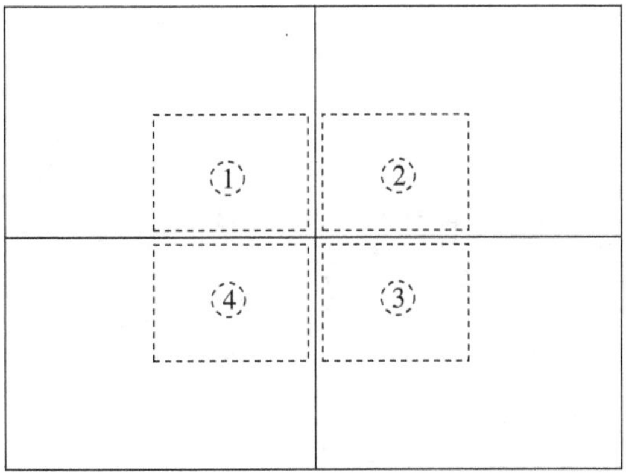

In Cartesian geometry, simple rectangular or diamond integration zones yield the same result, and are satisfactory. In cylindrical geometry, the values are different. However, if it is required to treat problems in spherical geometry and maintain the symmetry, a further sub-division into four sub-zones around the mesh point is advisable. In this case, the acceleration is written as

$$\frac{\partial \underline{\mathrm{v}}}{\partial t} = \sum_{i=1}^{4} \frac{f_i}{M_i} \tag{4.41}$$

The diamond sub-zone form is very similar to the Jacobian form. The integration procedure is similar to that described earlier for Jacobian methods. We note that these methods should have superior energy conservation properties compared with the Jacobian approach.

4.3.4 Artificial Viscosity

The role of artificial viscosity in two or more dimensions is more complex than in one. As before, the dissipation must provide the entropy generation to stabilise any shocks, and to avoid overshoot and the subsequent generation of ripples. In addition, we may require

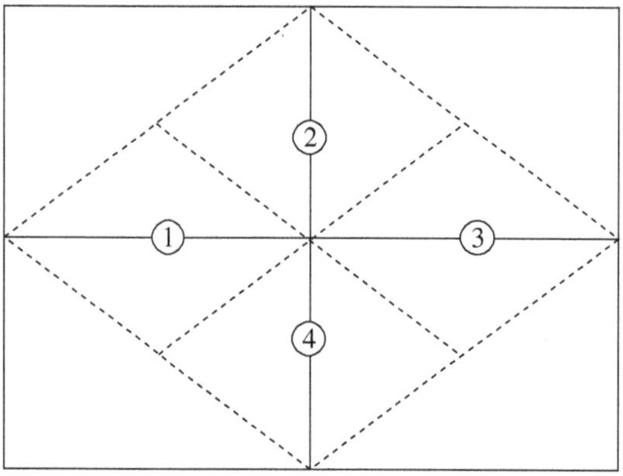

Figure 4.8 Sub-cell mesh with diamond sub-zones in (k, ℓ) space.

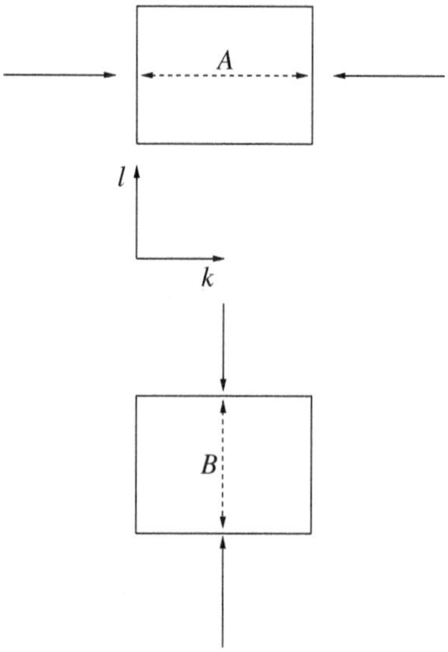

Figure 4.9 Components of the artificial viscosity produced into response to compression.

the dissipation to 'tidy up' spurious ripples (high spatial frequency noise) as in Eulerian schemes, and also to prevent severe mesh distortion.

Since there are two directions of mesh deformation, we require a two-component artificial viscosity, which must maintain the flow velocity parallel to any shock front unchanged.

Schultz' prescription (Schultz, 1964; Wilkins, 1980), which has been widely followed, is to introduce a tensor artificial viscosity in the form

$$\frac{\partial \underline{\mathrm{v}}}{\partial t} + \frac{\nabla p}{\rho} = -\frac{1}{M}\left[\frac{\partial}{\partial k}\left(q_A \frac{\partial \underline{\overline{R}}}{\partial \ell}\right) - \frac{\partial}{\partial \ell}\left(q_B \frac{\partial \underline{\overline{R}}}{\partial k}\right)\right] \tag{4.42}$$

and

$$\frac{\partial \epsilon}{\partial t} + p\frac{\partial V}{\partial t} = \frac{q_A}{M}\frac{\partial}{\partial k}\left(\frac{\partial \underline{\overline{R}}}{\partial \ell}\cdot \underline{\mathrm{v}}\right) + \frac{q_B}{M}\frac{\partial}{\partial \ell}\left(\frac{\partial \underline{\overline{R}}}{\partial k}\cdot \underline{\mathrm{v}}\right) \tag{4.43}$$

where $M = \rho j = const.$

To ensure that uniform compression does not increase entropy, we take the form for q_A and q_B as

$$q_A = -a^2\rho\left(\frac{\partial \underline{v}}{\partial k}\right)_A^- \cdot \left|\frac{\partial}{\partial k}\left(\frac{\partial \underline{v}}{\partial k}\right)_A^-\right|$$
$$q_B = -a^2\rho\left(\frac{\partial \underline{v}}{\partial \ell}\right)_B^- \cdot \left|\frac{\partial}{\partial \ell}\left(\frac{\partial \underline{v}}{\partial \ell}\right)_B^-\right| \quad (4.44)$$

where

$$\left(\frac{\partial \underline{v}}{\partial k}\right)_A = \frac{\partial\overline{\underline{R}}/\partial\ell \cdot \partial\underline{v}/\partial k}{|\partial\overline{\underline{R}}/\partial\ell|}$$
$$\left(\frac{\partial \underline{v}}{\partial \ell}\right)_B = -\frac{\partial\overline{\underline{R}}/\partial k \cdot \partial\underline{v}/\partial\ell}{|\partial\overline{\underline{R}}/\partial k|} \quad (4.45)$$

and $\gamma^- = \min(0, \gamma)$. q_A is produced by compression along the k direction and q_B along the ℓ.

This form is not satisfactory for long, thin zones, which may be compressing at one end but expanding at other. The remedy is introduce a different q for each side to avoid a pathological zone.

A further problem with such zones involves the energy change due to compression, which is added to the energy at the centre and leads to an increase in pressure over the whole zone. The net result is a spurious energy transfer along the mesh. The remedy for this problem

Figure 4.10 Response of artificial viscosity applied to long, thin zones.

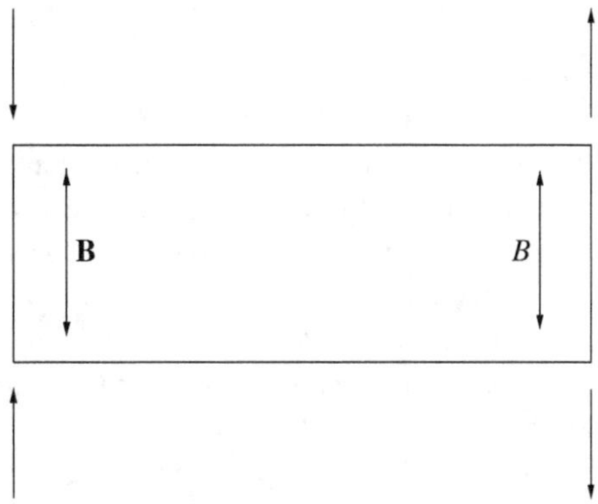

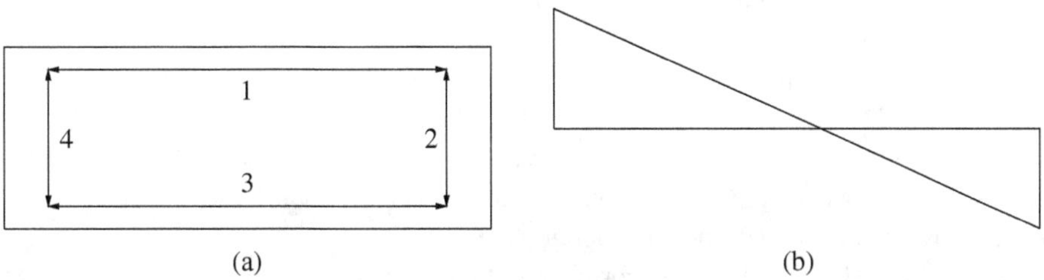

(a) (b)

Figure 4.11 Tensor viscosity components for each side of the cell prevents the creation of irregular zones. (a) Separate viscosity components for each side and (b) irregular or pathological zones.

may lie in the use of temporary triangular sub-zones (Browne and Wallick, 1973), but the programming is difficult and involves the sub-division of the mesh into retained structures, which may not be satisfactory.

4.4 Choice of Method

To a large extent, the choice of method is problem dependent. There is no best method, different methods handle different problems well. However, it is possible to identify some guidelines, which may help to make a sensible choice.

1) Composition differences are difficult to handle in Eulerian geometry due to inherent numerical smoothing, which cannot be avoided in the advection. Use a Lagrangian scheme, if possible.
2) Lagrangian schemes in one dimension are robust, easy to construct and use.
3) Lagrangian schemes are complicated and not robust to two or three dimensions.
4) Eulerian methods are generally robust, but not as easy to programme in one dimension. Their scope can be extended by adjusting the grid during the run.
5) Eulerian schemes are most appropriate for systems with complex boundary conditions, e.g. aerofoils.
6) Steady-state problems can be tackled with Eulerian schemes, by allowing them to time-march to a steady condition.

References

Browne, P L and K B Wallick. 1973. The reduction of mesh tangling in two-dimensional Lagrangian hydrodynamics by the use of viscosity, artificial viscosity and TTS (Temporary Triangular Subzoning for Long Thin Zones). Technical Report LA-4740-MS. Los Alamos National Laboratory.

Flanagan, D P and T Belytschko. 1981. "A uniform strain hexahedron and quadrilateral with orthogonal hourglass control." *Int J Num Meth Eng* 17:679–708.

Margolin, L G and J J Pyun. 1987. A method for treating hourglass patterns. In *Proceedings of the 5th International Conference on Numerical Methods in Laminar and Turbulent Flow, Montreal.* pp. 149–160.

von Neumann, J and R D Richtmyer. 1950. "A method for the numerical calculation of shocks." *J Appl Phys* 21:232–237.

Schultz, W D. 1964. Two dimensional Lagrangian hydrodynamic difference equations. In *Fundamental Methods in Hydrodynamics*, ed. B Alder, B Fernbach and M Rotenberg. Number 3 *in* "Methods in Computational Physics" New York: Academic Press pp. 1–45.

Wilkins, M L. 1969. Calculation of elastic-plastic flow. Technical Report UCRL-7322, Rev.1. Lawrence Livermore National Laboratory.

Wilkins, M L. 1980. "Use of artificial viscosity in multidimensional fluid dynamic calculations." *J Comput Phys* 36:2812–303.

5

Arbitrary Lagrangian–Eulerian Schemes

5.1 Introduction

As we have seen in Chapter 1 and in more detail in Chapter 3 there is no ideal method for tackling all problems. For certain problems, particularly those not involving material interfaces and fixed boundaries, Eulerian codes offer the best solution. On the other hand, changes of composition within the fluid or free expansion are better handled by a Lagrangian code. There remains a group of problems which do not conveniently fit into either category. These are multidimensional problems with time-varying boundaries and composition interfaces.

We may attempt to handle these types of problems by 'fixing' our original schemes to overcome their deficiencies. In the case of the Eulerian methods, this means allowing the boundaries to move and interfaces to remain fixed in the fluid. In the Lagrangian case, we adjust the cell boundaries to remove any of the damaging distortions, whilst retaining the interface surfaces. Both these two approaches lead to the same result, and the broad general class of codes so obtained is known as arbitrary Lagrangian–Eulerian (ALE) codes. A useful review of these methods is given by Margolin (1997), which highlights both the successes and failures of the methods prior to 1995. Subsequently, there has been extensive further development of ALE methods, much of it centred around the use of finite element methods (Donea et al., 2004).

To see the equivalence of these two approaches, we note that we may regard the standard development given by the equations of fluid mechanics as two steps. The first is the movement of the fluid element (convection) in a short time interval. This is followed by the second (interpolation) which re-assigns the values associated with the elements back to the co-ordinate points. The Lagrangian formulation treats a co-ordinate system fixed in the fluid, moving directly to the new state without interpolation. The Eulerian development follows the convection with interpolation back on to the original (laboratory) co-ordinate system. Thus, consider the Eulerian time derivative

$$\frac{\partial}{\partial t} = \frac{\mathrm{D}}{\mathrm{D}t} - \mathbf{v} \cdot \nabla \tag{5.1}$$

The term $\frac{\mathrm{D}}{\mathrm{D}t}$ is the Lagrangian time derivative and the term $(\mathbf{v} \cdot \nabla)$ the transport (re-zoning) back to the laboratory mesh.

Modelling and Simulation in Plasma Physics for Physicists and Mathematicians, First Edition. Geoffrey J. Pert.

We may consider these two developments separately and allow the co-ordinate system to move independently of the fluid by a suitable modification of the transport term to a co-ordinate system moving with velocity $\mathbf{v}'$, which may vary in space, with respect to the laboratory system, so that the time derivative in this system becomes

$$\frac{\partial}{\partial t} = \frac{\mathrm{D}}{\mathrm{D}t} - (\mathbf{v} - \mathbf{v}') \cdot \nabla \tag{5.2}$$

We note the expected limits:

1) If the co-ordinate system is fixed, i.e. Eulerian $\mathbf{v}' = 0$.
2) If the co-ordinate system moves with the fluid, i.e. Lagrangian, $\mathbf{v} = \mathbf{v}'$.

This involves two steps within our general formalism of a step on the old mesh followed by a transfer (re-zoning) on to the new mesh. A number of Eulerian schemes have been designed to work in this manner, for example, the FLIC scheme (Gentry, Martin and Daly, 1966). However, Eulerian schemes developed using the more general concepts of flux in which these steps are integrated, as in Chapter 3, cannot easily be adapted into the ALE approach.

It was very early appreciated that successful simulation using multidimensional Lagrangian codes demanded occasional re-zoning to overcome the mesh-tracking inherent in multidimensional Lagrangian codes, thereby to enable badly distorted Lagrangian meshes to be re-organised and the material interpolated from one mesh to the other. These methods identified problematic cells within the mesh and developed a re-organised mesh to cope with the problems. The material was generally moved as blocks, i.e. by a first-order interpolation scheme (essentially donor cell). However, these methods proved difficult to implement automatically due to problems of identifying when re-zoning was needed and how to adjust the cells (Wallick, 1987). Re-zoning generally took place at irregular intervals and was gross, essentially a conservative re-mapping from one mesh to the other (Dukowicz, 1984; Ramshaw, 1985). As re-zoning was relatively infrequent, errors associated with low-order routines were insignificant.

The original concept of ALE stems from the seminal paper of Hirt, Amsden and Cook (1974), which introduced an entirely new approach to these problems. Unlike the re-zoning techniques used previously, re-zoning was included on an equal footing with convection. Thus, both were carried out during each time step. This allowed much greater flexibility with a continuous controlled distortion of the computational mesh. This is the approach which has been followed subsequently. Hirt, Amsden and Cook (1974) also introduced a further important improvement for removing the Courant–Friedrichs–Lewy time-step constraint based on the sound speed (although there may be one associated with the re-zoning). They identified a series of steps to be followed in the programme development:

1) **The Lagrangian stage**: The flow element position, velocity and energy are updated taking into account pressure, viscosity and thermal conduction and any other external sources. This should be accomplished, if possible, using a conservative algorithm, e.g. the control volume method.
2) **The pressure method**: The optional inclusion of the pressure terms using an implicit scheme allows sound waves to propagate more than one cell width stably.

3) **Mesh generation**: The construction of a modified grid. Generally, the new mesh is constrained to follow the fluid motion, i.e. be Lagrangian in form, but without introducing badly structured cells.
4) **Re-zoning**: In their original paper, a first-order (donor cell) scheme was proposed. More modern schemes use either a second-order scheme such as a variant of flux corrected transport or van Leer's monotone scheme if the re-zoning is weak, or conservative re-mapping if extensive.

5.2 Step 1: The Lagrangian Stage

This step is basically uses a standard Lagrangian time-step, preserving as far as possible the essential flow symmetries, entropy and energy conservation. The integration of the Lagrangian methods with these conditions into an ALE scheme presents some new problems.

- The staggered grid is natural for Lagrangian systems which use the leap-frog approach, both for the spatial and temporal grids: the former being inherent in the definition of the nodes or grid points, which define the spatial structure of the cells. However, it is not the natural system for an Eulerian mesh where the density, velocities (momenta) and energies are defined at the cell centres.
- On a staggered spatial grid, kinetic and thermal energies are defined at separate locations, which makes exact energy conservation problematic.
- The staggered time integration can be avoided by using a predictor–corrector scheme, which will also introduce a measure of implicitness into the time integration. With a constant time step kinetic energy can be conserved in time by defining the kinetic energy as $\frac{1}{2}\mathbf{u}^{n+1/2} \cdot \mathbf{u}^{n-1/2}$.
- If the total energy (internal + kinetic) is conserved, this can lead to large errors in the internal energy, due to small errors in the kinetic energy.
- Entropy conservation is difficult to maintain, but use of the internal energy as the primary energy term will help to avoid large-scale growth.
- Care must be taken with the form of the finite difference operators to maintain spherical symmetry on a cylindrical grid, when the problem demands it (Whalen, 1996). This is a general problem with standard Lagrangian codes.
- The best form of artificial viscosity remains an unresolved problem. A useful background review of tensor viscosity is given by Wilkins (1980).

5.3 Step 2: The Iteration Stage

In materials where the flow is strongly sub-sonic, e.g. nearly solid materials, the integration would be severely constrained by the Courant–Friedrichs–Lewy condition based on the sound speed. We may avoid this limitation by treating the pressure as an implicit term (Hirt, Amsden and Cook, 1974). Thus, we perform a first predictor, Lagrangian step using the pressure at the old time step. The Lagrangian equations are linearised in the pressure and velocity differences between the first and the final corrector step, which uses the pressure at the end of the time step. The pressure difference is calculated by a Newton–Raphson iteration; the resulting simultaneous equations are typically five diagonal and can be solved by a preconditioned matrix method, e.g. ICCG (Kershaw, 1978).

5.4 Step 3: Mesh Generation

The mesh generation stage follows the general principles for mesh generation outlined earlier for Eulerian systems. The mesh generation usually takes place continuously at the same rate as the flow variables. For this approach, a standard mesh generation scheme such as, for example, Winslow's equipotential method (Winslow, 1963, 1981) is particularly valuable, a single iteration usually being sufficient to generate a suitable mesh on continuous adjustment. A modification (Brackbill and Saltzmann, 1982) to the equipotential method allows meshes to be generated adaptively to varying zone sizes and orthogonality in response to specific constraints.

Alternatively, we may decide to operate as a Lagrangian code with occasional re-zoning. Difficulties arise from the need for automatically identifying when to implement the rezoning algorithm and how the grid should be formed. There are few general rules for this step.

The several differencing advantages of an orthogonal grid point one in the direction of a re-zoning form which generates orthogonal grids automatically. This is simply achieved by moving the grid lines x or y as a block normally with a suitable prescription for the perpendicular grid point velocity. Moving each set of grid lines normally maintains the orthogonality automatically and is easy to implement the necessary 'advection'. The 'grid velocity' prescription of the mesh line i can be based on a suitable average of the flow velocity on i, to give a 'quasi-Lagrangian' form of re-zoning. The one-dimensional form of this method was examined earlier by the author Pert (1983).

5.5 Step 4: Re-zoning

Re-zoning follows the same general rules as for advection in Eulerian codes; namely, it should be conservative and avoid spurious oscillation. Since re-zoning is essentially the same operation as advection, it follows that the best methods discussed earlier, such as monotonicity preserving or flux-corrected transport are also appropriate in this case also, provided the re-zoning is small scale, i.e. no zone boundary moves across another cell. One of these methods is therefore used for continuous re-zoning.

Significant improvements in computational speed can be obtained, when re-zoning is not invoked every cycle, but relatively infrequently. However, in this case, the re-zoning is gross, and advection methods cannot be used: some form of direct mapping from one mesh to the other must then be used. The general conservative nonlocal re-mapping algorithm developed by Dukowicz, Ramshaw, Dukowicz and Kodis is particularly suitable for this purpose. In its simplest form (Dukowicz and Ramshaw, 1979) this assumes a uniform density within the cell which is then directly and conservatively mapped into the new cell using the overlap of volumes, the re-mapping is thus first order. However, by allowing the density distribution to be linear within the cell, the method may be made second order (Dukowicz and Kodis, 1987).

A major difficulty arises on the usual staggered grid, where the velocities are calculated at the cell nodes, not the centres. The normal method (Hirt, Amsden and Cook, 1974) is to define a dual mesh for the momentum about the nodes. This required extra computation as well as the inconsistent treatment of the node masses unless they also are re-mapped. Note that conservatism requires that momentum rather than velocity is re-mapped.

An alternative solution to average the nodal velocities into cell centres, advect and then redistribute the changes back to the grid points. This scheme is also unsatisfactory due to the averaging.

Material interfaces, including free surfaces, lead to problems in that conservation across the interface is difficult. The normal solution is to retain the interface along a grid line and allow the mesh to slide along the interface line (Wilkins, 1969). If advection across the interface cannot be avoided volume of fluid (VOF) methods can be used. These methods use a VOF variable, measuring the fraction of each fluid in the cell. They do not maintain the interface, but track it and then reconstruct it. They are strictly only applicable in incompressible flow, and lead to severe difficulties, for instance, in a shock.

In recent years ALE has become a general methodology rather than a specific computational method. ALE programmes are widely constructed for fluid mechanics as well as solid structural mechanics. The algorithms are based on both finite differences and finite elements, the latter having some advantages over the former for these applications. The general methodology is simply described and uses well-established numerical algorithms principally from fluid mechanics, which we have developed earlier: namely, Lagrangian mechanics, Sections 4.1 and 3.4.1 together with required peripheral modelling routines. The remaining problem concerns the re-zoning: both identifying a suitable criterion and implementing it. Additional physical behaviour may be added by including appropriate peripheral (to the core) routines. This latter unit is often both the most distinctive and difficult to satisfactorily construct.

References

Brackbill, J U and J S Saltzmann. 1982. "Adaptive zoning for singular problems in two dimensions." *J Comput Phys* 46:343–364.

Donea, J, A Huerta, J-Ph Ponthot and A Rodriguez-Ferran. 2004. *Arbitrary Lagrangian-Eulerian Methods.* Vol. 1. Chichester: Wiley pp. 413–437.

Dukowicz, J K. 1984. "Conservative rezoning (remapping) for general quadrilateral meshes." *J Comput Phys* 54:411–424.

Dukowicz, J K and J W Kodis. 1987. "Accurate conservative remapping (rezoning) for arbitrary Lagrangian-Eulerian computations." *SIAM J Sci Stat Comput* 8:305–321.

Dukowicz, J K and J D Ramshaw. 1979. "Tensor viscosity method for the control of convection in numerical fluid dynamics." *J Comput Phys* 32:71.

Gentry, R A, R E Martin and B J Daly. 1966. "An Eulerian differencing method for unsteady compressible flow." *J Comput Phys* 1:87–118.

Hirt, C W, A A Amsden and J L Cook. 1974. "An arbitrary Lagrangian-Eulerian computing method for all flow speeds." *J Comput Phys* 14:227–253.

Kershaw, D S. 1978. "The incomplete Cholesky-conjugate gradient method for the iterative solution of systems of linear equations." *J Comput Phys* 26:43–65.

Margolin, L G. 1997. "Introduction to "An arbitrary Lagrangian-Eulerian computing method for all flow speeds"." *J Comput Phys* 135:198–202.

Pert, G J. 1983. "Quasi-Lagrangain rezoning of fluid codes maintaining an orthogonal mesh." *J Comput Phys* 49:1–43.

Ramshaw, J D. 1985. "Conservative rezoning algorithm for generalised two dimensional meshes." *J Comput Phys* 59:193–199.

Wallick, K B. 1987. REZONE: A method for automatic rezoning in two-dimensional Lagrangian hydrodynamics problems. Technical Report LA-10829-MS. Los Alamos National Laboratory.

Whalen, P P. 1996. "Algebraic limitations on two-dimensional hydrodynamics simulations." *J Comput Phys* 124:46–54.

Wilkins, M L. 1969. Calculation of elastic-plastic flow. Technical Report UCRL-7322, Rev.1. Lawrence Livermore National Laboratory.

Wilkins, M L. 1980. "Use of artificial viscosity in multidimensional fluid dynamic calculations." *J Comput Phys* 36:2812–303.

Winslow, A M. 1963. Equipotential zoning of two-dimensional meshes. Technical Report UCRL-7312. Lawrence Livermore National Laboratory.

Winslow, A M. 1981. Adaptive-mesh rezoning by the equipotential method. Technical Report UCID-19062. Lawrence Livermore National Laboratory.

6

Hybrid or $1\,^1/_2$ d Schemes

6.1 Introduction

The numerical simulation of compressible gas flow is, as we have seen, a problem of severe complexity, demanding both computational time and large data storage. In addition to accurately integrating the plasma- dynamic equations, the integrations must be sufficiently simple to deal with other plasma effects, such as ionisation or recombination (Figure I). These constraints greatly limit the freedom to choose the best numerical scheme on which to attach the necessary algorithms. For example to treat ionisation or recombination, a Lagrangian scheme is almost essential: the individual particles are fixed in the mesh, whereas in an Eulerian scheme, there is a constant transfer of all the ionisation stage and excited state populations through the mesh – a time-consuming and error-prone process. On the other hand, Lagrangian codes in more than one dimension, do not maintain orthogonality and suffer from pathological instabilities. In consequence, where possible symmetry is used to reduce the number of dimensions treated in the problem.

In this section, an alternative approach is introduced for simplifying the hydrodynamics in a Lagrangian form. The flow is reduced to an analytic form in one or more dimensions by integrating the uniform self-similar form described in Section 2.1.1, with the flow in the remaining directions calculated by a cell method, such as the von Neumann–Richtmyer method (Section 4.2). Although only one such problem is considered here, it is likely that other such solutions exist and that the method is applicable to other problems. To this end, we first show that there exists an exact solution of the equations of fluid mechanics in which a body of gas possessing a similarity density profile in the (x, y)-plane, but varying in z, and of uniform temperature over that plane, expands into vacuum, maintaining its self-similar form in (x, y) and with a one-dimensional form of expansion in the z-direction. Such a flow is clearly a combination of a self-similar flow and a one-dimensional expansion, and we have therefore called it a *hybrid* flow.

The numerical integration of the hybrid model is readily accomplished by an extension of the well-known von Neumann–Richtmyer Lagrangian scheme for one-dimensional gas flow. The finite-difference forms of the governing differential equations are easily established, and their stability conditions examined. The hybrid model remains an academic exercise unless it can be applied to some useful physical problems. The expansion of non-uniformly heated cylinders of general cross section provides such a problem. If the initial length of the cylinder (in the z-direction) is long compared with the transverse

(z, y) dimensions, the flow will rapidly assume a similarity form in the (x, y)-plane. Furthermore, if the temperature is initially isothermal, or if the gas is uniformly heated, the temperature remains uniform across the (z, y)-plane. We have shown (Pert, 1980) that, for one-dimensional cylindrical and spherical flows, the self-similar solution gives an accurate representation of the flow, provided that the appropriate matching parameter between the initial scale lengths of the cylinder and the self-similar flow is used. In this work, we shall assume that this hypothesis can be applied to these flows and shall use matching parameters derived in our earlier publications (Pert, 1980, 1987).

6.1.1 The Hybrid Model

We consider the expansion (or contraction) of a body of gas subject to a velocity distribution within the gas of the following form. The velocities (u, v) in the (x, y)-plane are assumed to take a linear self-similar form separable in the coordinates (x, y) and (z, t), namely

$$u = \frac{dx}{dt} = \frac{x\,U(t)}{X(t)} \qquad v = \frac{dy}{dt} = \frac{y\,Vt}{Y(t)} \tag{6.1}$$

where the scale lengths $X(t)$ and $Y(t)$ are functions of time alone. The scale velocities

$$U = \frac{dX}{dt} \qquad V = \frac{dY}{dt} \tag{6.2}$$

The axial (z) velocity component w is assumed to be uniform over the (x, y)-plane and is a function of (z, t) only. Therefore, the transverse plane (x, y) is fixed in the flow. It follows from Eqs. (6.1) and (6.2) that the Lagrangian coordinates $(\xi_x = x/X,\ \xi_y = y/Y)$ remain constant for each fluid particle.

Consider a small cell fixed in the fluid of dimensions δx, δy and δz. Since w is a function of (z, t) alone, the spacing δz is independent of (x, y). The mass of the cell is constant:

$$\rho\,\delta x\,\delta y\,\delta z = \rho \cdot (X\,Y\,\delta z)\ \delta\xi_x\,\delta\xi_y)\delta z = const \tag{6.3}$$

Since $(X\,Y\,\delta z)$ is a function of (z, t) alone, the density distribution is separable:

$$\rho = \rho_0(z, t) f')\xi_x, \xi_y) \tag{6.4}$$

Furthermore, it follows from Eq. (6.1) that ρ must be even in (x, y), and therefore

$$\rho = \rho_0(z, t) \cdot f(\xi_x^2, \xi_y^2) \tag{6.5}$$

By consideration of Euler's equation in the (x, y)-plane, it follows that the pressure must be separable:

$$p = p_0(z, t) \cdot \phi(\xi_x^2, \xi_y^2) \tag{6.6}$$

where

$$p_0(z, t) = \frac{1}{2}\lambda_x X \frac{dU}{dt} = \frac{1}{2}\lambda_y Y \frac{dV}{dt} \tag{6.7a}$$

$$f(\xi_x^2, \xi_y^2) = -\lambda_x \frac{\partial \phi(\xi_x^2, \xi_y^2)}{\partial \xi_x^2} = -\lambda_y \frac{\partial \phi(\xi_x^2, \xi_y^2)}{\partial \xi_y^2} \tag{6.7b}$$

where λ_x and λ_y are separation constants.

The z-component of Euler's equation has the form

$$\frac{\partial w}{\partial t} + \frac{\partial w}{\partial z} = -\frac{\phi(\xi_x^2, \xi_y^2)}{f(\xi_x^2, \xi_y^2)} \cdot \frac{1}{\rho_0(z,t)} \cdot \frac{\partial p_0(z,t)}{\partial z} \tag{6.8}$$

which is only consistent with the condition $w = w(z,t)$ if the functions $f(\xi_x^2, \xi_y^2)$ and $\phi(\xi_x^2, \xi_y^2)$ are identical to within a constant factor, i.e. if the ratio p/ρ is dependent on (z,t) only. Furthermore, since X and Y are functions of time only, it follows from Eq. (6.6) that p/ρ must also be a function of time alone, i.e. be everywhere constant across the plane x, y. For an ideal gas, this condition, the *hybrid condition*, is upheld if the temperature is uniform everywhere. This requirement (*the hybrid condition*) is equivalent to the usual isothermal condition as applied to uniformly self-similar flows (Pert, 1974) and maintained either adiabatically or, in the case of external heating, by strong thermal conduction. In this case, it follows from Eq. (6.7b) that the self-similar distribution must take a Gaussian form

$$f(\xi_x^2, \xi_y^2) = \phi(\xi_x^2, \xi_y^2) = \exp(-(\xi_x^2 + \xi_y^2)) \tag{6.9}$$

for the case $\lambda_x = \lambda_y = 1$, where X and Y are the l/e widths of the distribution.

In view of this result, it is convenient to define the *'total' temperature* $T_A(z,t)$, independent (ξ_x, ξ_y), by the ratio of pressure to density according to the ideal gas law

$$\frac{p}{\rho} = \frac{R_g}{M} T_A \tag{6.10}$$

where M is the molecular mass number and R_g is the gas constant. For example in a plasma

$$T_A = T_i + ZT_e \tag{6.10$'$}$$

where T_i and T_e are the ion and electron temperatures, respectively, and Z is the ionic charge.

Similarly, it can be shown that a mass of gas with separable density distribution, Gaussian in (x, y), and whose 'total' temperature T_A is constant over the transverse plane (x, y) at rest will expand in a hybrid fashion.

The equations of motion follow from Eqs. (6.6) and (6.8)

$$X\frac{\mathrm{d}U}{\mathrm{d}t} = Y\frac{\mathrm{d}}{\mathrm{d}t} = \frac{2R_g}{M} T_A \tag{6.11a}$$

$$\frac{\partial w}{\partial t} + w\frac{\partial w}{\partial z} = -\frac{1}{\rho_)(z,t)} \frac{\partial p_0(z,t)}{\partial t} \tag{6.11b}$$

The 'total' temperature is determined from the overall energy balance of the gas. The average thermal energy per unit mass of a polytropic gas of adiabatic index γ

$$E_t = C_V T_A = \frac{1}{\gamma - 1} \frac{R_g}{M} T_A \tag{6.12}$$

and the average kinetic energy

$$E_k = \frac{1}{2}(U^2 + V^2) + \frac{1}{2}w^2 \tag{6.13}$$

averaged over the transverse plane (x, y). Thus, the total energy per unit mass

$$E = \frac{1}{2}\left[w^2 + \frac{1}{2}(U^2 + V^2)\right] + \frac{1}{2(\gamma - 1)} X \frac{\mathrm{d}U}{\mathrm{d}t} \tag{6.14}$$

6.1.1.1 The Local Approximation

The hybrid model provides a useful solution to problems for which its condition of validity is appropriate, namely that T_A be uniform everywhere. However, in many applications, we consider the flow of a system of initially large aspect ratio. In this case, we may imagine that an approximately self-similar flow will be rapidly established locally in the short dimensions of the system, the flow in the long dimension (z) being relatively slow. Provided the gradients in this long dimension are relatively small, we may relax the hybrid condition to apply in the (x, y)-plane only, with weak variations in X, Y and T_A with z permitted. Under this approximation, we consider $w(z, t)$ to be the velocity of flow in the z-direction averaged over the (x, y)-plane, so that there is no net flow through an (x, y)-plane moving with velocity w. We assume that the self-similar distribution is established sufficiently rapidly over the (x, y)-plane so that the changes in the distribution due to the variation of the z-component of velocity with x and y are always negligible. The assumption is thus made that the flow in the (x, y)-plane is locally self-similar everywhere with characteristic parameters X, Y, U and V, which vary from plane to plane, i.e. with z as well as t. The equations of motion are then most easily obtained by considering the flow of the fluid contained between two parallel (x, y)-planes separated by a distance δz. Equations (6.4) and (6.6) follow directly under the assumption of local similarity by a similar argument to that given earlier. The local hybrid approximation, however, unlike the hybrid model, is not exact, for in this case the analysis of Eq. (6.8) shows that the z-component of velocity varies across the (x, y)-plane due to the variation of X and Y with z. The appropriate equation of motion for w is obtained from consideration of the momentum balance of the line element δz, which after integration over the (x, y)-plane yields

$$\frac{\partial w}{\partial t} + w\frac{\partial w}{\partial z} = -\frac{1}{X Y \rho_0}\frac{\partial (X Y \rho_0)}{\partial z} \tag{6.11b$'$}$$

In this case, the 'total' temperature T_A varies with both z and t, but is constant across the plane (x, y). Hence, no thermal energy is convected across an (x, y)-plane moving with velocity w. The energy balance is constructed in a manner similar to that of the hybrid model itself, namely the average energy/unit mass across the plane (x, y) is

$$E = \frac{1}{2}\left(w^2 + \frac{1}{2}\left(U^2 + V^2\right)\right) + \frac{1}{2\,(\gamma - 1)} X \frac{\partial U}{\partial t} \tag{6.14a}$$

and is completed by including terms associated with energy deposition by external sources and thermal conduction. We may remark that the term w^2 in Eq. (6.14a) is identified by evaluating the work done by the pressure on the line element δz, namely $\partial(X\, Y\, \rho_0)/\partial z$ and using Eq. (6.11a). As noted earlier, the error in the local hybrid approximation arises from the variations of X and Y with z, which give rise to variations in the z-component of velocity across the plane (x, y). It is readily shown in cases of practical interest that the velocity varies monotonically with ξ_x and ξ_y, and furthermore that w given by Eq. (6.11a) is in fact the velocity at the l/e point. In contrast, the velocity w', calculated by Eq. (6.11b), is that at the symmetry axis of the similarity profile. Thus, $w - w'$ gives a measure of the total velocity variation across the (x, y)-plane. We may therefore assess the validity of the hybrid approximation during a calculation by requiring the variation $|(w - w')|/|w|$ to be less than some prescribed limit.

In most simulations of the type proposed for this model this condition is upheld. Using the numerical schemes to be described later we have calculated the error due to this local approximation during the expansion of uniform cylinders of various aspect ratios. As expected, the largest error occurs in the asymptotic, late-time motion. We find that generally the fractional error is reasonably constant over the bulk of the flow, but increases towards the end). These results are summarised in Pert (1983). For expansions involving aspect ratios greater than 1, the bulk error is $\approx 10^{-3}$ and only exceeds 10^{-2} in the endmost cell, whose differencing is in any case poor. For small aspect ratios, the end error is more extensive, but by no means severe, and in the bulk only of order 5×10^{-5}. For very large and very small aspect ratios, the error is nearly independent of the ratio; indeed at a ratio of 100, the error only exceeds 1% in the endmost cell. Results from other simulations involving heating, etc. show similar behaviour, but the error is in general smaller than that in the initially heated flow described here.

The nature of the simplification in the local approximation is comparable to that made in the representation of block (test) masses by a similarity model discussed in our earlier paper (Pert, 1980). Indeed, since we seek to model such structures here, it is clear that the inaccuracy introduced by the local approximation is no more severe than that due to the use of a similarity expansion in (x, y) itself.

In the absence of thermal conduction along the cylinders, the set of partial differential equations form a set of quasi-linear hyperbolic equations. The characteristic speeds can be easily shown to be ($w - c$, w and $w + c$), where c is the sound speed. The solutions to these equations can therefore be obtained by the method of characteristics. Such a procedure is, however, complex and we shall not pursue it here, preferring to develop simple finite-difference approaches. We treat in turn finite difference approaches in the Eulerian and Lagrangian formulations.

6.1.1.2 Eulerian Formulation

Let m be the fluid mass per unit length in the axial Z-direction. The equation of continuity has the form

$$\frac{\partial m}{\partial t} + \frac{\partial}{\partial z}(mw) = 0 \qquad \rho_0 = \frac{m}{XY} \tag{6.15a}$$

and the kinetic equations of motion

$$\frac{\partial X}{\partial t} + w\frac{\partial X}{\partial z} = U \qquad \frac{\partial Y}{\partial t} + w\frac{\partial Y}{\partial z} = V \tag{6.15b}$$

and the dynamic equations

$$\frac{\partial U}{\partial t} + w\frac{\partial U}{\partial z} = \frac{2}{X}\frac{p_0}{\rho_0} \qquad \frac{\partial V}{\partial t} + w\frac{\partial V}{\partial z} = \frac{2}{Y}\frac{p_0}{\rho_0}$$

$$\frac{\partial w}{\partial t} + w\frac{\partial w}{\partial z} = -\frac{1}{m}(XYp_0) \tag{6.15c}$$

The equation of energy balance

$$\frac{\partial mE}{\partial t} + w\frac{\partial}{\partial z}\left(m\left[E + \frac{p_0}{\rho_0}\right]\right) = m\left(W - \frac{\partial Q}{\partial z}\right) \tag{6.15d}$$

where W is the heat deposition rate per unit mass by external sources and Q is an averaged thermal conduction heat flux in the z-direction.

6.1.1.3 Lagrangian Form

The equations of motion take a simpler form more suitable for finite differencing in Lagrangian geometry. Consider the fluid between two transverse planes (x, y) separated by Δz at the point (x, y). Let the effective mass per unit area be m, then the mass of the corresponding cell is $M = m\Delta z$ per unit area. The equation of continuity is

$$\rho_0 = \frac{M}{XY\Delta z} \tag{6.16a}$$

with kinematic equations

$$\frac{\partial X}{\partial t} = U \qquad \frac{\partial Y}{\partial t} = V \qquad \frac{\partial z}{\partial t} = w \qquad \frac{\partial \Delta z}{\partial t} = \frac{\partial w}{\partial z} \tag{6.16b}$$

and the dynamic equations of motion

$$\frac{\partial U}{\partial t} = \frac{2}{X}\frac{p_0}{\rho_0} \qquad \frac{\partial V}{\partial t} = \frac{2}{Y}\frac{p_0}{\rho_0} \qquad \frac{\partial w}{\partial t} = -\frac{1}{M}\frac{1}{\partial z}\left(XYp_0\right) \tag{6.16c}$$

and the equation of energy balance

$$\frac{\partial E_t}{\partial t} + p_0\frac{\tau_0}{\partial t} = W - \frac{Q}{z} \tag{6.16d}$$

where $E - t$ is the thermal energy and $\tau_0 = 1/\rho_0$ is the specific volume on axis.

6.1.1.4 Shocks

The isothermal, linear self-similar form of the hybrid model is essentially a simple modification of one-dimensional hydrodynamics to include a transverse expansion. It therefore includes shocks lying in the (x, y)-plane and propagating in the z-direction. The upstream state, immediately ahead of the shock, has a uniform temperature and Gaussian distribution of pressure and density across the shock front. Consequently, the sound speed and therefore the Mach number is uniform over the $(2, y)$-plane in front of the shock. Since the gas is polytropic, the temperature, density and pressure ratios are uniform (Courant and Friedrichs, 1948, p. 148). The downstream temperature is therefore uniform, and the pressure and density Gaussian across the (x, y)-plane.

The Rankine–Hugoniot equations are modified to include a jump in the transverse velocities U and V to account for the corresponding transverse momentum changes, namely

$$\rho_l v_l U_l = \rho_r v_r U_r \qquad \text{and} \qquad \rho_l v_l V_l = \rho_r v_r V_r \tag{6.17}$$

As in the conventional von Neumann–Richtmyer, it is necessary to include artificial viscosity to provide the necessary entropy jump across the shock.

6.1.1.5 Artificial Viscosity

A variant of the von-Neumann–Richtmyer prescription is suitable to provide the artificial viscosity for the hybrid model. The flow in the z-direction is arbitrary and depends on the nature of the pressure distribution in z alone. Thus, compression flow in z may occur, which will develop shocks in extreme cases. More generally, a longitudinal nonlinear oscillatory growth may take place, rather similar in nature to the effects observed in two-dimensional Lagrangian codes. To overcome these problems, we introduce artificial viscosity with stress q into the Lagrangian dynamic equation of motion (6.16c)

$$\frac{\partial w}{\partial t} = -\frac{1}{M}\frac{\partial}{\partial z}\left\{X, Y\left(p_0 + q_0\right)\right\} \tag{6.16c$'$}$$

and the equation of energy balance

$$\frac{\partial E_t}{\partial t} + p_0 \frac{\partial \tau_0}{\partial t} + \frac{q_0}{\rho_0} \frac{\Delta z}{\partial t} = w - \frac{\partial Q}{\partial z} \tag{6.16d'}$$

where the stress

$$q = \begin{cases} a^2 \rho \dfrac{\partial (\Delta z)^2}{\partial t} & \left(\dfrac{\partial w}{\partial z} < 0\right) \\ 0 & \text{(otherwise)} \end{cases} \tag{6.18}$$

where a is a constant and $q\Delta z$ is the effective cell width. The artificial viscosity in this form satisfies the hybrid condition, namely q/ρ is a function of (z, t) alone. q_0 is the value on q on the axis.

6.1.1.6 Finite-Difference Form

The finite-difference form follows the von Neumann–Richtmyer form, section (Section 4.2.3) with the addition of the terms in the transverse velocities (U and V). The full algorithm can be found in Pert (1983).

We note that the stability has a sufficient condition (Pert, 1983, appendix B)

$$\Delta t \left\{ c \left[\frac{1}{2X^2} + \frac{1}{2Y^2} + \frac{1}{2\Delta z^2} \right]^{1/2} + 4a^2 \frac{\Delta w}{\Delta z} \right\} < 1 \tag{6.19}$$

which is a generalisation of the well-known Courant–Friedrichs–Lewy time-step constraint, Eq. (4.38).

6.1.1.7 Thermal Conduction

So far we have neglected thermal conduction along the z-direction, which leads to a redistribution of energy along the axial z. Since the gas is assumed isothermal in the x- and y-directions, the heat flux in the z-direction across a complete (x, y)-cross section is

$$Q = -\int K \frac{\partial T}{\partial z} \mathrm{d}x\,\mathrm{d}y \tag{6.20}$$

where K is the thermal conductivity and the integral is performed over the complete cross section. This procedure is unsatisfactory for the important case of a plasma, where the thermal conductivity is independent of density, and therefore (x, y), since the Gaussian density distribution extends to infinity in the (x, y)-plane. This unphysical behaviour is due to incorrect assumptions:

i) The density distribution is finite, so the assumed Gaussian distribution is in error near the plasma edge. We define the plasma surface by the edge factor Γ such that the boundary is described by the ellipse

$$\frac{x^2}{X^2} + \frac{y^2}{Y^2} = \frac{1}{\Gamma^2} \tag{6.21}$$

ii) Equation (6.20) assumes that the heat flow is described by Fourier's law. However, at low density, flux limitation occurs, which reduces the heat flux below that calculated using the classical thermal conductivity. A useful approximation for the flux-limited conductivity is the harmonic mean

$$K = \frac{K_0}{1 + \lambda|\nabla T|/T} \tag{6.22}$$

where K_0 is the classical conductivity. The empirical constant λ_0 is inversely proportional to density. Hence, if λ_0 is the value on the axis

$$\lambda = \lambda_0(z,t) \exp\left\{\frac{x^2}{X_2} + \frac{y^2}{Y^2}\right\} \tag{6.23}$$

Using Eqs. (6.21) and (6.22) we obtain

$$Q = -\pi XYK_0 \frac{\partial T}{\partial z}\left\{\Gamma^{-2} - \ln\frac{1 + \lambda_e|\nabla T|/T|}{1 + \lambda_0|\nabla T|/T}\right\} \tag{6.24}$$

where $\lambda_e = \lambda_0 \exp(\Gamma^{-2})$ is the value of λ at the plasma boundary.

In this calculation, we have specifically treated the case of a plasma; the generalisation to other media is straightforward and may be accomplished in a similar way.

The averaged thermal flux used in Eqs. (6.22) and (6.23) is simply obtained.

The heat Q may be added to W in Eqs. (6.15d) or (6.16d) in a finite-difference form. As usual, in these problems, stability requirements demand that the heat transport be treated by an implicit form if stringent time-step limitation is to be avoided (1967, p. 187). The problem then reduces to a tridiagonal matrix equation, whose solution is well known (Richtmyer and Morton, 1967, p. 199).

6.1.1.8 Concluding Remarks

The hybrid model was initially introduced in order to model the expansion and recombination in cylindrical plasmas with a small aspect ratio (width (R):length (L)) (Pert, 1979). The simple model of self-similar expansion model, integrated with a detailed ionisation calculation, was successfully used to investigate results from experiments in which laser gain was observed in laser-heated carbon fibres (Pert, 1979, 1981; Jacoby et al., 1982). Comparison against a two-dimensional Eulerian fluid code also showed good agreement with the self-similar model. The analysis of this fluid model in Pert (1983) indicated that the method was reasonably accurate for small aspect ratios.

The development of collisionally pumped soft X-ray lasers required modelling of the plasma from a slab target over fairly short pump pulse lengths, so that the aspect ratio may not have been necessarily small. However, simulation with a modified hybrid code proved remarkably accurate (Holden et al., 1994).

The reason for the accuracy of the similarity model as used by Holden et al., 1994 arises because:

- **either:** The aspect ratio is large ($L \gg R$), and the expansion is dominantly transverse; the transverse expansion being rapid, the longitudinal slow. The motion is therefore mainly transverse and well treated by the self-similar model with a longitudinal perturbation treated by the von Neumann–Richtmyer code.

- **or:** The aspect ratio is small ($L \ll R$), and the expansion is dominantly longitudinal; the axial expansion being rapid, the transverse slow. The motion is therefore mainly axial and well treated by the von Neumann–Richtmyer model with a transverse perturbation treated by the self-similar motion.

Numerical tests with both Eulerian and Lagrangian codes confirm that these two scenarios are applicable to collisionally pumped soft X-ray laser modelling at different times.

The hybrid model is particularly useful as a basic Lagrangian 'hydro' code to which additional physics may be linked (see Figure I) using the split operator principle (see Section 7.4), e.g. ionisation without the need for convective transport.

References

Courant, R and K O Friedrichs. 1948. *Supersonic Flow and Shock Waves*. New York: Interscience.

Holden, P B, S B Healy, M T M Lightbody, G J Pert, J A Plowes, A E Kingston, C L S Lewis and D Neely. 1994. "A computational investigation of the neon-like germanium collisionally pumped laser." *J Phys B: At Mol Opt Phys* 7:341–367.

Jacoby, D, G J Pert, L D Shorrock and G J Tallents. 1982. "Observations of gains in the extreme ultraviolet." *J Phys B: At Mol Phys* 15:3557–3580.

Pert, G J. 1974. "Some calculations of the heating of small pellets in rapidly expanding plasmas." *Plasma Phys* 16:1051–1068.

Pert, G J. 1979. "Model calculations of XUV gain in rapidly expanding cylindrical plasmas. II." *J Phys B: At Mol Phys* 12:2067–2079.

Pert, G J. 1980. "Self-similar flows with uniform velocity gradient and their use in modelling the free expansion of polyropic gases." *J Fluid Mech* 100:257–277.

Pert, G J. 1981. "Extreme ultraviolet laser action in plasmas." *Philos Trans R Soc Lond A* 300:631–640.

Pert, G J. 1983. "Quasi-Lagrangain rezoning of fluid codes maintaining an orthogonal mesh." *J Comput Phys* 49:1–43.

Pert, G J. 1987. "The use of flows with uniform velocity gradient in modelling free expansion of polyropic gas." *Lasers Part Beams* 5:643–658.

Richtmyer, R D and K W Morton. 1967. *Difference Methods for Initial Value Problems*. New York: Interscience.

7

Magneto-Hydrodynamics

7.1 Introduction

Thus far, we have assumed the working fluid of hydrodynamic equations to be a simple polytropic gas with a polytropic index γ. In reality, the actual working substance may be a general fluid with a complicated equation of state, or plasma with separated electrons and ions. We treat the former case elsewhere and examine the consequences of plasma behaviour in this chapter. Plasma is most simply treated as a two-fluid medium with interactions between the ions and electrons due to their different charges and masses. The presence of charges introduces currents and consequently magnetic fields, which interact directly with the particles. Plasma dynamics can be reduced to a series of equations closely resembling those of fluid dynamics (1.1) but including currents, and electric and magnetic fields: known as the magneto-hydrodynamics (MHD) equations.[1]

The MHD equations are derived from the Boltzmann distribution equations for each species (ions and electrons) by a series of steps, which are described in Pert (2021, §2.4 and §8.1–8.4). We briefly repeat these to give an outline of the details to enable the approximation made in their derivation to be appreciated.

Magneto-hydrodynamics generalises fluid behaviour beyond that seen in conventional fluid flow. This is a consequence of 'fluid' stiffness and the ability to carry and to support tensor transport coefficients, to generate magnetic fields internally and to allow more complex wave behaviour. In particular, shocks generated in plasma are much more complicated to describe than in fluids, and indeed may lack the clear uniqueness of those in fluids, which are determined by a simple form of the Rankine–Hugoniot relations as described in some detail in Chapter 1. As this will lead us into diverse areas of complexity which lie beyond the scope and aims of this book, we will refer the interested reader to specific texts devoted to this subject such as Anderson (1963), Landau and Lifshitz (1984), and Liberman and Velikhovich (1986).

The resulting equations of the flow of conducting flow contain those of a simple fluid as a limiting case when the conductivity is zero. As a result the resulting equations may be expected to be an extension of those we have described for fluid flow. Indeed,

1 Magneto-hydrodynamics is the study of a single conducting fluid subject to electromagnetic fields.

Modelling and Simulation in Plasma Physics for Physicists and Mathematicians, First Edition. Geoffrey J. Pert.

much of the methodology described for fluids carries over to MHD. Unfortunately, however, the additional degrees of freedom introduced by the conductivity greatly increase the complexity of modelling any realistic problem. Problems involving only one spatial dimension are relatively simple, see the work of Anderson (1963), Landau and Lifshitz (1984), and Liberman and Velikhovich (1986). In two dimensions, magnetic fields and their effect on the plasma motion become an important factor, for example, when the plasma motion is two-dimensional cylindrical (r, z) and the field is azimuthal (ϕ). A number of practical flows of this type occur, for example in: z and θ pinches, plasma focus devices and self-generated magnetic fields in laser-produced plasmas (Stamper et al., 1971). General methods of treating magnetohydrodynamic simulations are presented in this Chapter with particular emphasis on energy conservation within the hydrodynamic equations.

Most problems in MHD are three dimensional due to the axial nature of the magnetic field. As a result, hydro-codes for dealing with fusion devices such as tokamaks are large making them impractical for use on relatively smal computers due to the memory and speed requirements and are therefore outside the aims of this book. However, we note that most of the essential numerical techniques are simply extensions of those of lower dimensions and do not introduce new numerical algorithms or techniques.

In this chapter, we will examine the most general problem in MHD when the plasma and magnetic field are intimately coupled self-consistently, i.e. the case of self-generated fields. The coupling of the fields and fluid introduces new difficulties involving the conservation of various quantities. The development of this methodology is the underlying motivation of this chapter.

7.2 The MHD Equations

1) Take moments of the Boltzmann equation. Proceeding as before (Pert, 2021, §8.1), we set the

   electron/ion number conservation $Q(\mathbf{v}) = 1$ which means:

$$\frac{\partial n_i}{\partial t} + \nabla \cdot (n_i \mathbf{u}_i) = 0 \qquad \frac{\partial n_e}{\partial t} + \nabla \cdot (n_e \mathbf{u}_e) = 0 \tag{7.1a}$$

   and the momentum conservation $Q(\mathbf{v}) = m\mathbf{v}$:

$$\frac{\partial}{\partial t}(n_i\, m_i\, u_{i_i}) + \frac{\partial}{\partial x_j}\left(n_i\, m_i\, u_{i_i}\, u_{i_j} + \mathsf{P}_{i_{ij}}\right) - Z e n_i\,(E_i + \epsilon_{ijk}\, u_{i_j}\, B_k) - n_i m_i \frac{\partial \phi}{\partial x_i} = \mathcal{P}^{ie}_i$$

$$\frac{\partial}{\partial t}(n_e\, m_e\, u_{ei}) + \frac{\partial}{\partial x_j}\left(n_e\, m_e\, u_{e_i}\, u_{e_j} + \mathsf{P}_{e_{ij}}\right) + e n_e\,(E_i + \epsilon_{ijk}\, u_{e_j}\, B_k) - n_e m_e \frac{\partial \phi}{\partial x_i} = -\mathcal{P}^{ie}_i$$

(7.1b)

   where ϕ is the gravitational potential (which will be zeroed henceforward), $\mathbf{P}$ the stress tensor and $\mathcal{P}^{ie}$ the momentum exchange rate from electrons to ions known as the *ion-electron drag force*.[2]

2 The first subscripts i and e represent the particle, ions i and electrons e, respectively; subsequent subscripts i, j, etc. represent directional components, e.g. u_{i_i} is the ith component of the ion mean velocity $\mathbf{u}_i$.

and the energy conservation $Q(\mathbf{v}) = \frac{1}{2}mv^2$:

$$\frac{\partial}{\partial t}\left[n_i m_i \left(\epsilon_i + \frac{1}{2}u_i^2 + \phi\right)\right] + \frac{\partial}{\partial x_i}\left[n_i m_i \left(\epsilon_i + \frac{1}{2}u_i^2 + \phi\right) \; u_{ii} + \mathbf{P}_{i_{ij}} u_{i_j} + q_{i_i}\right]$$
$$= Z e n_i E_i u_{ii} + Q^{ie}$$
$$\frac{\partial}{\partial t}\left[n_e m_e \left(\epsilon_e + \frac{1}{2}u_e^2 + \phi\right)\right] + \frac{\partial}{\partial x_i}\left[n_e m_e \left(\epsilon_e + \frac{1}{2}u_e^2 + \phi\right) \; u_{e_i} + \mathbf{P}_{e_{ij}} u_{ej} + q_{e_i}\right]$$
$$= -e n_e E_i u_{ei} - Q^{ie} \tag{7.1c}$$

where ϵ_i and ϵ_e are the individual specific energies of the ion and electrons, respectively, i.e. the energy per unit mass of either ion or electron as appropriate. $\mathbf{q}_i$ and $\mathbf{q}_e$ are the ion and electron heat fluxes, respectively. Q^{ie} is the *collisional energy transfer rate* from electrons to ions, which we treat in Chapter 11.

2) If the Debye length λ_D is small compared to the characteristic lengths of the problem $\ell \gg \lambda_D$, charge differences are smoothed out and the space charge density $q = Z e n_i - e n_e \approx 0$. This arises due to the characteristic role of the Debye length as separator, the scale length over which incoherent electric fields can be supported against the large neutralising field.[3] The plasma is *quasi-neutral* so that $n_e \approx Z n_i$. Thus we have $\nabla \cdot \mathbf{u}_i = \nabla \cdot \mathbf{u}_e$, but $Z\nabla \wedge \mathbf{v}_i \neq \nabla \wedge \mathbf{v}_e$. The plasma behaves as a two-component fluid having independent mass flow velocity and current with properties

Mass density	$\rho = n_i m_i + n_e m_e$	Flow velocity	$\overline{\mathbf{u}} = \left(n_i m_i \mathbf{u}_i + n_e m_e \mathbf{u}_e\right)/\rho$
Charge density	$q = Z e n_i - e n_e$	Current density	$\mathbf{j} = Z e n_i \mathbf{u}_i - e n_e \mathbf{u}_e$
Specific internal energy			$\epsilon = \left(n_i m_i \epsilon_i + n_e m_e \epsilon_e\right) / \left(n_i m_i + n_e m_e\right)$
Total stress tensor			$\mathbf{P} = \mathbf{P}_i + \mathbf{P}_e$

(7.2)

Adding the electron and ion conservation equations and taking the divergence gives the equation of continuity for current:

$$\frac{\partial q}{\partial t} + \nabla \cdot \mathbf{j} = 0 \tag{7.3}$$

(the $\partial q/\partial t$ term is struck through on the page)

Therefore, in quasi-neutral flow the current paths form closed loops.

The total velocity of the particles thus follows the *Helmholtz' decomposition* into an *irrotational* ($\nabla \wedge \mathbf{u} = \mathbf{0}$) and *solenoidal* ($\nabla \cdot \mathbf{j} = \mathbf{0}$) (Sommerfeld, 1964, §20).

Unfortunately, the moment equations cannot be expressed simply in the single fluid variable set (7.2). It is therefore necessary to simplify them by taking advantage of the large difference in mass between the ions and electrons, $m_i \gg m_e$. Furthermore, since a small charge imbalance leads to a large neutralising electric field, the space charge q is approximately zero (*quasi-neutral approximation*) and $Z n_i \approx n_e$. These two

3 Coherent (wave) fields of wavelength greater than the Debye length of course exist, but are strongly damped for shorter wavelengths.

approximations allow us to re-write in a simpler form

$$\begin{aligned}
n_i &= \frac{e\rho - m_e q}{e\,(m_i + Z m_e)} \approx \frac{\rho}{m_i} \\
n_e &= \frac{Z e\rho - m_i q}{e\,(m_i + Z m_e)} \approx \frac{Z\rho}{n_i} - \frac{q}{e} \approx \frac{Z\rho}{n_i} \\
\mathbf{u}_i &= \frac{e\rho\mathbf{u} + m_e\mathbf{j}}{e\rho + m_e q} \approx \mathbf{u} \\
\mathbf{u}_e &= \frac{Z e\rho\bar{\mathbf{u}} - m_i\mathbf{j}}{Z e\rho - m_i q} \approx \mathbf{u} - \frac{m_i}{Z e\rho}\mathbf{j} \approx \mathbf{u} - \frac{\mathbf{j}}{e n_e}
\end{aligned} \tag{7.4}$$

3) Since the electric field is small, it is dominated by the induced field $\nabla \wedge \mathbf{B}$, and thus, for many purposes, we replace Maxwell's equations by the simpler form

$$-\frac{\partial \mathbf{B}}{\partial t} = \nabla \wedge \mathbf{E} \qquad \mathbf{j} = \frac{1}{\mu_0}\nabla \wedge \mathbf{B} \tag{7.5}$$

4) These two fluid equations still suffer from the problem of closure. To overcome this, we apply the Chapman–Enskog method to deal with closure (Chapman and Cowling, 1952; Braginskii, 1965). Namely, we consider the slowly varying in time behaviour of a perturbation about the equilibrium state. Since the zero-order term is the equilibrium (Maxwellian) distribution it avoids the closure problem, the small perturbations giving rise to the difficulty being of lower order. This introduces the transport terms, diffusion, viscosity and thermal conduction terms, which depend directly on the respective gradients and are thereby introduced into the resulting equations (Chapman and Cowling, 1952; Shkarofsky, Johnston and Bachynski, 1966; Lifshitz and Pitaevskii, 1981; Pert, 2021). The MHD equations become as follows.

Equation of mass conservation:

$$\frac{\partial \rho}{\partial t} + \nabla \cdot (\rho \mathbf{u}) = 0 \tag{7.6}$$

Equation of momentum conservation:

In the general case where the plasma is non-quasi-neutral, the interaction between the plasma and the field must be taken into account using Maxwell's equation:

$$\begin{aligned}
\nabla \wedge \mathbf{E} &= -\frac{\partial \mathbf{B}}{\partial t} \quad \text{and} \quad \nabla \wedge \mathbf{B} = \mu_0\left(\mathbf{j} + \epsilon_0 \frac{\partial \mathbf{E}}{\partial t}\right) \\
\nabla \cdot \mathbf{E} &= \frac{q}{\epsilon_0} \quad \text{and} \quad \nabla \cdot \mathbf{B} = 0
\end{aligned} \tag{7.7}$$

and using the identity

$$(\nabla \wedge \mathbf{A}) \wedge \mathbf{A} = \frac{1}{2}\nabla A^2 - (\mathbf{A}\cdot\nabla)\mathbf{A} = \frac{1}{2}\nabla A^2 - \nabla \cdot AA + A\nabla \cdot \mathbf{A} \tag{7.8}$$

we obtain the force per unit volume

$$\begin{aligned} q\mathbf{E}+\mathbf{j}\wedge\mathbf{B} &= \epsilon_0\,\mathbf{E}\,\nabla\cdot\mathbf{E}-\epsilon_0\,(\nabla\wedge\mathbf{E})\wedge\mathbf{E}-\frac{1}{\mu_0}\,(\nabla\wedge\mathbf{B})\wedge\mathbf{B}-\epsilon_0\,\frac{\partial\,(\mathbf{E}\wedge\mathbf{B})}{\partial t} \\ &= \nabla\cdot\left(\epsilon_0\,\mathbf{EE}+\frac{1}{\mu_0}\,\mathbf{BB}\right)-\frac{1}{2}\nabla\left(\epsilon_0\,E^2+\frac{1}{\mu_0}B^2\right)-\epsilon_0\,\frac{\partial\,(\mathbf{E}\wedge\mathbf{B})}{\partial t} \\ &= \nabla\cdot\boldsymbol{\Phi}-\frac{\partial\mathbf{G}}{\partial t} \end{aligned} \tag{7.9}$$

where $\boldsymbol{\Phi}=\left[\frac{1}{2}\,(\mathbf{E}\cdot\mathbf{D}+\mathbf{B}\cdot\mathbf{H})\,\mathbf{I}-(\mathbf{ED}+\mathbf{BH})\right]$ is Maxwell's stress tensor written in dyadic form (Pert2021, p152) and $\mathbf{I}$ the identity dyadic. The force per unit volume is therefore the sum of two terms: the gradient of the stress tensor and the rate of change of the momentum density. The momentum of the electromagnetic field per unit volume is

$$\mathbf{G}=\frac{1}{c^2}\mathbf{S}=\frac{1}{c^2}\mathbf{E}\wedge\mathbf{H}=\mathbf{D}\wedge\mathbf{B}$$

$c\,\mathbf{G}$ is the momentum flux and $\mathbf{S}=c^2\,\mathbf{G}$ the energy flux (Sommerfeld, 1952, p. 260), where $\mathbf{S}=\frac{1}{\mu_0}\,\mathbf{E}\wedge\mathbf{B}$ is Poyntings vector and $\mathbf{T}=\mathbf{P}+\boldsymbol{\Phi}$ the total stress tensor. The total momentum per unit volume is $\rho\mathbf{u}+\mathbf{G}$ and the momentum $\mathbf{T}+c\,\mathbf{S}$.
Neglecting the small momentum of the electrons due to the current in view of the small electron mass, but including that of the field the change in the momentum per unit volume per unit time is

$$\begin{aligned} &\frac{\partial}{\partial t}\,(\rho\,\mathbf{u})+\nabla\cdot\left[n_i\,\mathbf{u}_i\,\mathbf{u}_i+n_e\,\mathbf{u}_e\,\mathbf{u}_e+\mathbf{P}\right]-\mathbf{E}-\mathbf{j}\wedge\mathbf{B}=0 \\ &\approx\frac{\partial}{\partial t}\,(\rho\,\mathbf{u}+\mathbf{G})+\nabla\cdot\left[(\rho\,\mathbf{u}\,\mathbf{u})+(\mathbf{P}+\boldsymbol{\Phi})\right]\approx 0 \end{aligned} \tag{7.10}$$

where $\mathsf{I}\equiv\delta_{ij}$ is the identity dyadic. Expressed in a conservative form, we have

$$\frac{\mathrm{d}(\rho\mathbf{u})}{\mathrm{d}t}+\frac{\partial\mathbf{G}}{\partial t}=-\nabla\cdot\mathbf{T} \tag{7.11}$$

Fortunately, in practice, this complex equation involving the momentum the electromagnetic field is not normally needed. As we have already discussed, the plasma in the 'continuum' regime is usually quasi-neutral so that the space charge density is zero and the bulk force due to the electric field is zero. It this case, the terms involving the electromagnetic stress vanish and the only force due to the field is the Lorentz force due to the current and the magnetic field. The equation of momentum conservation reduces to

$$\frac{\partial\,(\rho\mathbf{u})}{\partial t}+\nabla\cdot(\rho\mathbf{uu})=-\nabla\mathbf{P}+\mathbf{j}\wedge\mathbf{B}=\nabla\cdot\boldsymbol{\Phi} \tag{7.12}$$

where the stress tensor $\boldsymbol{\Phi}$ takes the reduced form $\boldsymbol{\Phi}\approx\frac{1}{2}\,(\mathbf{B}\cdot\mathbf{H}-\mathbf{BH})$, which is conservative. In the quasi-neutral approximation, fulfilled by most practical plasma systems, the electric field $E\approx 0$ and the electromagnetic field density $\mathbf{G}=\mathbf{D}\wedge\mathbf{B}\approx 0$.

The MHD momentum equation therefore reduces to that for a simple fluid with particle mass $(m_i + Zm_e)$ but with the additional term for the Lorentz force. We may, therefore, use the well-established CFD routines we discussed earlier to treat such problems provided the magnetic field is known.

5) **Equation of current flow**: The current density $\mathbf{j}$ is given by the product of the electron density and the relative velocity of the electrons with respect to the ions, namely, $\mathbf{j} = -e\,n_e(\mathbf{u_e} - \mathbf{u_i})$. Performing closure, using the Chapman–Enskog method we obtain the result from the momentum Eq. (7.1b),[4]

$$\frac{m_e}{n_e e^2}\left[\frac{\partial \mathbf{j}}{\partial t} + \nabla\cdot\left(\mathbf{uj}+\mathbf{ju} - \frac{m_i}{Ze\rho}\mathbf{jj}\right)\right]$$
$$= \frac{m_e}{n_e e^2}\left\{\left[\frac{\partial}{\partial t}\left(Zen_i\mathbf{u}_i\right) + \nabla\cdot\left(Zen_i\mathbf{u}_i\mathbf{u}_i\right)\right] - \left[\frac{\partial}{\partial t}\left(en_e\mathbf{u}_e\right) + \nabla\cdot\left(Zen_i\mathbf{u}_i\mathbf{u}_i\right)\right]\right\}$$
$$\approx \mathbf{E} + \overline{\mathbf{u}}\wedge\mathbf{B} - \frac{\mathbf{j}\wedge\mathbf{B}}{n_e e} + \frac{\nabla\cdot\mathbf{P}_e}{n_e e} - \frac{\mathcal{P}^{ie}}{n_e e} \approx 0 \tag{7.13}$$

The first term on the left-hand side of Eq. (7.13) is normally neglected in view of the relative smallness of the electron mass. In contrast to the mean velocity, we retain the momentum exchange term, which contributes the Ohmic resistance of the plasma.
The current flow is obtained from the difference between the electron and ion flow. The convective and time-dependent terms on the left-hand side of Eq. (7.14) can normally be neglected to give the generalised form of Ohm's law

$$en_e\,\mathbf{E} + \mathbf{u}\wedge\mathbf{B} = \mathbf{j}\wedge\mathbf{B} - \nabla\cdot\mathbf{P}_e \approx \mathcal{P}^{ie} = \mathbf{R}_e \tag{7.14}$$

This equation is easily interpreted by considering the momentum exchange $\mathcal{P}^{ie} = \mathbf{R}_e$ from the electrons to the ions as a result of collisions. The electrons are accelerated by the generalised driving force $-e(\mathbf{E} + \mathbf{u}\wedge\mathbf{B}) + \left[\mathbf{j}\wedge\mathbf{B} - \nabla\cdot\mathbf{P}_e\right]/n_e$. In a collision with an ion, the electrons transfer momentum to the ions at a rate determined by the relative velocity. It may, therefore, be expected that the momentum exchange $\mathcal{P}^{ie}$ is parallel to the electron velocity relative to ions.
More generally, the momentum exchange term $\mathcal{P}^{ie}$ is the frictional resistance term and includes both Ohmic resistive and thermo-electric terms. It can be expressed in a number of ways depending on the choice of the conjugate pair of force/flux terms (see Pert, 2021, §8.4.2). In Braginskii's (Braginskii, 1965) form

$$\mathbf{R}_e = en_e\boldsymbol{\alpha}\cdot\mathbf{j} - n_e\boldsymbol{\beta}^0\cdot\nabla(\kappa T_e) \tag{7.15}$$

where $\boldsymbol{\alpha}$ and $\boldsymbol{\beta}^0$ are the resistivity and thermo-electric tensors, respectively, values for which are given by Braginskii (1965) and Epperlein and Haines (1986) (see also Shkarofsky, Johnston and Bachynski, 1966).

6) **Magnetic field generation:**
In early studies of the interaction of intense laser fields with solid targets, it was found that relatively strong magnetic fields were generated. These were ascribed to the currents generated by electron pressure gradients in the plasma. Although the assignation remains uncertain as other particle mechanisms are also possible, there is

4 This procedure thus introduces expressions for the collisional exchange terms $\mathcal{P}^{ie}$ and $\mathcal{Q}^{ie}$ introduced empirically in Eq. (7.1).

little doubt that a purely MHD source exists. This has been best explored by simulation using finite-difference forms of the MHD equations, which we outline in this chapter. Since these are quite complicated, they can be used as a model to exemplify methods of calculation of time-dependent MHD problems.

The generation of a magnetic field by electron currents driven by gradients in the electron pressure (or density and temperature) is given by the simplified Maxwell equation (neglecting the displacement current), Eq. (7.5) and the equation of current flow Eq. (7.13), namely

$$\frac{\partial \mathbf{B}}{\partial t} - \nabla \wedge (\mathbf{u} \wedge \mathbf{B}) + \nabla \wedge [\boldsymbol{\eta} \cdot (\nabla \wedge \mathbf{B})] = \nabla \wedge \mathbf{S} \tag{7.16}$$

where the diffusivity tensor $\boldsymbol{\eta}$ and the source vector $\mathbf{S}$ may be written in two alternative forms depending on whether the Hall term is included as a magnetic stress, or not included:

- **Explicit Magnetic stress:**

$$\mathbf{S} = \frac{1}{e}\left\{\left[\nabla p_e + \nabla\left(\frac{B^2}{\mu_0}\right) - \frac{1}{\mu_0}(\mathbf{B}\cdot\nabla)\,\mathbf{B}\right] + \boldsymbol{\beta}^0 \cdot \nabla\left(\mathcal{k}T_e\right)\right\} \tag{7.17}$$

  and the magnetic diffusivity

$$\eta = \alpha/\mu_0 \tag{7.18}$$

- **Tensor resistivity component (without magnetic stress)**

$$\mathbf{S} = \frac{1}{e}\left\{\frac{1}{n_e}\nabla p_e + \boldsymbol{\beta}^0 \cdot \nabla\left(\mathcal{k}T_e\right)\right\} \tag{7.18a}$$

  and

$$\eta = \rho/\mu_0, \qquad \text{when} \quad \rho'_{\parallel} = \rho_{\parallel}, \quad \rho'_{\perp} = \rho_{\perp} \quad \text{and} \quad \rho'_{\wedge} = \rho_{\wedge} + B/en_e$$

  where the subscripts $\parallel$, $\perp$ and $\wedge$ refer to field components parallel, perpendicular to the magnetic field and a component perpendicular to both electric and magnetic fields in the usual way.

The source vector $\mathbf{S}$ is only defined through its *curl* and is undefined to a term which is the *gradient* of a scalar point function. Thus, we may equivalently write

$$\frac{1}{n_e}\nabla p_e \longrightarrow \mathcal{k}T_e \nabla \ln(n_e) \longrightarrow -\ln n_e\, \nabla(\mathcal{k}T_e) \tag{7.19}$$

and

$$\mathbf{S} = \frac{1}{e n_e}\boldsymbol{\beta} \cdot \nabla \ln(\mathcal{k}T_e)$$

If the thermo-electric term $\boldsymbol{\beta}^0$ is small, we obtain the usual form

$$\frac{\partial \mathbf{B}}{\partial t} = \nabla \wedge \mathrm{S} = \frac{1}{e}\,\nabla(\mathcal{k}T_e) \wedge \nabla(\ln n_e) \tag{7.20}$$

7) **Equation of energy conservation:**
The general equation for the total conservation of energy summed over both ions and electrons is rarely used due to the slow equilibration rate between the particles. As a consequence, the electron and ion temperatures are generally unequal, and the equilibration between them must be treated explicitly as part of the time development of each population. Nonetheless, the general equation, which is valid if the system time development is sufficiently slow, is of interest and, therefore, we develop it here.

Summing the equivalent equations for the electrons and ions, we obtain

$$\frac{\partial}{\partial t}\left\{\begin{array}{c} n_i m_i \left(\epsilon_i + \frac{1}{2} u_i^2\right) \\ + n_e m_e \left(\epsilon_e + \frac{1}{2} u_e^2\right) \end{array}\right\} + \frac{\partial}{\partial \mathbf{r}} \cdot \left\{\begin{array}{cc} n_i m_i \mathbf{u_i} \left(\epsilon_i + \frac{1}{2} u_i^2\right) & + \mathbf{p_i} \cdot \mathbf{u_i} + \mathbf{q_i} \\ + n_e m_e \mathbf{u_e} \left(\epsilon_e + \frac{1}{2} u_e^2\right) & + \mathbf{p_e} \cdot \mathbf{u_e} + \mathbf{q_e} \end{array}\right\} - \mathbf{E} \cdot \mathbf{j} = 0 \tag{7.21}$$

The current density

$$\mathbf{j} = n_i q_i \mathbf{u}_i + n_e q_e \mathbf{u}_e = \nabla \wedge \mathbf{H} - \frac{\partial \mathbf{D}}{\partial t} = \frac{1}{\mu_0} \nabla \wedge \mathbf{B} - \epsilon_0 \frac{\partial \mathbf{E}}{\partial t} \tag{7.22}$$

and the Ohmic work term (including the work done by the electromagnetic stress)

$$\begin{aligned} \mathbf{E} \cdot \mathbf{j} &= \frac{1}{\mu_0} \mathbf{E} \cdot \nabla \wedge \mathbf{B} - \epsilon_0 \mathbf{E} \cdot \frac{\partial \mathbf{E}}{\partial t} \\ &= -\frac{1}{\mu_0} \nabla \cdot (\mathbf{E} \wedge \mathbf{B}) + \frac{1}{\mu_0} \mathbf{B} \cdot \nabla \wedge \mathbf{E} - \epsilon_0 \mathbf{E} \cdot \frac{\partial \mathbf{E}}{\partial t} \\ &= -\frac{1}{\mu_0} \nabla \cdot (\mathbf{E} \wedge \mathbf{B}) - \frac{1}{2\mu_0} \frac{\partial (B^2)}{\partial t} - \frac{\epsilon_0}{2} \frac{\partial (E^2)}{\partial t} \end{aligned} \tag{7.23}$$

Hence, replacing the Ohmic work term, we finally obtain an equation in the conservation law form

$$\frac{\partial}{\partial t}\left\{\begin{array}{l} n_i m_i \left(\epsilon_i + \frac{1}{2} u_i^2\right) + \\ n_e m_e \left(\epsilon_e + \frac{1}{2} u_e^2\right) + \\ \epsilon_0 E^2/2 + B^2/2\mu_0 \end{array}\right\} + \frac{\partial}{\partial \mathbf{r}} \cdot \left\{\begin{array}{l} n_i m_i \mathbf{u}_i \left(\epsilon_i + \frac{1}{2} u_i^2\right) + \mathbf{P}_i \cdot \mathbf{u}_i + \mathbf{q}_i + \\ n_e m_e \mathbf{u}_e \left(\epsilon_e + \frac{1}{2} u_e^2\right) + \mathbf{P}_e \cdot \mathbf{u}_e + \mathbf{q}_e + \\ \mathbf{E} \wedge \mathbf{B}/\mu_0 \end{array}\right\} = 0 \tag{7.24}$$

Note the introduction of the electromagnetic energy flux determined by the Poynting vector $\mathbf{E} \wedge \mathbf{B}/\mu_0$. This equation can be cast into the more familiar fluid dynamics form by replacing the total stress $\mathbf{P}$ separating the pressure p from the 'viscous' stress $\mathbf{T}$, namely $\mathbf{P} = p\mathbf{I} - \mathbf{T}$, where $\mathbf{I}$ is the identity tensor, and introducing the enthalpies $h = \epsilon + p/nm$.

$$\frac{\partial}{\partial t}\left\{\begin{array}{l} n_i m_i \left(\epsilon_i + \frac{1}{2} u_i^2\right) + \\ n_e m_e \left(\epsilon_e + \frac{1}{2} u_e^2\right) + \\ \epsilon_0 E^2/2 + B^2/2\mu_0 \end{array}\right\} + \frac{\partial}{\partial \mathbf{r}} \cdot \left\{\begin{array}{l} n_i m_i \mathbf{u}_i \left(h_i + \frac{1}{2} u_i^2\right) - \mathbf{T}_i \cdot \mathbf{u}_i + \mathbf{q}_i + \\ n_e m_e \mathbf{u}_e \left(h_e + \frac{1}{2} u_e^2\right) - \mathbf{T}_e \cdot \mathbf{u}_e + \mathbf{q}_e + \\ \mathbf{E} \wedge \mathbf{B}/\mu_0 \end{array}\right\} = 0 \tag{7.25}$$

7.3 MHD Equations

The MHD equations are a slightly simplified set of the general equations derived earlier, which are valid for a large set of problems in plasma dynamics. The approximations used that the ion mass is much larger than that of the electrons ($m_i \gg m_e$), that the plasma is quasi-neutral $\sigma = \nabla \cdot \mathbf{E} \approx 0$ and that the displacement current is small in Maxwell's equations $\partial D/\partial t \ll j$. Provided the Debye length is small compared to characteristic lengths in the physical system, the equation of mass continuity and the equation of momentum conservation are those of a single fluid in conservation law form. The relevant equation are summarised below:

1) **Equation of mass continuity:**

$$\boxed{\frac{\partial \rho}{\partial t} + \nabla \cdot (\rho \mathbf{u}) = 0} \tag{A}$$

2) **Momentum conservation (Euler's equation):**

$$\boxed{\frac{\partial}{\partial t}(\rho \mathbf{u}) + \nabla \cdot (\rho \mathbf{u}\mathbf{u}) = \frac{1}{\mu_0}(\mathbf{B} \cdot \nabla)\mathbf{B} - \frac{1}{2\mu_0}\nabla\left(B^2\right) - \nabla p} \tag{B}$$

Turning now to energy balance, we can no longer use the single-fluid approximation for the total energy Eq. (7.24). Due to the large mass difference between the ions and electrons, energy transfer from the electrons to ions is relatively slow (see Pert, 2021, §2.1.4) and may be comparable with the other characteristic times in the physical problem. It is therefore necessary to return to separate electron and ion energy balances.

3) **Electron energy balance:**
The magnetic field is generated by electron currents driven by the electron pressure gradient. The ions are not involved due to their large mass and play little additional role in the energy balance. Therefore, we need only consider the modification to the electron energy equation as a result of the magnetic field generation. Closing Eq. (7.1c) for the electrons, we obtain

$$\boxed{\begin{aligned} &\underbrace{\frac{\partial \epsilon_e}{\partial t} + \nabla \cdot \left(\epsilon_e \mathbf{u}\right)}_{\substack{\text{Electron}\\ \text{Convection}}} + \underbrace{p_e \nabla \cdot \mathbf{u}}_{\substack{\text{Electron}\\ \text{pressure work}}} + \underbrace{\nabla \cdot \mathbf{q}_e}_{\substack{\text{Thermal}\\ \text{diffusion}}} \\ &\quad = \underbrace{W_e}_{\substack{\text{External}\\ \text{heating}}} - \underbrace{G - J}_{\substack{\text{Electron/ion}\\ \text{exchange}}} + \underbrace{\mathbf{j} \cdot \nabla\left(\frac{3}{2}kT_e/e\right)}_{\text{Current transport}} + \underbrace{\rho j^2}_{\substack{\text{Ohmic}\\ \text{heating}}} - \underbrace{\mathbf{S} \cdot \mathbf{j}}_{\substack{\text{Magnetic}\\ \text{source}}} \end{aligned}} \tag{C}$$

4) **Ion energy balance:**
Similarly for the ion energy balance

$$\boxed{\begin{aligned} &\underbrace{\frac{\partial}{\partial t}\left[n_i m_i \left(\epsilon_i + \frac{1}{2}u^2\right)\right] + \frac{\partial}{\partial x_j}\left[n_i m_i \left(\epsilon_i + \frac{1}{2}u^2\right) u_j\right]}_{\substack{\text{Ion}\\ \text{convection}}} - \underbrace{\mathbf{P}_{ij} u_j}_{\substack{\text{Ion pressure}\\ \text{work}}} - \underbrace{\frac{\partial q_{ij}}{\partial x_j}}_{\substack{\text{Ion thermal}\\ \text{conduction}}} \\ &\quad = \underbrace{W_i}_{\substack{\text{Ion}\\ \text{heating}}} + \underbrace{G + J}_{\substack{\text{Ion/electron}\\ \text{exchange}}} \end{aligned}} \tag{D}$$

5) **Magnetic field generation:**
In the early studies of the interaction of intense laser fields with solid targets, it was found that relatively strong magnetic fields were generated. These were ascribed to the currents generated by electron pressure gradients in the plasma. Although the assignation remains uncertain as other particle mechanisms are also possible, there is little doubt that a purely MHD source exists. This has been best explored by simulation using finite-difference forms of the MHD equations, which we outline in this chapter. Since these are quite complicated, they can be used as a model to exemplify methods for solving time-dependent MHD problems.
The generation of a magnetic field by electron currents driven by gradients in the electron pressure (or density and temperature) is given by the simplified Maxwell equation (neglecting the displacement current) (Eq. (7.5)) and the equation of current flow (Eq. (7.13)), namely

$$\boxed{\frac{\partial \mathbf{B}}{\partial t} - \underbrace{\nabla \wedge (\mathbf{u} \wedge \mathbf{B})}_{Advection} + \underbrace{\nabla \wedge [\boldsymbol{\eta} \cdot (\nabla \wedge \mathbf{B})]}_{Diffusion} = \underbrace{\nabla \wedge \mathbf{S}}_{Source}} \tag{E}$$

where the diffusivity tensor $\boldsymbol{\eta}$ and the source vector $\mathbf{S}$ may be written in two alternative forms depending on whether the Hall term is included as a magnetic stress (see Equations 7.17 to 7.20).

$$\eta = \rho/\mu_0, \qquad \text{when} \quad \rho'_{\parallel} = \rho_{\parallel}, \quad \rho'_{\perp} = \rho_{\perp} \quad \text{and} \quad \rho'_{\wedge} = \rho_{\wedge} + B/en_e \tag{7.26}$$

$$\frac{1}{n_e}\nabla p_e \longrightarrow \mathcal{k}T_e \nabla \ln(n_e) \longrightarrow -\ln n_e\, \nabla(\mathcal{k}T_e) \tag{7.27}$$

$$\mathbf{S} = \frac{1}{e n_e}\boldsymbol{\beta} \cdot \nabla \ln(\mathcal{k}T_e) \tag{7.28}$$

$$\frac{\partial \mathbf{B}}{\partial t} = \nabla \wedge \mathrm{S} = \frac{1}{e}\nabla(\mathcal{k}T_e) \wedge \nabla(\ln n_e) \tag{7.29}$$

7.4 Conjugate Conservative Operators

The underlying MHD equations have been listed as Eqs. (A)–(E) above. Since the underlying fluid dynamics obey a set of conservation relations, there is an underlying structure to the equations, which may be exploited to improve the accuracy of the results.

The solution of the equations of magneto-hydrodynamics requires the integration of several sets of coupled differential equations. This is accomplished by the technique of operator splitting, discussed in Supplement M.4 where the general multi-variable equations are split into a series of simpler ones:

$$\frac{\mathrm{d}u}{\mathrm{d}t} = \mathcal{L}u = \sum \mathcal{L}_1 u + \mathcal{L}_2 u + \cdots + \mathcal{L}_n u \tag{7.30}$$

where the set of operators $\mathcal{L}_n u$ are those in Eqs. (A)–(E) and the variable u the set ρ, $\rho\mathbf{u}$, ϵ_e, ϵ_i and $\mathbf{B}$.

Although $\mathcal{L}$ must be conservative, the individual terms $\mathcal{L}_\nu$ may not be. It therefore follows that there must be conjugate pairs of elements in the split operators $\mathcal{L}_m u^\alpha$ and $\mathcal{L}_\nu^\beta$ such that their sum is conservative.

Simple conservation occurs when each quantity in Eqn. (A) to (E) is separately conserved. Such simple conservation is exemplified by thermal diffusion in both electron (C) and ion (D) energy equations, and electron current energy transport in Eq. (C).

More commonly, conservation is only achieved by two or more operators in the governing equations (*conjugate conservation*). Within the MHD equations, the following are conservative conjugates:

Conservative conjugates

Eq. (E)	Source	:	Eq. (C)	Source
Eq. (E)	Advection	:	Eq. (C)	Magnetic stress
Eq. (D)	Work	:	Eq. (E)	Particle pressure
Eq. (E)	Work	:	Eq. (E)	Particle pressure
Eq. (D)	Ion-electron exchange	:	Eq. (E)	Ion-electron exchange

It may be expected that, in view of the conservative energy exchange between different terms, there will be several different oscillatory speeds in the system in particular: hydrodynamic sound waves, magneto-sonic waves and thermal magnetic waves. As a result, there are three characteristic speeds, namely: flow speed, Alfvén speed and thermal magnetic wave speed. On an Eulerian grid, these must satisfy the Courant-Friedrichs-Levy condition for stability. When possible explicit time steps should be avoided and centred difference implicit ones used.

7.5 Finite-Difference Approximations

The introduction of currents and magnetic fields greatly increases the complexity of a problem. Although the basic form of the governing equations are those met in the study of fluids, namely second-order differential equations of hyperbolic and parabolic form, the increase in the number of variables and the interaction between them results a considerable increase in complexity. In particular, the interaction between the various terms in the overall equation set means that care should be taken with the finite-difference development to avoid instabilities and cumulative errors. The method we have identified separates the individual equations through their interactions in energy using mutual conservation as the identifier using assembling the resulting equations via an operator splitting techniques (see Supplement M.4).

The solution of the equations of magneto-hydrodynamics requires the simulation integration of several sets of coupled differential equations. This is the problem discussed in Supplement M.4, where the general multi-variable equations are split into a series of simpler ones

$$\mathcal{L}u = \sum \mathcal{L}_1 u + \mathcal{L}_2 u + \cdots + \mathcal{L}_n u \tag{7.31}$$

where the set of operators $\mathcal{L}_n u$ are those in Eqs. (A)–(E) and the variable u the set ρ, $\rho\mathbf{u}$, ϵ_e, ϵ_i and $\mathbf{B}$.

Although $\mathcal{L}$ may be conservative, the individual terms $\mathcal{L}_v$ may not be. It therefore follows that there must be conjugate pairs of elements in the split operators $\mathcal{L}_m u^\alpha$ and $\mathcal{L}_v^\beta$ such that their sum is conservative. For the equation set (A)–(E), we have the set identified earlier, namely:

Conservative conjugates

Eq. (E)	Source	:	Eq. (C)	Source
Eq. (E)	Advection	:	Eq. (C)	Magnetic stress
Eq. (D)	Work	:	Eq. (E)	Particle pressure
Eq. (E)	Work	:	Eq. (E)	Particle pressure
Eq. (D)	Ion-electron exchange	:	Eq. (E)	Ion-electron exchange

By splitting the operators in this way it is possible to ensure that the resulting computer models show at least weak conservation. In particular, the requirement to demonstrate conservation requires the variables on a mesh cantered grid.

7.6 Cylindrical Geometry – Self-Generated Magnetic Fields

The simplest geometrical problem is the two-dimensional plasma problem, with the magnetic field B_ϕ is orthogonal to the two spatial directions (z, r) – typically a cylindrical geometry with an azimuthal magnetic field. Typically, this is the case of a cylindrically focussed incident laser on a solid surface. The finite-difference approximations to model this problem are complex, but treated in some detail by the author in Pert (1981). We therefore refer the interested reader to that work rather than detailing the lengthy analysis of that work, The finite- difference equations derived therein are stable and conservative. They have proved successful in practical use.

This relatively simple case which is analysed in detail in Pert (1981) will serve as an exemplar of how a stable, well-behaved computer code to treat the time-dependent magneto-hydrodynamic flow of a plasma may be constructed (see also Brackbill and Goldman, 1983). The general problem is considered, but simplification involving an external field rather than a self-generated one is easily obtained by setting the source term (**S**) to zero. Three-dimensional diffusion is treated by Douglas' ADI method (Section 10.18).

References

Anderson, J E. 1963. *Magnetohydrodynamic Shock Waves*. Cambridge, MA: M I T Press.

Brackbill, J U and S R Goldman. 1983. "Magnetohydrodynamics in laser fusion: fluid modelling of energy transport in laser targets." *J Comput Phys* 36:415–436.

Braginskii, S I. 1965. Transport processes in plasma. In *Reviews of Plasma Physics*, ed. M A Leontovich. Vol. 1. New York: Consultants Bureau pp. 205–311.

Chapman, S and T G Cowling. 1952. *The Mathematical Theory of Non-Uniform Gases*. Second ed. Cambridge: Cambridge University Press.

Epperlein, E M and M G Haines. 1986. "Plasma transport coefficients in a magnetic field by direct numerical solution of the Fokker–Planck equation." *Phys Fluids* 29:1029–1041.

Landau, L D and E M Lifshitz. 1984. *Electrodynamics of Continuous Media*. Vol. 8 of *Course of Theoretical Physics* Second ed. Oxford: Butterworth-Heinemann.

Liberman, M A and A L Velikhovich. 1986. *Physics of Shock Waves in Gases and Plasmas*. New York: Springer.

Lifshitz, E M and L P Pitaevskii. 1981. *Physical kinetics*. Vol. 10 of *Course of Theoretical Physics* Second ed. Oxford: Butterworth-Heinemann.

Pert, G J. 1981. "Algorithms for the self-consistent generation of magnetic fields in plasmas." *J Comput Phys* 43:111–163.

Pert, G J. 2021. *Introduction to Plasma Physics*. London: Wiley.

Shkarofsky, I P, T W Johnston and M P Bachynski. 1966. *The Particle Kinetics of Plasmas*. Reading, MA: Addison-Wesley.

Sommerfeld, A. 1952. *Electrodynamics*. Vol. 3 of *Lectures on Theoretical Physics* New York: Academic Press.

Sommerfeld, A. 1964. *Mechanics of Deformable Bodies*. Vol. 2 of *Lectures on Theoretical Physics* New York: Academic Press.

Stamper, J A, K Papdopulous, R N Sudan, S O Dean, E A McLean and J A Dawson. 1971. "Spontaneous magnetic fields in laser-produced plasmas." *Phys Rev Lett* 26:1012.

Part II

Particle Methods

8

Particle in Cell Simulations

8.1 Introduction

A large number of problems in applied physics and engineering involve the integration of the trajectory of a single or group of particles along a path subject to externally applied forces and inter-particle interactions. These embrace astronomical orbits for planets and stars, electron optics and molecular dynamics – the latter including the calculation of thermodynamic functions and chemical structures requiring a knowledge of the short-range inter-molecular force. Probably the most important application of particle methods has been in plasmas where the force is the long-range Coulomb interaction between charged particles (electrons and ions). Similar methods are also required for the astronomical calculation of the structure of galaxies under gravitational inverse square laws of force.

An elementary introduction to particle methods was given in our earlier work (Woolfson and Pert, 1999, ch. 3) and it is unnecessary to repeat this discussion of these methods, but we refer the reader to that earlier text. Short-range inter-particle forces are relatively simple to treat with particle dynamics. The particles are enclosed in a cell whose dimensions are large compared to the range of the force. Forces between particles outside and inside the box are satisfactorily treated by introducing a periodic box arrangement as in Section 9.5 (see Figure 9.3).

The particle-in-cell (PIC) method was originally developed for fluid mechanics by Harlow (1955) (see also Cooper, 1989, pagel 44) and his collaborators at Los Alamos (see Amsden, 1966). However, it was quickly realised by Harlow at Los Alamos (Harlow, 1964), Dawson at Princeton and Buneman at Cambridge that the method was particularly suited for the simulation of dilute plasma. where the palasma responds to the long range co-operative electro-magnetic fields rather than short range collisions. Using particle-in-cell methods allows the treatment of long-range inter-particle forces due to the charge on the particle. In the simplest case, these are purely electrostatic, but more generally involve electromagnetic waves. These models have now seen significant development in both two and three dimensions since the introductory discussion in our earlier work, and we therefore give a more up-to-date explanation of the method. A nice succinct introduction to more modern methods is given by Evans (1986) and comprehensive descriptions by Hockney and Eastwood (1988) and Birdsall and Langdon (1991). A thorough description of a working code was given by Langdon and Lasinski (1976) for studying laser–plasma interactions and should be consulted if such a model is required.

The behaviour of plasma is described in terms of two characteristic quantities (see Pert 2021), which are associated with the collective behaviour of plasma, where the particles move together in coherent motion.

1) **Plasma frequency** is the characteristic frequency at which plasma electrons oscillate under the influence of their mutual electric field. Its value is given by

$$\omega_p^2 = \frac{n_e e^2}{\epsilon_0 m_e} \tag{8.1}$$

where n_e is the electron density and e and m_e the electron charge and mass, respectively. This term represents the natural frequency of oscillation of a plasma block.

2) **Debye length** is the screening distance over which the field due to an isolated charge is screened by a charge cloud around it:

$$\lambda_D = \sqrt{\frac{\epsilon_0 \not{k} T_e}{n_e e^2}} \tag{8.2}$$

where T_e is the electron temperature.

We note the important relation

$$\lambda_D \omega_p = \sqrt{\not{k} T_e / m_e} = \bar{v}_e \tag{8.3}$$

The inter-particle interaction in plasma physics takes two distinct forms:

a) **Collisional interaction:** If the inter-particle separation is small ($< \lambda_D$) then the interaction is basically two particle, as in a classical collision.

b) **Co-operative interaction:** If the interaction distance is large ($> \lambda_D$) then the interaction is many body and the particles respond co-operatively.

PIC calculations generally take place in co-operative mode only, since collisions are difficult to include.

The density of electrons in a typical experiment is too large to allow us to calculate the motion of individual particles. However, we may consider a limited number of representative particles whose dynamic behaviour is identical to the real ones. Thus, consider *super-particles* whose mass and charge are N times that of an electron, Nm and Nq, respectively, and whose density is n/N; then, the charge-to-mass ratio $Q/M = e/m$, the charge density $\mathscr{P} = n/NqN = nq = \rho$ and similarly current density $J = n/NqNv = nqv = j$ are the same; therefore, the dynamics and self-generated electromagnetic fields are identical.

The dynamical equations are Newton's law for particle motion and Maxwell's equations for electrodynamics: Eq. (8.4). Thus, consider a particle ℓ of mass m_ℓ and charge q_ℓ with position $\mathbf{r}_\ell \equiv (x_\ell, y_\ell, z_\ell)$ and velocity $\mathbf{v}_\ell \equiv (u_\ell, v_\ell, w_\ell)$. The charge and current densities are obtained by a sum over the particles ℓ in the system. Thus, suppose the particle ℓ is spatially spread out by a distribution function $w_\ell(|\mathbf{r} - \mathbf{r}_\ell|)\mathbf{dr}$ at $\mathbf{r}$ known as the *shape function*. The dynamic equations are:

$$\frac{\mathrm{d}\mathbf{r}_\ell}{\mathrm{d}t} = \mathbf{v}_\ell \qquad \frac{\mathrm{d}}{\mathrm{d}t}\left(\gamma_\ell \mathbf{v}\right) = \frac{q_\ell}{m_\ell}(\mathbf{E} + \mathbf{v}_\ell \wedge \mathbf{B}) \tag{8.4a}$$

$$\nabla \wedge \mathbf{E} = -\frac{\mathrm{d}\mathbf{B}}{\mathrm{d}t} \qquad \frac{1}{\mu_0}\nabla \wedge \mathbf{B} = \mathbf{j} + \epsilon_0 \frac{\mathrm{d}\mathbf{E}}{\mathrm{d}t} \tag{8.4b}$$

$$\epsilon_0 \nabla \cdot \mathbf{E} = \rho \qquad \nabla \cdot \mathbf{B} = 0 \tag{8.4c}$$

where $\gamma_\ell = \sqrt{1 - v_\ell{}^2/c^2}$ is the relativistic correction for particle ℓ.

The charge and current densities are calculated by summing over all the particles taking account of their distributions

$$\rho(\mathbf{r}) = \sum_\ell \int q_\ell \, w_\ell(|\mathbf{r} - \mathbf{r}_\ell|) \, \mathrm{d}\mathbf{r}/\Delta V \quad \mathbf{j}(\mathbf{r}) = \sum_\ell \int q_\ell \mathbf{v}_\ell \, w_\ell(|\mathbf{r} - \mathbf{r}_\ell|) \, \mathrm{d}\mathbf{r}/\Delta V \tag{8.5}$$

We define a Cartesian grid ($x_i = i\Delta x$, $y_j = j\Delta y$, $z_k = k\Delta z$) with spacing Δx, Δy, Δz and with cells of volume $\Delta V = \Delta x \, \Delta y \, \Delta z$. The mesh fills a simply connected space of $X = N_x\Delta x$, $Y = N_y\Delta y$, $Z = N_z\Delta z$, where $x \in (0 : X)$, $y \in (0 : Y)$, $z \in (0 : Z)$. The particle ℓ is found in the cell centred at ($i = \lfloor x_\ell/\Delta X \rfloor, j = \lfloor y_\ell/\Delta y \rfloor, k = \lfloor z_\ell/\Delta z \rfloor$).[1]

In principle, the mesh may be three-dimensional with particles moving in the x-, y-, z-direction. In practice, such systems are computationally very expensive and smaller sub-sets are usually used. These are:

$$\begin{array}{llll}
1\mathrm{D} & : \partial/\partial z = 0 & (0 \;\; 0 \;\; 0) & (E_x \;\; 0 \;\; 0) \\
 & \;\; \partial/\partial y = 0 & (x \;\; 0 \;\; 0) & (v_x \;\; 0 \;\; 0) \\
1\frac{1}{2}\mathrm{D} & : \partial/\partial z = 0 & (0 \;\; 0 \;\; B_z) & (E_x \;\; E_y \;\; 0) \\
 & \;\; \partial/\partial y = 0 & (x \;\; 0 \;\; 0) & (v_x \;\; v_y \;\; 0) \\
2\mathrm{D} & : \partial/\partial z = 0 & (B_x \;\; B_y \;\; B_z) & (E_x \;\; E_y \;\; 0) \\
 & & (x \;\; y \;\; 0) & (v_x \;\; v_y \;\; 0) \\
2\frac{1}{2}\mathrm{D} & : \partial/\partial z = 0 & (B_x \;\; B_y \;\; B_z) & (E_x \;\; E_y \;\; E_z) \\
 & & (x \;\; y \;\; 0) & (v_x \;\; v_y \;\; v_z)
\end{array} \tag{8.6}$$

The 1D and 2D cases are electrostatic, and the $1\frac{1}{2}$D and $2\frac{1}{2}$D cases electromagnetic. The latter two are particularly useful for studying the interaction of laser beams with plasma. As the dimension of phase space increases, the number of particles required for the simulation increases and therefore both the storage and runtime requirements. Fortunately, the nature of the algorithms are well suited to modern parallel processing machines. Typically, one uses in excess of 100 particles per cell.

There is a clear difference between a PIC simulation and the actual physical environment associated with the difference between the finite charge/mass of the simulation particles and the nearly continuous distribution of the real ones. This will introduce error in the same way as finite-difference models from continuum solutions. The method works by assigning a charge/mass distribution (*shape function*) to a finite mesh, calculating the charge/mass in a set of finite volume cells (c.f. Section 1.11 for fluid dynamics) [Hockney and Eastwood (1988, ch. 5) give a thorough account of various weighting methods]. The charge and mass of individual particles are assigned to the cells by their overlap with the cell volume. Thus, remembering that a super-particle is an assembly of many real particles, we may split them to obtain a smoother distribution. Given a particle position (x_i, y_i, z_i) on a grid with mesh spacings (Δx, Δy, Δz) the particle ℓ lies in the cell ($i_\ell = \lfloor x_\ell/\Delta x \rfloor, j_1 = \lfloor y_\ell/\Delta y \rfloor, k_\ell = \lfloor z_\ell/\Delta z \rfloor$). Although some higher-order interpolations are available, two simple methods are popularly used, which we first illustrate in one dimension

i) **Nearest grid point** (NIP): Assigns the full weight of the element ℓ to the cell containing the nearest grid point (x_i, y_j, z_k)

$$W(x_\ell; x_i) = \begin{cases} 1 & \text{if } \; |x_\ell - x_i| < \frac{1}{2}\Delta x \\ 0 & \text{otherwise} \end{cases} \tag{8.7}$$

1 The floor function $f(x)$ takes a real number x as an input and returns the greatest integer less than or equal to x.

ii) **Cloud in cell** (CIC) [also known as area weighting (Harlow, 1964; Morse, 1970)]: Assigns the weight of the particle amongst the nearest neighbours of the nearest grid point.

$$W(x_\ell; x_i) = \begin{cases} 1 - |x_\ell - x_i|/\Delta X & \text{if } |x_\ell - x_i| < \Delta x \\ 0 & \text{otherwise} \end{cases} \tag{8.8}$$

In three dimensions the particle is assumed to have a volume $\Delta x \, \Delta y \, \Delta z$ centred at x_ℓ, y_ℓ, z_ℓ. The shape function is the overlap of the particle shape function on the mesh. With a simple Cartesian mesh, the overall weight is the product of the three one-dimensional mesh weights

$$W(x_\ell, y_\ell, z_\ell; x_i, y_j, z_k) = W^x(x_\ell; x_i)\, W^y(y_\ell; y_j)\, W^z(z_\ell; z_k) \tag{8.9}$$

Integrating Eq. (8.5) we obtain

$$\begin{aligned} \rho(\mathbf{r}) &= \sum_\ell W(x_\ell, y_\ell, z_\ell; x, y, z) q_\ell / \Delta V \\ \mathbf{j}(\mathbf{r}) &= \sum_\ell W(x_\ell, y_\ell, z_\ell; x_i, y_j, z_k) q_\ell \mathbf{v}_\ell / \Delta V \end{aligned} \tag{8.10}$$

There is a further important application of the weights, namely in assigning the force $\mathbf{F}_\ell = q_\ell(\mathbf{E} + \mathbf{v}_\ell \wedge \mathbf{B})$ in that both electric ($\mathbf{E}$) and magnetic ($\mathbf{B}$) fields are calculated on the configuration space and must be interpolated to the particle position:

$$\mathbf{E}_\ell = \sum_{i,j,k} W(x_\ell, y_\ell, z_\ell; x_i, y_j, z_k)\, \mathbf{E}_{i,j,k} \quad \text{and} \quad \mathbf{B}_\ell = \sum_{i,j,k} W(x_\ell, y_\ell, z_\ell; x_i, y_j, z_k)\, \mathbf{B}_{i,j,k} \tag{8.11}$$

Note that it is essential that the *same* shape function is used in both these operations.

It will be convenient in dealing with the electromagnetic field cases in one or two dimensions to define a set of centred difference operators in time and space, thus:

$$\mathfrak{d}_{t\,i,j}^{\,n+1/2} A \equiv \frac{A^{n+1}_{i,j} - A^n_{i,j}}{Dt} \tag{8.12a}$$

$$\mathfrak{d}_{x\,i,j}^{\,n} A \equiv \frac{A^n_{i+1/2,j} - A^n_{i-1/2,j}}{\Delta x} \tag{8.12b}$$

$$\mathfrak{d}_{y\,i,j}^{\,n} A \equiv \frac{A^n_{i,j+1/2} - A^n_{i,j-1/2}}{\Delta y} \tag{8.12c}$$

It is obvious that these are the centred differences of the derivatives; thus, $\nabla \equiv \mathfrak{d}_x$ for example. These difference operators have the useful property that they all commute.

8.1.1 One-Dimensional (1D) Electrostatic Model

The simplest example of PIC dynamics describes the one-dimensional motion of plasma in a steady electric field. Despite its simplicity, it exemplifies many of the essential features of PIC simulation, and enables us to draw some characteristic limiting conditions. The only interaction fields are the electrostatic described by Poisson's Eq. (8.4c) and the particle dynamics by Eq. (8.4a); there is no contribution from time-varying fields or current and the magnetic field is zero (Eq. 8.4b).

Firstly, we map the electric charge density from the particles to the grid (Eq. 8.10) at integer time (ρ_i^n)

$$\rho_i^n = \sum_\ell q_\ell W(x_\ell^n; x_i)/\Delta V \tag{8.13}$$

The potential $\phi(x)$ is obtained from Poisson's equation

$$\epsilon_0 \nabla^2 \phi = -\rho \tag{8.14}$$

or most simply (see Appendix M.6.i) from the Gauss integral form in one dimension

$$E_{i+1/2}^n = E_{i-1/2}^n + \rho_i^n \,\Delta x/\epsilon_0 \tag{8.15}$$

to obtain the electric field on the grid at half-intervals $E_{i+1/2}^n$. The total potential across the plasma with N intervals is 0 if the net charge is 0.

Setting the particle position at integer time x_ℓ^n and the velocity at half-integer time $u_\ell^{n+1/2}$ Eq. (8.4a) can be integrated by a leap-frog algorithm over a time-step Dt,

$$u_\ell^{n+1/2} = u_\ell^{n-1/2} + q_\ell \, E_\ell^n \, Dt \qquad \text{and} \qquad x_\ell^{n+1} = x_\ell^n + u_\ell^{n+1/2} \, Dt \tag{8.16}$$

In the 1D electrostatic model, we note that the equation of current continuity:

$$\frac{\partial \rho}{\partial t} + \nabla(\mathbf{j}) = 0 \rightarrow \mathfrak{d}_t\rho + \mathfrak{d}_x(j_x) \tag{8.17}$$

is automatically satisfied in the finite-difference form since

$$\mathfrak{d}_t(\rho) = -\sum_\ell \dot{W}(x - x_\ell) q_\ell \mathfrak{d}_t x_\ell = -\sum_\ell \dot{W}(x - x_\ell)\, q_\ell \, v_{x\ell} \tag{8.18}$$

$$\text{and} \quad \mathfrak{d}_x(j_x) = \sum_\ell \dot{W}(x - x_\ell)\, q_\ell \, v_{x\ell} \tag{8.19}$$

8.1.2 One-Dimensional ($1\frac{1}{2}$D) Electromagnetic Model

To treat one-dimensional electromagnetic waves, we must add the transverse wave components E_y and B_z from Eq. (8.4b) see Figure 8.1

$$\frac{\partial E_y}{\partial x} = -\frac{\partial B_z}{\partial t} \tag{8.20a}$$

$$\frac{1}{\mu_0}\frac{\partial B_z}{\partial x} = -j_y - \epsilon_0 \frac{\partial E_y}{\partial t} \tag{8.20b}$$

or for the alternative independent polarisation E_z and B_y,

$$\frac{\partial E_z}{\partial x} = \frac{\partial B_y}{\partial t} \tag{8.21a}$$

$$\frac{1}{\mu_0}\frac{\partial B_y}{\partial x} = j_z + \epsilon_0 \frac{\partial E_z}{\partial t} \tag{8.21b}$$

We consider plane electromagnetic waves propagating along the x-direction, so that either $E_y \neq 0$ and $B_z \neq 0$ or $B_y \neq 0$ and $E_z \neq 0$; the direction of propagation $\hat{\mathbf{x}}$ depending on the Poynting vector: product $E_y B_z$ or $E_z B_y$. It is easy to show that the two modes E_z, B_y and

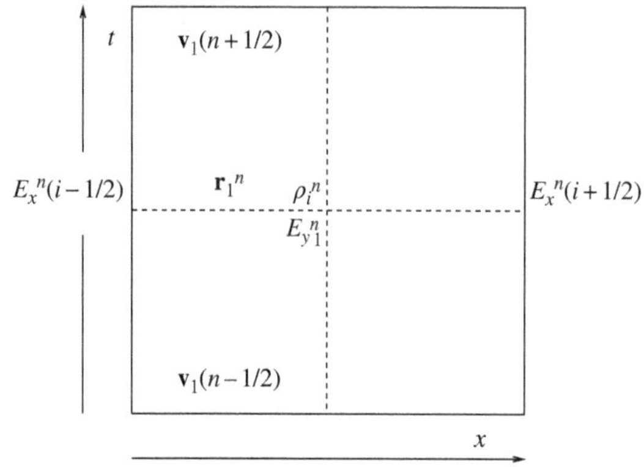

Figure 8.1 Staggered grid for $1\frac{1}{2}$ calculation.

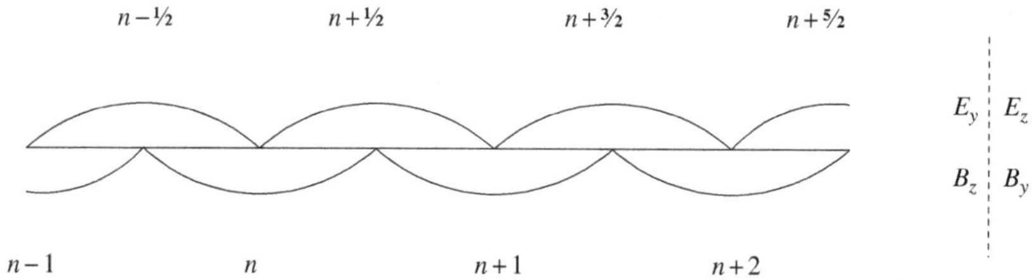

Figure 8.2 Sketch to show the two time level mesh employed for integrating the two modes of the electromagnetic wave.

E_y, B_z are not coupled, i.e. independent as is well known from standard electromagnetic theory. It is easy to show that this result holds after finite differencing. The sets E_zB_y and E_yB_z can therefore be calculated independently.

Two sets of spatial grids are introduced along the x-axis (Figure 8.2). One is a full-integer grid system, and the other is a half-integer grid system. E_y, B_y, J_y and ρ are defined on full-integer grids, while E_x, E_z, B_z and J_z are defined on the half-integer grids. Spatial and time derivatives in Maxwell's equations are replaced by centred differences by Δx and Dt.

Using a linked mesh for the electromagnetic wave terms E_y and B_z,

$$\left(E_{y_i}^{n+1} - E_{y_i}^{n}\right)/Dt = -c^2\left(B_{z_{i+1/2}}^{n+1/2} - B_{z_{i-1/2}}^{n+1/2}\right)/\Delta x \tag{8.22a}$$

$$\left(B_{z_{i-1/2}}^{n+1/2} - B_{z_{i-1/2}}^{n-1/2}\right)/Dt = -\left(E_{y_i}^{n} - E_{y_{i-1}}^{n}\right)/\Delta x \tag{8.22b}$$

where $c = \sqrt{1/\epsilon_0\,\mu_0}$ is the velocity of light in free space. We require the magnetic field at integer steps to treat the Lorentz force in the particle mover

$$\mathbf{B}_i^n = \frac{1}{2}\left(\mathbf{B}_{\mathbf{z}_i}^{n-1/2} + \mathbf{B}_{\mathbf{z}_i}^{n+1/2}\right) \tag{8.23}$$

The Courant condition for the time step and grid spacing is used. $cDt < \Delta x$. We need to push particles in the electromagnetic field by solving the Lorentz equation for $\mathrm{d}v/\mathrm{d}t$

$$\frac{\mathrm{d}v}{\mathrm{d}t} = \frac{q}{m}(\mathbf{E} + \mathbf{v} \wedge \mathbf{B}) \quad \text{and} \quad \frac{\mathrm{d}x}{\mathrm{d}t} = v_x \tag{8.24}$$

by the Buneman–Boris method. The advantage of the method is that it is strictly conservative of the kinetic energy in the calculation of cyclotron motion. The Buneman–Boris algorithm has a quasi-implicit form

$$\frac{\mathbf{v}^{n+1/2} - \mathbf{v}^{n-1/2}}{Dt} = \frac{q}{m}\left(\mathbf{E}^n + \frac{\left(\mathbf{v}^{(n+1/2)} + \mathbf{v}^{(n-1/2)}\right)}{2} \wedge \mathbf{B}^n\right) \tag{8.25}$$

The algorithmic form in one dimension is:

i) Advance velocity due to an electric field only to the mid-point:

$$\mathbf{v}^- = \mathbf{v}^{(n-1/2)} + \frac{q}{m}\mathbf{E}^n\frac{\Delta t}{2} \tag{8.26a}$$

$$\mathbf{v}^+ = \mathbf{v}^{(n+1/2)} - \frac{q}{m}\mathbf{E}^n\frac{\Delta t}{2} \tag{8.26b}$$

ii) Rotate velocity vector from $\mathbf{v}^-$ to $\mathbf{v}^+$ in a magnetic field:

$$\frac{\mathbf{v}^+ - \mathbf{v}^-}{Dt} = \frac{1}{2}\frac{q}{m}\left(\mathbf{v}^+ + \mathbf{v}^-\right) \wedge \mathbf{B}^n \tag{8.26c}$$

iii) Solving for $\mathbf{v}^+$

$$\mathbf{v}^+ = \mathbf{v}^- + \frac{2}{(1+T^2)}(\mathbf{v}^- + \mathbf{v}^- \wedge \mathbf{T}) \wedge \mathbf{T} \tag{8.26d}$$

where $\mathbf{T} = \dfrac{q}{2m}\,\mathbf{B}^n\,Dt$.

iv) The final velocity advance:

$$\mathbf{v}^{n+1/2} = \mathbf{v}^+ + \frac{q}{2m}\,\mathbf{E}^n\,Dt \tag{8.26e}$$

8.1.2.1 Boundary Conditions

- **Electrostatic codes:** The field boundary conditions in a 1D electrostatic model are easily established, namely that the potential across the plasma must be equal to the applied potential supplied by some external source (capacitor model), typically zero as in an infinitely periodic system; see Appendix M.6.
  For the particle in a finite system with a defined boundary there are three possibilities:
  a) Reflect the particles at the boundary.
  b) Reflect but with a random (Gaussian) normal velocity, v_x.
  c) Reflect with a random velocity, but with a random delay.
- **Electromagnetic codes:** We may set the boundary conditions as for the electro-static case, but must take account of the fact that periodic boundary conditions do not allow for propagating electromagnetic waves. There is therefore no mechanism to introduce the important case of laser radiation interaction with the plasma.
  When the field is electromagnetic, we must take account of its travelling wave nature. Examining Eq. (8.22) we see that propagation by a full time step requires knowledge of (E_y, B_z) or (E_z, B_y), depending on the mode over the full mesh.
  Consider the E_y, B_z case. If the boundary is set at an integer space point, we require a knowledge of either E_{y_ν} or of $B_{z\nu+1/2}$ across the mesh at the boundary surface. For example, if the mesh extends so that E_{y_1} and E_{y_N} are known, we require $B_{z1/2}$ and $B_{zN+1/2}$ to update E_y.

There are three important cases:

i) **Reflection:** Suppose the wave is reflected at the surface $i = 1$, then the field is zero at the surface:

$$E_{y_1}^{\ n} = 0 \qquad \text{for all} \qquad n \geq 1 \tag{8.27}$$

and similarly at the surface N if reflecting.

ii) **Transmission:** A plane wave is transmitted through the boundary with no loss. Consider the one-dimensional case ($1\frac{1}{2}$ code). A plane wave travelling in the x-direction has $E_y = +cB_z$, and vice versa ($E_y = -cB_z$) for one in the negative x-direction. Therefore, the right-hand wave travelling in the x-direction at the right-hand boundary (N) and exiting without reflection has

$$E_{yN}^{\ n+1} + E_{yN}^{\ n} = c\left(B_{zN-1/2}^{\ n+1/2} + B_{zN+1/2}^{\ n+1/2}\right) \tag{8.28}$$

Maxwell's equation yields

$$E_{yN}^{\ n+1} - E_{yN}^{\ n} = -\frac{c^2\,Dt}{\Delta x}\left(B_{zN+1/2}^{\ n+1/2} - B_{zN-1/2}^{\ n+1/2}\right) \tag{8.29}$$

Eliminating $E_{yN}^{\ n+1}$ gives

$$cB_{zN+1/2}^{\ n+1/2} = \frac{2E_{yN}^{\ n} - (1 - cDt/\Delta x)\,cB_{zN-1/2}^{\ n+1/2}}{1 + cDt/\Delta x} \tag{8.30}$$

A similar result is obtained at the left-hand boundary ($i = 1$) for exiting radiation propagating in the negative x-direction with $E_y = -cB_z$.

$$E_{y_1}^{\ n+1} + E_{y_1}^{\ n} = c\left(B_{z1/2}^{\ 1/2} + B_{z3/2}^{\ 1/2}\right) \tag{8.31}$$

$$E_{y_1}^{\ n+1} - E_{y_1}^{\ n} = -\frac{c^2Dt}{\Delta x}\left(B_{z3/2}^{\ n+1/2} - B_{z1/2}^{\ n+1/2}\right) \tag{8.32}$$

Eliminating $E_{y_1}^{\ n+1}$ we obtain

$$cB_{z1/2}^{\ n+1/2} = \frac{2E_{y_1}^{\ n} - (1 - cDt/\Delta x)\,cB_{z3/2}^{\ n+1/2}}{1 + cDt/\Delta x} \tag{8.33}$$

iii) **Injection:** A wave incident on the plasma is introduced by adding the incident field to B_z and E_y at the boundary. Thus, the values of $B_{zN+1/2}^{\ i+1/2}$ and $E_{y_i}^{\ n}$ are set directly. The inward propagation of the wave is set by the sign of E_yB_z opposite to that of transmission treated above.

This corresponds to the important case of a laser beam incident on the plasma.

The alternative mode is treated identically.

In two dimensions the boundary conditions are more difficult to set since waves at arbitrary angles to the mesh are transmitted. For appropriate algorithms see Lindman (1975).

8.1.2.2 Initialisation

There are two characteristic quantities, which determine the performance of the model, namely the time step Dt and the spatial cell sizes Δx, Δy, Δz.

If the problem has corresponding characteristic parameters τ_0, X_0, Y_0, Z_0, clearly the modelling parameters must be much smaller than those of the physical problems, which thereby give rise to the constraints on the time step.

The one-dimensional electrostatic model identifies stability constraints (see Appendix 8.B.1). These must also be applied in the more general cases with additional dimensions and electromagnetic fields. The time step must therefore be a small fraction of the plasma period, i.e. the inverse of the plasma frequency, typically $Dt < 0.15\,\omega_p{}^{-1}$. In the electromagnetic case, this must be supplemented by the Courant–Friedrichs–Lewy condition for electromagnetic waves $Dt < \Delta x/c$ but this is rarely the limiting condition (Richtmyer and Morton, 1967, §10.2).

The space step is determined by the constraint to avoid aliasing by Landau damping to resolve the spurious spatial oscillation associated with plasma oscillations, discussed in Appendix 8.B.2, namely $\Delta x < \lambda_D$.

The principal source of noise in the PIC code is due to the limited number of particles in each cell. The initial number of particles included in each cell is therefore an important parameter, but is limited by the fact that the CPU time expended increases approximately linearly with particle number and the number of cells.

The number of particles in each cell is determined by the density distribution in the problem to be solved. The number is proportional to the relative density at the cell centre. This is set by the requirement of a reasonable number of calculational 'super-particles' in each cell such that the distribution function can be determined over the mesh. The total number of particles may be several million in a large multidimensional code. As a minimum each cell must contain at least 100 particles. The ratio of the plasma density to the particle density determines the scaling between real and super particles.

The initial particle distribution may take one of three forms:

i) **Cold start**: Particles initially at rest. Debye length initially undefined. Particles gain energy from the field.
ii) **Thermal plasma**: Particles have a Maxwell distribution in the appropriate number of dimensions set by a random number of the appropriate distribution
   1) In one dimension. Particles have a Gaussian given by Eq. (9.A.8).
   2) In two dimensions. Particles have a 2D Maxwellian given by Eq. (9.A.9).
   3) In three dimensions. Particles have a 3D Maxwellian.
iii) **Quiet start**: Due to relatively few particles per cell compared to the real plasma, fluctuations from the mean can be significant. These can be suppressed by forcing the mean velocity to be zero, $\overline{v} = 0$ and the variance $\overline{v^2} = \not{k}T$, and thereby reducing random noise.

8.1.3 Two-Dimensional (2D) Electrostatic Model

This is a relatively simple direct extension of the 1D case treated earlier involving a two-dimensional solution of Poisson's equation. This is usually accomplished via the electro-static potential $\mathbf{E} = -\nabla\phi$, which satisfies the Poisson equation

$$\nabla^2\phi = \frac{\partial^2\phi}{\partial x^2} + \frac{\partial^2\phi}{\partial y^2} = -\frac{\rho}{\epsilon_0} \tag{8.34}$$

subject to the boundary conditions, which are typically either:

- Specified value at the boundary, e.g. $\phi = 0$.
- Zero field normal to the boundary $\frac{\partial \phi}{\partial n} = 0$.
- Mixed of either type on each surface.

This equation utilizes finite differences similar to the standard five-point schemes discussed earlier. The potential may be obtained by a number of standard methods (see Hockney and Eastwood, 1988, ch. 6). The fields $E_x = \frac{\partial \phi}{\partial x}$ and $E_y = \frac{\partial \phi}{\partial x}$ are then obtained by finite differences.

8.1.4 Two-Dimensional ($2\frac{1}{2}$D) Electromagnetic Model

Let us now consider two dimensional meshes with propagating electromagnetic fields. The variables are assigned to the mesh as shown in Figure 8.3, which is a logical generalisation of the one-dimensional case treated above. The time is assigned to the variables in a leap-frog manner as in 1D. The field equations and particle dynamics are finite differenced in the same manner as in the $1\frac{1}{2}$D case. Taking the co-ordinates (x, y) in the plane of the **E** and **B** fields, Maxwell's equations take the form of Eqs. (8.20b) and (8.21b), using finite differences as in Eq. (8.35),

$$\left(E_{yi}^{n+1} - E_{yi}^{n}\right)/Dt = -c^2\left(B_{zi+1/2}^{n+1/2} - B_{zi-1/2}^{n+1/2}\right)/\Delta x \tag{8.35a}$$

$$\left(B_{zi-1/2}^{n+1/2} - B_{zi-1/2}^{n-1/2}\right)/Dt = -\left(E_{yi}^{n} - E_{yi-1}^{n}\right)/\Delta x \tag{8.35b}$$

and for the other polarisation

$$\left(E_{zi}^{n+1} - E_{zi}^{n}\right)/Dt = c^2\left(B_{yi+1/2}^{n+1/2} - B_{yi-1/2}^{n+1/2}\right)/\Delta x \tag{8.36a}$$

$$\left(B_{yi-1/2}^{n+1/2} - B_{yi-1/2}^{n-1/2}\right)/Dt = \left(E_{zi}^{n} - E_{zi-1}^{n}\right)/\Delta x \tag{8.36b}$$

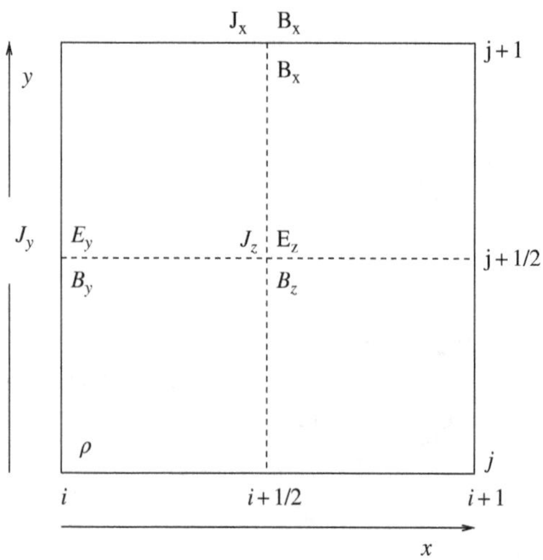

Figure 8.3 Spatial arrangement of variables on a two-dimensional grid.

As before, we require the magnetic field at integer steps for the particle mover

$$\mathbf{B}_i^n = \frac{1}{2}\left(\mathbf{B}_{\mathbf{z}_i}^{n-1/2} + \mathbf{B}_{\mathbf{z}_i}^{n+1/2}\right) \tag{8.37}$$

There are two problems with the code at this stage:

- **Charge continuity:** In analytic theory with continuous functions for **E**, **B**, **j** and ρ, Maxwell's equations require that the equation of charge continuity

$$\frac{1}{\mu_0}\nabla \wedge \mathbf{B} = \mathbf{j} + \epsilon_0 \frac{\partial \mathbf{E}}{\partial t} \tag{8.38}$$

is consistent with

$$\epsilon_0 \nabla \mathbf{E} = \rho \qquad \text{and} \qquad \frac{\partial \rho}{\partial t} + \nabla \cdot \mathbf{j} = 0 \tag{8.39}$$

This is no longer true when the equations are discretised and the current densities are calculated from the individual particle currents and their weights as per Eq. (8.5).
The usual way this is handled is to calculate a correction to the electrostatic potential such that the current density satisfies the continuity equation. Writing this correction to the current density **j** as $-\nabla\xi$ where[2]

$$\frac{1}{\epsilon_0}\nabla^2 \xi = \nabla \cdot \mathbf{j} + \partial\rho/\partial t \tag{8.40}$$

The algorithm based on this form is time-centred and therefore second-order. The term in ∇^2 is just the simple five-point operator given by Poisson's equation, whose solution is discussed in Supplement M.6.
Computationally, it is simpler (as is normally done) to advance **E** using the uncorrected **j** and apply a correction $-\nabla\delta\phi$ to **E** afterwards

$$-\frac{1}{\epsilon_0}\nabla^2 \delta\phi = \rho/\epsilon_0 = \nabla \cdot \mathbf{E} \tag{8.41}$$

where $\xi = \partial_t^2 \delta\phi$.
- **Boundary conditions** are treated in detail by Langdon and Lasinski (1976, §D) and Birdsall and Langdon (1991, §15.1) to which we refer the interested reader and treat them briefly here. Systems with periodic boundary conditions are easily solved using a Fourier decomposition of the fields, but are generally not applicable. Practical problems generally involve incoming and outgoing waves across at least one boundary. Using a two-dimensional Cartesian grid, this typically requires an incoming wave in the $-x$ and outgoing waves at arbitrary directions at the boundary x_N and a reflection at x_1.
For many problems, the simple one-dimensional method described for $1\frac{1}{2}$D meshes in item 2 on p. 128 is readily adapted to two dimensions and easily integrated into the field calculation. The more general case where the exit rays are not parallel but inclined to the mesh lines x, treated by Lindman (1975), is more complicated to program but works well.

2 Since the operation $\nabla \cdot \nabla \wedge \equiv 0$ and using the *fundamental theorem of vector calculus* (Helmholtz' theorem), we may simply write the correction to **j** as a scalar potential.

Appendix 8.A Buneman–Boris Algorithm

The Buneman–Boris algorithm advances the particle velocity by a second-order time development but with marked advantages over the direct leap-frog type difference, namely being relativistic

$$\frac{\mathbf{v}^{n+1/2} - \mathbf{v}^{n-1/2}}{Dt} = \frac{q}{m}\left(\mathbf{E}^n + \frac{\left(\mathbf{v}^{(n+1/2)} + v^{(n-1/2)}\right)}{2} \wedge \mathbf{B}^n\right) \tag{8.A.1}$$

Using the Buneman–Boris method usually takes five steps to update the position and velocity information of the particles as follows: Defining the relativistic factor $\gamma = 1/\sqrt{1 - v^2/c^2}$ and $u = \gamma v$

a) E field acceleration over time $Dt/2$,

$$\mathbf{u}' = \mathbf{u}^{n-1/2} + \frac{q}{m}\mathbf{E}^n \frac{Dt}{2} \tag{8.A.2a}$$

b) Cyclotron rotation over time Dt through the angle $qBDt/\gamma m$,

$$\mathbf{u}^0 = \mathbf{u}' + \mathbf{v}' \wedge \mathbf{T}\mathbf{v}^+ = \mathbf{v}^- + \mathbf{v}^0 \wedge \mathbf{S} \tag{8.A.2b}$$

where $\mathbf{S} = 2\mathbf{T}/(1 + \mathbf{T}^2)$.

c) E field acceleration over time $Dt/2$,

$$\mathbf{u}^{n+1/2} = \mathbf{u}^+ + \frac{q}{m}\mathbf{E}^n \frac{Dt}{2} \tag{8.A.2c}$$

d) Move particle over time Dt,

$$\mathbf{v}^{n+1/2} = \frac{\mathbf{u}_{n+1/2}}{\sqrt{1 + u^2/c''}} \tag{8.A.2d}$$

$$\mathbf{r}^{n+1} = \mathbf{r}^n + \mathbf{v}^{n+1/2} Dt \tag{8.A.2e}$$

Appendix 8.B Stability of the 1D Electrostatic Model

8.B.1 Time-Step Limitation

Let us first examine the nature of the solution expected from the 1D electrostatic model using differentials instead of finite differences. Consider the simplest case of an ambient plasma with zero space charge (ion density = charge no × electron density) and consider a small displacement of the electron. The equations of motion are

$$\frac{\partial x_\ell}{\partial t} = v_\ell \quad \frac{\partial v_\ell}{\partial t} = q_\ell E(x_\ell) W(x_\ell - x)/m_\ell \approx qE(x_\ell)/m \quad \frac{\partial E}{\partial x} = \rho(x)/\epsilon_0 \tag{8.B.1}$$

$$\rho(x) = \sum_\ell \rho_\ell = \sum_\ell q_\ell (W(x_\ell - x)) \tag{8.B.2}$$

Noting that the particle shape function must satisfy the equation of continuity

$$\frac{\partial W_\ell}{\partial t} = -\frac{\partial}{\partial x}\left[v_\ell W_\ell (x_\ell - x)\right] \tag{8.B.3}$$

If the number of particles is large, the sum in Eq. (8.B.2) may be averaged over particles ℓ, to be replaced by an equation of continuity for the net charge density $\rho = (n_i - zn_e)$:

$$\frac{\partial \rho}{\partial t} \approx -\sum_{\ell} \frac{\partial}{\partial x} \left(\bar{v}_\ell \, \rho_\ell \right) \tag{8.B.4}$$

If due to their large marge mass difference, the ions are stationary and the ions and the electrons mobile, we may consider only the case $\ell = 1$.

$$\frac{\partial^2 \rho_\ell}{\partial t^2} = -\rho_{\ell 0} \frac{\partial}{\partial x} \frac{\partial v_\ell}{\partial t} = -\rho_{\ell 0} q_\ell \frac{\partial E}{\partial x} = -\left(\frac{q_{\ell 0} \rho_{\ell 0}}{\epsilon_0 m_\ell} \right) \rho_\ell = -\omega_p^2 \, \rho_\ell \tag{8.B.5}$$

The electrons therefore oscillate with frequency $\omega_p = \sqrt{n_e e^2/\epsilon_0 \, m_e}$ known as the *plasma frequency* (n_e is the electron number density).

Turning now to the finite-difference form, we expect that a form of Eq. (8.B.5) will be generated namely

$$\left[\rho_{i+1} - 2\rho_i + \rho_{i-1} \right] / \Delta x^2 = -\omega_p^2 \rho \tag{8.B.6}$$

Setting $\rho = \rho_0 \mathrm{e}^{\imath \omega t}$ as a trial solution, we obtain

$$\exp(\imath \omega Dt) - 2 + \exp(-\imath \omega Dt) = 2\left[-\cos(\omega Dt) - 1 \right] = 4 \sin^2(\omega Dt/2) = -\omega_p^2 Dt^2 \tag{8.B.7}$$

Hence,

$$\omega Dt = \pm 2 \arcsin(\omega_p Dt/2) \tag{8.B.8}$$

which has real solutions if ($0 < \omega Dt < 2$), otherwise the solution is complex. For small time-steps $\omega_p Dt \ll 1$, $\omega = \omega_p$ as required; however, as ωDt becomes larger the solutions diverge leading to increasing phase errors in the solution. For small $\omega_P Dt$

$$\omega Dt \approx \omega_p \frac{Dt}{2} \left[1 + \frac{1}{6} \left(\omega_p \frac{Dt}{2} \right)^2 \right] + \cdots \tag{8.B.9}$$

and the phase error is $\frac{1}{24}(\omega_p Dt)^3$ per time step. To keep this within reasonable bounds, one generally limits the time step $Dt < 0.15\,\omega_p^{-1}$

8.B.2 Space Step Limitation

During the calculation the spatial values are stored as a grid of N points. If these are Fourier transformed, we get another set of N values at the points $1 \leq k \leq 2\pi/N$ which can, in turn, be inverted back to the original set. However, due to errors in the finite-difference calculation, additional terms of high spatial frequency $k > 2\pi/N$, not relevant to the problem, may be generated. On inversion, these transform to a frequency in the original set $k' = k - 2pi/N$, introducing additional terms into solution, usually at a low frequency. This effect, which occurs even if the terms are not *explicitly* Fourier transformed, is known as *aliasing*; it can be extremely serious in PIC codes where wave dynamics are being studied.

To overcome this effect, we may use the natural damping in the plasma due to its thermal motion, namely Landau damping (see Pert, 2021, §12.4). The damping rate for a spatial wave of wave number k is

$$\gamma = \sqrt{\frac{\pi}{8}} \frac{\omega_p}{k^3 \lambda_D^3} \exp\left[-\frac{1}{2k^2 \lambda_D^2} - \frac{3}{2} \right] \tag{8.B.10}$$

The Landau damping rate approximately equals the plasma frequency if $k\lambda_D \approx 3.4$. To prevent aliasing it is usual to limit the mesh size Δx, which defines the spatial frequencies in the Fourier expansion so that we require that $\Delta x < \lambda_D$ to avoid aliasing problems.

8.B.3 Stability of the 1D Electromagnetic Model

Neglecting the current in Eq. (8.20b) the equations reduce to the set of Eq. (8.22), which are equivalent to those studied by Richtmyer and Morton (1967, p. 262), who show that they are stable provided a Courant–Friedrichs–Lewy condition $cDt \leq \Delta x$ is obeyed.

References

Amsden, A A. 1966. The particle-in-cell method for the calculation of the dynamics of compressible fluids. Technical Report LA-3466. Los Alamos National Laboratory.

Birdsall, C K and A B Langdon. 1991. *Plasma Physics Via Computer Simulation*. Bristol, U.K.: Adam Hilger.

Cooper, N G, ed. 1989. *From Cardinals to Chaos*. Cambridge University Press. Originally published as: Stanislaw Ulam 1909–1984 in *Los Alamos Science No.15 1987*.

Evans, R G. 1986. Plasma simulation using fluid and particle models. In *Laser-Plasma Interactions 3*, ed. M B Hooper. Proceedings of the 29th Scottish Universities Summer School in Physics Scottish Universities Summer School in Physics pp. 247–278.

Harlow, F H. 1955. A machine calculation method for hydrodynamic problems. Technical Report LAMS-1956. Los Alamos National Laboratory.

Harlow, F H. 1964. The particle-in-cell computing method for fluid dynamics. In *Fundamental Methods in Hydrodynamics*, ed. B Alder S Fernbach and M Rotenberg. Number 3 *in* "Methods in Computational Physics" New York: Academic Press pp. 319–345.

Hockney, R W and J W Eastwood. 1988. *Computer Simulation Using Particles*. Abingdon, U.K.: Taylor and Francis.

Langdon, A B and B F Lasinski. 1976. Electromagnetic and relativistic plasma simulation models. In *Controlled Fusion*, ed. B Alder, S Fernbach, M Rotenberg and J Kileen. Number 16 *in* "Methods in Computational Physics" New York: Academic Press pp. 327–366.

Lindman, E L. 1975. ""Free-space" boundary conditions for the time dependent wave equation." *J Comput Phys* 18:66–78.

Morse, R L. 1970. Multidimensionalplasma simulation by the particle-in-cell method. In *Plasma Physics*, ed. B Alder S Fernbach and M Rotenberg. Number 9 *in* "Methods in Computational Physics" New York: Academic Press pp. 213–240.

Pert, G J. 2021. *Introduction to Plasma Physics*. London: Wiley.

Richtmyer, R D and K W Morton. 1967. *Difference Methods for Initial Value Problems*. New York: Interscience.

Woolfson, M M and G J Pert. 1999. *An Introduction to Computer Simulation*. Oxford: Oxford University Press.

Part III

Stochastic Methods

9

Monte Carlo Schemes

9.1 Monte Carlo Methods

Monte Carlo was the earliest technique of numerical simulation to use the power of electronic computers to rapidly perform large numerical calculations. Developed in the late 1940s by Ulam and von Neumann and their co-workers[1] to treat problems of neutron diffusion through reactor masses (see for example Spanier and Gelbard, 2008), it has subsequently broadened into a method of wide application with applications in radiative transfer, solid state and quantum mechanics (see for example MacKeown, 1997). An elementary introduction to the Monte Carlo technique is given by Woolfson and Pert (1999, ch 3). In recent years, there has been a great improvement in the quality of generators available. Press et al. (2007) discuss this in some detail, building on the work by Knuth (1998), and provide a set of working routines in C^{++}.

Monte Carlo techniques are based on sampling theory whereby a statistically distributed group of elements is examined and the properties of the ensemble deduced by averaging.

The problems studied may either be deterministic or probabilistic; the technique is most evident in the study of probabilistic events, for example the diffusion and scattering of neutrons within a nuclear reactor. The actual process here involves the motion of individual particles through the medium as they undergo collisions and scatter with changes in the velocity direction and energy. In this form, the method follows single 'particles' which may take several forms – neutrons, photons, electrons, etc. – but the general process is more abstract.

Monte Carlo methods follow the trajectory of a system based on the individual probabilities of the events which, taken sequentially, individually determine the probability of reaching a particular final state. The random probability of each possible event is used to calculate a possible individual result usually as a random deviate of the appropriate probability. For example, if a particle travelling through a background medium suffers collisions with a mean free path λ, then the probability of an element of free path of length $\mathrm{d}x$ is $\mathrm{d}x/\lambda$ or the probability of a collision within a path less than x is $\exp(-x/\lambda)$. Therefore, we may construct a random free path of length ζ where ζ is a random number from a probability distribution $p(\zeta) = \lambda \exp(-\zeta)$.

1 An excellent history of the development of Monte Carlo methods at Los Alamos is given in a collection of memoirs (Cooper, 1989).

Modelling and Simulation in Plasma Physics for Physicists and Mathematicians, First Edition. Geoffrey J. Pert.

The same problem may be approached at a second deterministic level, in which the particles are imagined as elements of a stream, and the averages performed implicitly in terms of distribution-averaged terms, for example fluxes. The statistical fluctuations inherent in a particle model are absorbed into thermal motion, defined by a temperature. If both methods are carried out to sufficient accuracy, they will give identical results within their inherent levels of accuracy.

The probabilistic approach, whose name Monte Carlo was suggested by Nicholas Metropolis, reflects the interest in the theory of games of the originators Ulam and von Neumann.

Monte Carlo has a reduced applicability in much of plasma physics as finite-difference methods are preferred, which do not suffer from statistical noise. However, it is a simple, easily programmed methodology, which should be part of any computational physicist tool-kit. We have used it often for quick simulations in a preliminary investigation of complex problems. A most useful application has been to solve the Fokker–Planck equation in three dimensions in a study of plasma relaxation described in some detail in Pert (2021, Ch. 7). We have therefore included a brief introduction of the approach and treated two classic examples to indicate the versatility of the method.

An introduction to basic probability theory is given in Supplementary Material M.5.

9.1.1 Random Number Generators

Monte Carlo programmes are based on random variables which are replaced by numerical values. It therefore becomes essential to be able to generate a sequence of numbers which are statistically uncorrelated. This is a difficult task for any sequence generated by a machine is produced by a programme of operations which are inherently predictable and hence not truly random.

Truly random events occur relatively infrequently in nature, electrical noise, radioactive decay, photon emission, etc. are examples which may, in principle, be used as the generator of random events in Monte Carlo. In practice, they are usually too slow to operate within a working programme. One solution is to generate a large table of random numbers, but this is too cumbersome and limited in its range to be satisfactory for large programmes. We therefore seek to generate a set of numbers with the appropriate statistical properties to be identified as behaving similarly to a set of real random numbers. These are *psuedo-random numbers* or *quasi-random numbers*. Exhaustive studies of random number generation are given by Knuth (1998, Ch. 3) and Press et al. (2007, Ch. 7), and should be studied by all workers in this field. Useful introductions and working FORTRAN programmes are to be found in Press et al. (1986, ch. 7).

However, practical random number generators do exist, which satisfy statistical conditions approximating to random numbers. The sets of numbers so generated are called *psuedo-random numbers*, as opposed to physical sets. These are key to rapid and successful Monte Carlo calculation.

From the programming point of view, we identify randomness as the requirement that the deterministic programme generating the random sequence must be different from and statistically uncorrelated with the programme using the results. In other words two different random number generators should produce statistically the same results.

This puts the onus for suitability on the application, for example quasi-random numbers are used for specific applications (MacKeown, 1997). For general purposes the exist a set of statistical tests which should be used to test any such programmes (Knuth, 1998, ch. 3).

It is clear from the above illustration in the introduction that the key to successful Monte Carlo simulation is the ability to generate random numbers to a particular distribution determined by the physical event described. This falls into two parts:

- The generation of a set of numbers, which are uncorrelated and have a uniform distribution in the range 0–1, i.e. for probability intervals dy wide, the probability density

$$f(y) = \begin{cases} dy & \text{if} \quad 0 < y \leq 1 \\ 0 & \text{otherwise} \end{cases} \tag{9.1}$$

- Using the set of uniform random numbers construct a set of uncorrelated deviates to the required probability density distribution $\phi(y)$.

9.1.1.1 Uniform Deviates

Uniform deviates are a set of random numbers with a uniform probability distribution, so that any number is equally probable, usually continuous in the range 0–1. Their distribution function is therefore given by Eq. (9.1). Uncorrelated sequences of uniform deviates with a long periods are not easy to construct; care must be taken to ensure that the algorithm used for their generation is reliable (see Knuth, 1998, §3.2) for an extensive discussion and model programs. Since the deviate to any other distribution can be obtained from these, they form the essential building blocks of Monte Carlo.

Since uniform deviates form the essential building block of any Monte Carlo simulation, it is clearly important that they do not suffer from any defect (principally correlation); to that end considerable effort has been put into creating algorithms that are well behaved, and to creating a set of tests, which must be satisfied. As neither are straightforward it is proposed that, if required, the routines in Press et al. (2007, §7.1) provide a useful set already programmed and tested. The DIEHARD and NIST sets of tests provide measures against which any new routine should be evaluated. We will now outline some of the more popular basic routines. Those using integers directly are treated in detail by Knuth (1998), who should be referred to for details of the underlying mathematics.

1) *Linear congruential generators* generate a sequence of integers $I_1, I_2 \cdots$ by the recurrence

$$I_{j+1} = (aI_j + c) \mod m \tag{9.2}$$

where m is the modulus, a the multiplier and c the increment, all positive integers. If $c = 0$ the generator is known as a *multiplicative linear congruential generator*. Clearly, the recurrence will repeat with a period not exceeding m. If a, c and m are well chosen (Knuth, 1998, §3.2) the full period is obtained and all the integers $0 - (m - 1)$ generated.[2] Any seed (initial value $0 \leq I_0 \leq (m - 1)$) is as good as any other to start the sequence. The random number is simply I_{j+1}/m.

2 Knuth (1998, §3.2.1.2) gives a necessary and sufficient condition on b and c for the full period to be obtained.

The generator has the advantage that it is fast and easily programmed in a high-level language. However, it is not free of sequential correlation on successive calls. In addition, the low-order bits are much less random than high order ones. Thus, you should be careful in your use of the random numbers generated and never try to split them up. Careful choice of the constants a, m and c can mitigate these effects. Press et al. (1986, §7.1) give a table of suitable values for different integer lengths.

As a result, neither linear congruential nor multiplicative linear congruential generators can be recommended for general use on their own. However, they have a role to play as elements in a compound generator where the correlations are eliminated in a second randomising independent process, and the period also increased.

System-supported generators were originally often of the linear congruential type. In some cases, the values of m, a and c were badly chosen, and it is sensible to check them through the tests outlined in Knuth (1998). Modern routines are generally reliable but it is sensible to be aware of potential problems.

2) *Shift register generators* also known as lagged Fibonacci generator use an algorithm of the form

$$u_{n+1} = u_{n-p} \odot u_{n-q} \pmod m \qquad 0 < p < ql \tag{9.3}$$

where u_n is the set of integers generated by the algorithm, and the operation $\odot$ is one of *plus* ('conjunction' *'and'* $\wedge$), *minus* ('disjunction' *'or'* $\vee$) or *exclusive or* (XOR). The new term is therefore a combination of two previous terms, m is usually a power of $2: m = 2^M$, making the mod operation simple in binary arithmetic. They require k words (the last k values are remembered), and therefore require careful initialisation.

If the operation is additive the generator is called an *additive Fibonacci generator* and if the exclusive or operation a *generalised feedback shift generator*. The maximum period is $(2^q - 1) \times (2^M - 1)$ if addition or subtraction is used and $2^{q-1} \times q$ if XOR is used. If multiplication is used, the maximum period is reduced to $(2^q - 1) \times (2^M - 3)$ or $1/4$ that of an additive generator. The maximum period is only obtained if the polynomial $y = x^p + x^q + 1$ is primitive over integers mod(2).

A very successful form of the shift register generator using subtraction and may be used with real numbers as well as integers is given by Knuth (1998, p. 39, 186–188). It is easily programmed and has proved reliable in practice, the only problem being the need for careful initiation as a set of p initial approximately random values are needed.

3) *XOR shift random number generators*: A major step forward was taken by Margalia (2013), who introduced the idea of *bit* shifts on long (32 or 64 bit) integers to identify a very simple routine involving 3 bit shifts and 3 exclusive *XOR* bit operations and most simply expressed using the C++ computer language.[3] A psuedo-code version is

state:	x	(unsigned long integer)
initialise:	$x \neq 0$	
update:	$x \longleftarrow x \wedge (x \gg a_1)$	
	$x \longleftarrow x \wedge (x \ll a_2)$	
	$x \longleftarrow x \wedge (x \gg a_3)$	

3 This is most easily utilised in a FORTRAN program by compiling the C++ element in an object file, which is incorporated with the FORTRAN at the linking stage. Care must be taken to ensure the types of integers defined in the two languages match.

OR

state:	x	(unsigned long integer)
initialise:	$x \neq 0$	
update:	$x \longleftarrow x \wedge (x \ll a_1)$	
	$x \longleftarrow x \wedge (x \gg a_2)$	
	$x \longleftarrow x \wedge (x \ll a_3)$	

where a_1, a_2 and a_3 are the number of bit shifts in each of the three steps. The algorithm has two forms, depending whether the first shift is forwards or backwards.

4) *Shuffling*: If it is suspected that the generator suffers from correlation problems, the easiest way to overcome this is to break up any sequential correlations by shuffling the values in the set.

9.1.2 Generating Random Numbers to a Specific Distribution

Crucial to the development of successful Monte Carlo simulation is the ability to follow the trajectory of an element through a series of random events, each specified by a known probability distribution. These use a uniform deviate as their basis, thereby introducing the essential underlying random element. In view of their importance in the trajectory of the system, it is essential that they be as efficient as possible to avoid unnecessary wastage of CPU effort. To this end, a number of strategies have been devised.

As we have noted: the distribution of random numbers is specified through the probability distribution in one of two ways as:

Density distribution	$p(x)$	i.e. probability of value $x \rightarrow x + \mathrm{d}x$	is	$p(x)\,\mathrm{d}x$
Cumulative distribution	$P(x)$	i.e. probability of a value $< x$	is	$P(x)$

Clearly, $P(x) = \int_{-\infty}^{x} p(x')\mathrm{d}x'$.

The probability is usually normalised: $P(\infty) = \int_{-\infty}^{\infty} p(x')\mathrm{d}x' = 1$ so that the total probability of an event is unity. In practice, the probability distribution is often specified either as an analytic function or through a data table.

Random numbers of a specified form are generally generated from a uniform deviate in the range $0 \leq x < 1$ using one of five methods:

1) **Transformation method**: If the normalised cumulative distribution can be easily inverted i.e. given a uniform deviate x, the solution y to the equation $x = P(y)$ may be calculated, the set of random numbers y is obtained from the set x of uniform deviates, by using the inverse cumulative distribution function $y = P^{-1}(x)$.
Since the cumulative probability of a uniform distribution is

$$P(x) = \begin{cases} x & 0 \leq x < 1 \\ 0 & \text{otherwise} \end{cases} \tag{9.4}$$

the number of random events $\eta < y$ therefore equals the number of uniform random events $\zeta < x$ and the random numbers y so generated have the correct distribution. Provided the cumulative distribution function can be easily inverted, this should always be the method of choice.

2) **Rejection method:** Given a trial function $\phi(y)$, such that $p(y) \leq \phi(y)$ for all y; for this probability deviates may be easily calculated for the density distribution $\phi(y')$. Then, the required set of deviates y to $p(y)$ are obtained from a trial deviate to ϕ, namely y' and a separate uniform deviate. We accept y', i.e. $y = y'$, if $[x < p(y')/\phi(y')]$; otherwise, we reject the trial and repeat with a second pair of x and y'. The value y is accepted $p(y')/\phi(y')$ and the value used. Noting $\phi(y')/\phi(y')p(y')$ gives the required probability distribution for the set y. The efficiency, namely the total fraction of accepted calls, is clearly $\int_{-\infty}^{\infty} p(y)\,\mathrm{d}y / \int_{-\infty}^{\infty} \phi(y)\,\mathrm{d}y$.
The function $\phi(y)$ is not normalised but must be chosen to always be larger than the probability $p(y)$. It needs not to be calculated outside the range of $p(y)$.
The comparison distribution $\phi(y)$ need not be continuous. Significant improvements in efficiency can be made by selecting a discontinuous function to closely fit the required distribution. Which branch of the comparison function is to be used to determine the trial deviate is decided by its relative probabilities, i.e. the fractional integral under the branch, known as the branching ratio. Thus, if there are n branches and the corresponding integrals are $I_1 \cdots I_n$ the branching ratio for branch m is $\Gamma_m = I_m / \sum I_n$. The appropriate branch m is found by comparing a uniform deviate x with the cumulative branching probability, i.e. $\sum_{l=1}^{m} \Gamma_l \leq x < \sum_{l=1}^{m+1} \Gamma_l$.
This is a very powerful technique; it is the fall-back if all other methods are not practicable. The method can clearly be made more efficient by a judicious choice of the test function $\phi(y)$ to match the probability distribution $p(y)$ as closely as possible, subject to $\phi(y) > p(y)$ for all values of y.
There are a few rather special methods applicable to specific probabilities. They are generally quite old and better methods are usually available, but they may be of value in special problems and so are outlined here:
3) **Cumulative method:** Suppose the particular event can be broken down into a series of events each with its own specific probability, then the overall event can be regarded as the culmination of the individual events, i.e. as the sum on the individual probabilities. Thus, if we can generate a series of random numbers from the constituent events $(\eta_1, \eta_2, \ldots)$, we may associate these in an appropriate manner to generate the required number $y = f(y_1, y_2, \ldots)$. This method is used relatively infrequently but examples will be found widely in the cases described forthwith.
4) **Composition method:** Suppose the density distribution can be written as either a sum $p(y) = \sum f_n g_n(y)$ or an integral $p(y) = \int g_z(y) dh(z)$ where $g_n(y)$ and $g_z(y)$ are a set of functions depending on both y and n [or z]; and f_n are a set of specified numbers [or $h(z)$ a known function]. If we regard these equations in terms of probabilities, we interpret them as a series of **OR** distinct events, each with a discrete probability f_- [or cumulative probability g_z]. Thus, an event of type n occurs with probability f_n, whose value prescribes the form of the distribution $g_n(y)$ and determines the random number y. We must first generate a random number n [or z] to give the appropriate distribution and the required deviate y n z.
5) **Markov (Metropolis) method:** Suppose we take a random walk through a set of values y generating the values $y_1, y_2 \cdots$ to the required probability density $p(y)$. From a point y_n we make a trial step to y_t, the step length being randomly selected by multiplying the

prescribed step by a uniform deviate ζ: $y_t = y_n + \zeta d$. The trial step is either accepted or rejected depending on the ratio $r = p(y_t)/p(y_n)$. If $r \geq 1$ the point is accepted. Otherwise, if $x \in (0,1)$ is a uniform deviate, we accept the point if $r \geq x$ and set $y_{n+1} = y_t$: otherwise, we reject the trial and either repeat the step or set $y_{n+1} = y_n$ either yielding the same distribution. This method yields a correlated set of deviates, which may not be suitable for many applications. The correlation may be reduced by shuffling, but is not recommended.

6) **Miscellaneous methods:** There are a number of methods based on the cumulative rule for probabilities M.5.1 and multiple random deviates. These may be useful in special cases:
   - Given two uniform deviates $x_1 \in (-1,1)$ and $x_2 \in (-1,1)$, the point (x_1, x_2) lies in the unit circle $x_1^2 + x_2^2 \leq 1$. Since uniform deviates (x_1, x_2) are uniformly distributed over the area of the unit circle, the probability of a value lying in the sector $(\theta, d\theta)$ is $d\theta/2\pi$. Hence, if $\theta \in (0,2\pi)$ is a uniform deviate, $\tan\theta = y = x_1/x_2$ and $p(y) = 1/(1+y^2)$. Similarly, $\sin\theta = x_1/\sqrt{x_2 + x_2^2}$ and $\cos\theta = x_2/\sqrt{x_1^2 + x_2^2}$. Reject values outside the unit circle.
     Note the pair $r = \sqrt{x_1^2 + x_2^2}$ and $\theta = \arctan(x_1/x_2)$ are a pair of independent random variables, being the polar co-ordinates (r, θ) of the point (x_1, x_2).
   - An alternative method not requiring the surds is: given two uniform deviates $x_1 \in (-1,1)$ and $x_2 \in (-1,1)$ lying in the unit circle, applying the double angle trigonometric formulae $\tan\theta = 2x_1x_2/(x_1^2 + x_2^2)$, $\sin\theta = 2x_1x_2/(x_1^2 + x_2^2)$, and $\cos\theta = (x_1^2 - x_2^2)/(x_1^2 + x_2^2)$ gives the values of the trigonometric functions of a random deviate.

9.2 Monte Carlo Integration

A simple, but practically very useful Monte Carlo application is the evaluation of integrals in multi-dimensions, where conventional methods are clumsy and calculation times lengthy. The method is most simply described in one dimension, but the generalisation to many dimensions is obvious.

Consider the integral of the function $f(x)$ from $x = a$ to $x = b$

$$I = \int_a^b f(x)dx = \sum_{i=1}^{M} f(x_i)\Delta x \tag{9.5}$$

where

$$x_i = a + (i - 1/2)\Delta x \qquad \text{and} \qquad \Delta x = (b - a)/M \tag{9.6}$$

for a large number M of intervals.

Suppose we take a large number N of uniform deviates x_i in the range (a, b) and form the sum:

$$S = \sum_{i=1}^{N} f(x_i) \tag{9.7}$$

The number of deviates with values $x_i - \Delta x/2$ to $x_i + \delta x/2$ is $N\Delta x/(b-a)$. Hence, for large values of N

$$S = N\sum_{i=1}^{M} f(x_i)\,\Delta x_i/(b-a) = \int_a^b f(x)\,\mathrm{d}x/(b-a) \tag{9.8}$$

Therefore,

$$I \approx \frac{1}{N}\sum_{i=1}^{N} f(x_i)/(b-a) = \frac{1}{N} I/(b-a) \tag{9.9}$$

which is the desired result.

More generally, if the deviates have a density distribution $p(x)$, then the number of deviates in the element $(x_i, \Delta x)$ is $N\,p(x_i)\,\Delta x_i$ and given a set of N deviates y_i and a function $\phi(y)$, we form the sum $f(y_i)/p(y_i)$:

$$S = \sum_{i=1}^{N} \phi(y_i) = N\sum_{i=1}^{M} \phi(x_i)p(x_i)\Delta x \approx N\int_a^b \phi(x)\Delta x \tag{9.10}$$

Letting $f(x) = \phi(x)p(x)$, we obtain the general result

$$I = \int_a^b f(x)\mathrm{d}x = \frac{1}{N}\sum_{i=1}^{N} \frac{f(y_i)}{p(y_i)} \tag{9.11}$$

where y_i are a set N deviates with probability density $p(y)\mathrm{d}y$ in the $y \in (a, b)$. Since the range of the probability function can extend to infinity, this generalisation allows the calculation of integrals with infinite limits.[4]

We may view this result from a different perspective, namely the *mean value theorem*. In this case

$$\int_a^b f(x)\mathrm{d}x/(b-a) = \langle f \rangle \tag{9.12}$$

is just the mean value of $f(x)$ over the range (a, b). Similarly, $\frac{1}{N}\sum_{i=1}^{N} f(y_i)$ is the mean value of the function $f(y)$ of the uniform deviate and $\frac{1}{N}\sum_{i=1}^{N} \frac{f(y_i)}{p(y_i)}$ the weighted mean of the function $f(y)/p(y)$ weighted by the probability density $p(y)$ of the set of values y_i.

The value of the integral is directly proportional to the mean value of the function at a set of uniformly distributed points:

$$I \approx \langle f(y)/p(y) \rangle \tag{9.13}$$

The variance of this set of values is given by:

$$\begin{aligned}\sigma^2 &= \frac{1}{(N-1)}\sum_{i=1}^{N}\left\{\frac{f(y_i)}{p(y_1)} - \left\langle \frac{f(y)}{p(y)} \right\rangle\right\}^2 \\ &\approx \int_a^b \left\{\frac{f(y)}{p(y)} - I\right\}^2 p(y)\mathrm{d}y\end{aligned} \tag{9.14}$$

---

4 Note that if the set y_i are uniform deviates, then $p(y)1/(b-a)$ and we recover our earlier result.

The error in the result (see Section M.5.vi) is determined by the mean square root of the variance in the usual way for the variance of the mean of a limited set of values, i.e.

$$\text{Error} \sim \frac{1}{\sqrt{N}}\sigma \tag{9.15}$$

We can reduce the variance and therefore the error in an integration result by making the ratio $f(y)/p(y)$ as nearly constant as possible, i.e. the random number distribution should be chosen to match the integrand $f(y)$ as closely as possible.[5] This is very similar to the requirement on the comparison function in the rejection method.

The idea of sampling (i.e. $p(y)$ large) when the function $f(y)$ is large to make most use of the available deviates is known as *importance sampling*, and the general procedure is known as *variance reduction*.

9.3 Random Walks

Most physical problems involve a series of progressive effects, each of which is independent of the preceding ones, known as a *Markov process*. In this case, the development of a particular sequence of events – a trajectory – takes the form of a random walk in a general parameter space. Each step moves to another point in the generalised space with a specified probability.

The simplest example of such a process is the diffusion of a particular set of particles through space subject to regular, random collisions in which the direction of travel is changed according to a known probability. This type of behaviour accounts for simple diffusion, Brownian motion, molecular orientation in polymer chains, etc.

Although random walks are usually Markovian (i.e. independent of previous effects), they are not always so. Non-crossing walks, in which the path of the walk may *not* cross any previous part of the walk, occur in polymer chains for example. These are clearly non-Markovian. They are also much more difficult to treat as a record of the entire walk must be retained and checked as the walk proceeds.

Since a Markovian random walk takes a random walk through the system domain, we may use it to construct a set of random deviates to a prescribed distribution, using the distribution to assign the probability of a particular step (see Section 9.1.2 item 5).

9.4 Nuclear Reactor Criticality

One of the earliest problems to which the Monte Carlo method was successfully applied was the design of reaction cores for nuclear reactors (Davis, 1963; Spanier and Gelbard, 2008) using uranium as fuel. These methods were extensively developed into the 1960–70s as computer power increased, but have more lately been superseded by discrete methods,

---

5 If $f(y)/p(y) = const = I$, the variance is zero; but to achieve this result requires that we integrate $f(y)$!

which are more efficient, but have more difficulty in treating the fine structure in cross sections. For the basic principles underlying the design of nuclear reactors, see Hill (1957). Nuclear reactor modelling can be used to demonstrate the strengths and weaknesses of the Monte Carlo approach. We give a brief outline of the method, much simplified from a practical system. We wish to calculate the behaviour of neutrons as they diffuse through a block of fissile, absorbing and scattering material. In a real nuclear reactor, the fissile material is uranium, both $^{235}\mathbf{U}$ and $^{238}\mathbf{U}$ isotopes, normally in the from of enriched uranium and absorbing control material (usually boron). Neutrons are released following fission held separately in narrow fuel rods inside a matrix of scattering material (carbon or heavy water). As a pedagogic exercise we consider a simple geometry using graphite as moderator, and consider fuel, absorber and scatterers intimately mixed, although the extension to real systems is straightforward, the basic principles are unchanged, but the detail is much more complicated.

Approximately 2.5 neutrons are released following fission.[6] The cross section for fission in $^{235}\mathbf{U}$ is very small unless the energy is approximately thermal (0.025 eV). The neutrons must therefore loose a large fraction of their energy before they can be absorbed by a $^{235}\mathbf{U}$ nucleus and cause fission. This is accomplished in a series of elastic scatterings as the neutrons diffuse through the moderator, a low absorption, light nuclei, either hydrogen, $^{1}\mathbf{H}$ (pressurised water reactors), deuterium $^{2}\mathbf{D}$ (heavy water reactors) or graphite, $^{12}\mathbf{C}$. Due to the small mass ratio $2\,M^{n}/\,M^{12C} \sim 1/6$ approximately 100 collisions to thermalise a neutron during which it may be absorbed by a $^{238}\mathbf{U}$ nucleus or exit from the reactor.

Because $^{238}\mathbf{U}$ has a high absorption for fast neutrons, it is necessary to separate the fuel from the moderator. This done by separating the fuel in rods of small diameter immersed in the moderator matrix. The adjustable control rods containing absorber are also fitted into the moderator.

The trajectory of a typical neutron may be split into the following decision points, which form the framework of a working programme.

i) Random initial velocity, direction and position of neutron from known distributions. Add to neutron number counter.
ii) Random length of path to collision from known cross sections and material.
iii) Does path cross a material boundary, or exit from the reactor? If exit, update loss counter. Go to step (i) with a new particle.
iv) Identify type of collision from known branching ratios, namely absorption, fission or elastic scatter
v) a) Scattering – calculate new energy and direction, energy loss, etc.
   b) Fission – calculate number of new fission neutrons and go to step (i) update fission counter
   c) Absorption path terminates. Update loss counter. Go to step (i) for a new neutron.
vi) Go to the step (ii) or terminate after a set number of iterations.

6 We ignore the technically important delayed neutrons, which are emitted by 'daughter' nuclei after a delay following fission with lower energy.

At each decision there are several different outcomes determined by the individual probability:

a) **Elastic scattering:** The neutron may scatter elastically without change in the total kinetic energy (see the Kinematics of Elastic Scattering treatment, eqns.) but, on average, the neutron loses $2/A$ of its kinetic energy per collision, where A is the mass number of the principal scattering particle. If n_s is the density of scatterers and σ_s the scattering cross section, the probability of a scattering per unit path:

$$\Sigma_s = n_m \sigma_{s\,m} + n_{235}\,\sigma_{s\,235} + n_{238}\,\sigma_{s\,238} \tag{9.16}$$

Note, elastic scattering is the principal moderating process.

b) **Absorption:** ^{238}U is strong absorber of fast neutrons in resonance processes with a complicated spectrum. To follow this in detail requires a large number of paths. To avoid this, it is possible to average over this distribution over the interval $(E_0 - E_1)$

$$\overline{\sigma}_{a38} = \int_{E_0}^{E_1} \sigma_a \frac{\mathrm{d}E}{E} \Big/ \int_{E_0}^{E_1} \frac{\mathrm{d}E}{E} \tag{9.17}$$

The absorption of ^{235}U is small for fast neutron but large for thermal ones. A fraction of these events yields fission. These total absorption probability per unit path is

$$\Sigma_a = n_{238}\,\sigma_{a\,238} + n_{235}\,\sigma_{a\,235} \tag{9.18}$$

c) **Boundary loss:** The total probability of a collision per unit path length is therefore $\Sigma = \Sigma_s + \Sigma_a = 1/\lambda$, where λ is the mean free path. The next free path is given by the exponential distribution $\ell_1 = \lambda x$, where x is an exponential deviate $(x \in 0, \infty)$ and determines the next event point (x_1, ψ_1) from Eq. (9.23).
If the reactor is a sphere of radius R, the next path is limited to $r_1 < R$, i.e

$$\ell_1 < \ell_1' = (R^2 - r_0^2) / (2r_0 \cos\psi_0) \tag{9.19}$$

or the neutron is lost and the path terminated. The probability of escape is therefore $W_l = \mathrm{e}^{-(\ell_1'/\lambda)}$.

There are therefore three possible outcomes of the event: scattering, absorption or loss with probabilities: $W_s = \Sigma_s\ell_0$, $W_a = \Sigma_a\ell_0$, W_l, where ℓ_0 is the length of the preceding step, two of which ((b) and (c)) terminate the path. Which occurs is determined by the branching ratios

$$\Phi_i = W_i / \sum_i W_i \tag{9.20}$$

and a decision obtained from a uniform deviate such that $(W_{i-1} \leq x < W_i)$.

Let us examine these individual steps in more detail, identifying the corresponding random processes and their distribution. As a pedagogical exercise, we will consider a particularly simple case where the fuel, moderator, control absorber, etc. are all mixed a spherical system of radius R. Although this does not correspond to a practical system, it illustrates the application of Monte Carlo principles.

Step (i): At fission, a fast neutron is produced with energy given by a deviate from the Watt distribution, Eq. (9.A.27) moving in a random direction with respect to the radius vector **r** defined by a single random angle ψ defined with respect to an appropriate co-ordinate system where $\cos\psi = (2x - 1)$, and x is a single uniform random deviate. The radial position, in principle, is determined by the distribution of fission events during the calculation. In principle, the best approach is to initiate one or more fission-generated neutrons when a fission event occurs. In this case, the initial distribution is relatively unimportant as a neutron steady state distribution is generated reasonably quickly if the system is near-critical. However, a good approximation can be obtained from the density of neutrons determined by diffusion; since the probability of a fission is proportional to the density, we may obtain an approximate distribution of the neutron density at critically when the number of neutrons diffusing out of an element volume just balances the number generated by fission, i.e.

$$D\nabla^2 n = \frac{1}{r^2}\frac{\partial}{\partial r}\left(r^2\frac{\partial n}{\partial r}\right) = \gamma\, n \tag{9.21}$$

where γ is the number of neutrons generated per neutron per unit volume per unit time.

Since the probability of fission is only large if the neutrons are thermalised, we may concentrate on the density of thermal neutrons alone to generate a distribution of fission events. The constant of proportionality $\gamma \sim \alpha\, n^{^{235}\mathrm{U}}\, \sigma_{\mathrm{fis}}\, v_{therm}$ where $\alpha \approx 2.47$ is the number of neutrons released per fission, $n^{^{235}\mathrm{U}}$ is the density of $^{235}\mathbf{U}$ nuclei, σ_{fis} the fission cross section and v_{therm} the thermal speed. Furthermore, since the diffusion coefficient $D \sim \frac{1}{3}n_0$, where n_0 is the density of scatterers, the ratio $\gamma/D \sim \left(\sigma_{fis} n^{\mathrm{U}^{235}}\right)/(\sigma n_0) \ll 1$. We may neglect the term on the right-hand side of Eq. (9.21) and seek a solution to $\nabla^2 n = 0$. A simple approximation with $n = 0$ at the boundary $r = R$ is $n \sim \frac{1}{r}\sin\left(\frac{\pi r}{R}\right)$. A further correction may be added to take account of the failure of the diffusion approximation near the boundary,[7] namely $n \sim \frac{1}{r}\sin\left(\frac{\pi r}{(R + 0.71\,\lambda)}\right)$. Deviates to this distribution are easily calculated by the rejection method.

Step (ii): Given the total scattering cross section, calculate a random scattering path length from the exponential distribution, Eq. (9.A.4). The neutron moves in an environment containing fuel, moderator (graphite) and/or absorber(boron) for control. The cross sections for each material at the neutron energy are known and hence the overall cross section in the medium for elastic scattering, absorption, moderation and loss (fast fission).

Step (iii): Update the neutron position. The spatial position r is determined by the free path ℓ, Eq. (9.23) (see Figure 9.1). In spherical geometry, the spatial position is **r** with respect to the centre. The angle ψ defines the direction of motion with

7 Within about 1 m.f.p. the flux to the wall changes from collision limited (diffusion) $\phi = -D\nabla n$ to free particle $\phi = \frac{1}{4}n\bar{v}$, see Hill (1957).

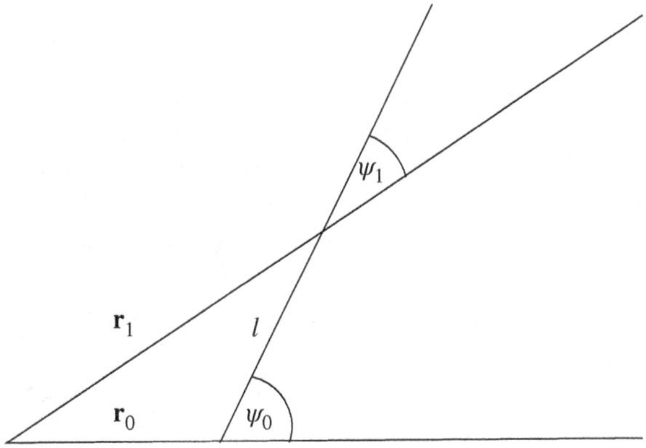

Figure 9.1 The particle following elastic scattering at $\mathbf{r}_0$ leaving at an angle ψ_0 has a second elastic scatter at $\mathbf{r}_1$ after a free path ℓ leaving at radial angle ψ_1.

respect to radial direction from the centre $\mathbf{r}$ and the angle of the path to the radius vector changes from χ_0 to χ_1 where

$$r_1^2 = r_0^2 + 2\ell\, r_0 \cos\psi_0 \tag{9.23a}$$

$$\psi_1 = \arccos\{\frac{r_1^2 + \ell^2 - r_0^2}{2\ell\, r_1}\} \tag{9.23b}$$

If the path length exceeds the distance to the boundary, the neutron is lost.

Step (iv): Otherwise, the nature of the event is determined by the branching ratios given the probability of an individual event j: $n_j\sigma_j/\sum n_i\sigma_i$ tested against a uniform deviate.

Step (v): If the neutron is scattered calculate the new energy and direction from the standard results for elastic scattering for a free path ℓ.

9.4.1 Statistical Properties

The Monte Carlo method is inherently statistical. The accuracy of any simulation is therefore determined by the variance of the final result, in addition to any simplification in the physics.

In the reactor criticality problem described in the previous section, we measure the final result as a probability in terms of the fraction of trials that yield a successful outcome, for example the probability that the neutron will thermalise (i.e. its energy is less than 0.1 eV)

$$p_T = n_t/N \tag{9.24}$$

where n_T is the number of thermal neutrons in N trials.

The inherently statistical nature of the calculation means that n_T is subject to random variation. Since the number must be integral, we conclude that the variation is Poisson distributed with mean $\overline{n_T}$ and variance $\overline{n_T}$ also. Thus, as the number of trials is increased, the error scales as $1/\sqrt{N}$.

Taking the average over the neutron distribution generated in each trial. The error associated with the final result (the mean) Eq. (M.5.35) is given by its variance $\sqrt{V/N}$ where V is the variance in energy.

This statistical variation is inherent in the Monte Carlo. It is clearly important to appreciate the uncertainty in any result in its interpretation.

It is obvious that we may reduce the variance by increasing the number of trials. However, as the error only decreases as $1/\sqrt{N}$, this can be prohibitively expensive in computer time if accurate results are essential, e.g. the design of nuclear reactors, and/or the event is comparatively rare, e.g. penetration of a barrier. As a result, although the Monte Carlo method was a popular technique in the early days of simulation, being comparatively easy to programme, it fell into disfavour in the 1960s in comparison with finite-difference methods. In this case, errors scale at least as $1/N$, but the programming is more complex and the code-less flexible.

The realisation that the Monte Carlo method could be made more efficient by variance reduction techniques has led to a re-examination of the method for specific problems for which it is well suited. A suitable problem for the Monte Carlo method should generate repeatable expectation values, with small variance. Monte Carlo techniques can often increase the amount of physics content of modelling and reduce the variance of determined quantities by increasing the number of data points of the variable of interest. One then tries to obtain results over a range of trajectories. The general aim is for the trajectories to cover a maximum of phase space with a minimum of trials.

Since variance reduction frequently results from the inclusion of additional physics into the problem, it is to some extent problem specific. We now examine some general methods, which can be identified and are particularly appropriate to transport problems, such as in a nuclear reactor.

9.4.1.1 Weighted Sampling

A Monte Carlo calculation consists of a calculation along a known trajectory of the system. Averages are formed as ensemble averages over the trajectories. Each trajectory can be assigned a weight to represent its probability in the ensemble, so each calculation can use several trajectories simultaneously. The weight can change during the calculation as trajectories are added. We give a brief introduction to weighted trajectory techniques. MacKeown (1997) gives a fairly comprehensive survey of variance reduction techniques.

i) **Survival biasing**: The trajectory is assigned a weight at the start of the calculation to account for those events which would terminate the path. Thus, the weight is decreased in proportion to the branching ratio for termination, i.e. in the neutron diffusion case the branching ratio for absorption, namely the ratio of the absorption probability to the total scattering probability.[8] Any other loss processes are similarly included.
   At each decision point, we may similarly add the fission probability over that step to obtain the value for the whole path and the weight correspondingly reduced as the path is terminated.
   The final result is the average over all such trajectories.
ii) **Terminating events**: If the event under investigation is one which terminates the chain, it may be advantageous to calculate its weighted probability at each decision point in the path and to continue the calculation with an appropriate reduction in weight. The total probability for the trajectory is then the sum of the weighted contributions from each point. This method is particularly useful if the event has low probability, e.g. fast fission. The known value of the parameter x_t (in this case, unity) is multiplied by its probability and by the total weight to give the contribution.

8 Note that escape to the wall is easily included in the total probability of loss.

iii) **Splitting**: Each trajectory can be split at some decision point into ν trajectories each assigned a weight $1/\nu$ of the incoming trajectory. The splitting technique may be useful to account for an initial loss in regions of little contribution to the final result. For example, the transmission through a barrier of high-energy particles, which suffer a loss of $\exp(-\mu_{abs}x)$ in a distance x. We can maintain their number approximately constant by increasing them every distance x by ν, such that $\nu\exp(-\mu_{abs}x) \sim 1$.

iv) **Russian roulette**: This, rather colourfully named technique, discards trajectories which make little contribution to the final result. The philosophy is that, at any stage of the calculation, the number of trajectories of any specific type is proportional to their expected contribution. At some level, trajectories are discarded with a probability α by random sampling, the remainder having their weight increased by $1/(1-\alpha)$. This method can be particularly useful for deciding when to discard trajectories of low weight. For example, if the weight is less than some specified minimum W_{min}, compare $w_{min}/0.5$ against a uniform variate η. If $\eta > w_{min}/0.5$, discard the trajectory, otherwise re-assign its weight to 0.5 and continue.

All these methods ensure that the final Monte Carlo evaluations cover a range of trajectories with fewer runs. In addition, a smaller number of the runs contribute negligibly small scores. As a result, for the same number of trials, the variance will be less. When combined in an optimum way, the methods can lead to a significant reduction in the error necessary to establish a result.

9.5 Thermodynamic Properties and Equation of State

The Monte Carlo method has been used to calculate the equation of state of equilibrium liquids and gases for which the intermolecular force-law is known.

For example, suppose we have a liquid non-polar gas such as argon. The force between two molecules (in this case atoms) is given approximate by the Lennard–Jones (6–12) potential (see Figure 9.2),

$$\phi = \underbrace{A/r^{12}}_{\substack{\textit{Attractive}\\ \textit{Short range}}} - \underbrace{B/r^{6}}_{\substack{\textit{Repulsive}\\ \textit{Long range}}} \tag{9.25}$$

If we know the position of all the molecules, we may calculate the total potential energy of the interaction between the molecules within a few scale lengths of the force separation, i.e. we consider a box of appropriate dimensions. The box may without loss of generality be considered as a unit of a periodic system, i.e. surrounded by identical boxes (see Figure 9.3). The total potential energy of the configuration Φ is easily calculated by summing over all pairs.

In thermal equilibrium, the probability of such a configuration is proportional to $\exp(-\Phi/\kappa T)$. We may therefore calculate the thermodynamic properties by generating a series of such configurations and averaging over the probability distribution, for example the potential part of the internal energy

$$U_{pot} = \overline{\Phi} = \sum \Phi\, exp(-\Phi/\kappa T) \Big/ \sum \exp(-\Phi/\kappa T) \tag{9.26}$$

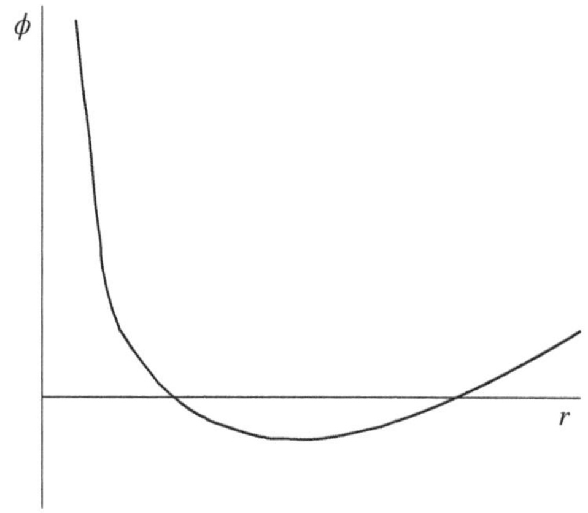

Figure 9.2 Sketch of the Lennard-Jones interparticle potential.

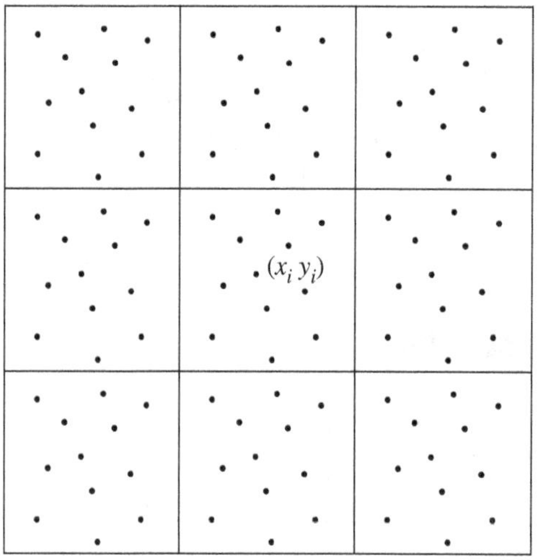

Figure 9.3 Periodic grid with randomly positioned molecules.

The summation must be taken over a large number of configurations, great enough to include all real ones of reasonable probability.

We may generate these configurations by the Monte Carlo method during a random walk through the configuration space. Starting from the initial arrangement, we displace one of the molecules by a random shift

$$x_i \to x_i + \alpha\Delta \qquad y_i \to y_i + \beta\Delta \tag{9.27}$$

where α and β are uniform deviates and Δ the maximum shift. The molecule may be selected at random from the set by a third random number.

The method as outlined is very inefficient in that it calculates any configuration which may contribute to the sum, even if its probability is small. This is due to the fact that as a result of the close packing an arbitrary displacement may bring the molecule close to a neighbour where strong repulsive forces lead to a large positive potential energy.

To avoid this problem, a variance reduction technique due to Metropolis (item(5), Section 9.1.2), based on the rejection method, is used. Starting from the old configuration, we form a new one and calculate the energy change $\Delta\Phi$.

- If $\Delta\Phi \leq 0$, we accept the change.
- If $\Delta\Phi > 0$, we test $\exp(-\Delta\Phi/kT)$ against a uniform deviate. If the exponential is larger the new configuration is accepted, otherwise return to the old.

By analogy with the rejection method, it is clear that the probability of individual configurations is $\exp(-\Phi/kT)$, i.e. the Boltzmann probability.

Supplementary Material

Appendix 9.A Deviates to Specific Distributions

In this section, we consider methods of constructing deviates to specific distributions which fall outside the general methods outlined earlier.

9.A.1 Power Law Distribution

In a power law distribution the probability of an event is given by the cumulative probability distribution $P(y) = y^{\upsilon}$ where $y \in (0,1)$, with probability density $p(y) \propto y^{\nu-1}$. Applying the transformation method, a random number to this distribution is

$$y = x^{\nu+1} \tag{9.A.1}$$

where x is a unit deviate $x \in (0,1)$.

An alternative older method may be used for when $\upsilon = n$ and n is an integer. If the cumulative probability $P(y) = y^n$ then

$$y = \max(x_1, x_2, \ldots, x_n) \tag{9.A.2}$$

yields a deviate to the cumulative distribution y^n; the required deviate is given by the largest value of a set of uniform deviates $x \in (0,1)$. This follows from the cumulative distribution of a uniform deviate $P_1(x) = x$, thus the cumulative probability of a set of random deviates $x_1, x_2, \ldots x_n$, all being less than a value $x_{max} = \max(x_1 \ldots x_n) \leq (x_{max})$ is

$$P_1(x_1)P_1(x_2)\cdots P_1(x_n) = x^n \leq P_1(x_{max}) = x_{max} \tag{9.A.3}$$

This method is normally less efficient than the transformation result unless $n \approx 2$.

9.A.2 Exponential Distribution

An important set of random events are those with an exponential distribution, such as waiting times or free path events. In these, the probability of an event in the interval (y, dy) is directly proportional to the duration of the interval dy, the constant of proportionality being either the mean free path (λ) or mean free time (τ). Thus, normalising the duration, if there is unit probability of an event occurring in an interval of unit duration, the probability of one in dy is just dy. The probability of an event NOT having occurred before y and occurring in (y, dy) is therefore

$$\exp(-y)\,\mathrm{d}y \tag{9.A.4}$$

where y is normalised to the mean free path or mean free time.

The distribution has unit mean and variance. The cumulative distribution is $P(y) = \exp(-y)$. Using the transformation method the desired random number to an exponential distribution is $y = \ln(x)$ where x is a uniform deviate. In practical usage, this must be multiplied by the normalising factor, i.e. mean free path or mean free time. The random deviate is the *waiting time* for the event.

9.A.3 Lorentz Distribution

The Lorentz (or Cauchy) distribution represents the spectral profile of homogeneously broadened spectral lines. It is the profile of a suddenly terminated oscillatory pulse. Strictly speaking, it has neither mean nor variance, although being symmetric about its zero point, which is conventionally taken as its mean. This distribution is particularly useful as a comparison function for the rejection method having a 'bell-shaped' profile. The density and cumulative distributions are

$$p(y) = \frac{1}{\pi(1+y^2)} \tag{9.A.5a}$$

$$P(y) = \{\arctan(y) + \pi/2\}/\pi \tag{9.A.5b}$$

The deviates are therefore

$$y = \tan[\pi(x - 1/2)] = -\cot(\pi x) \tag{9.A.6}$$

In this case also, there is an older, alternative method not requiring the use of machine arithmetic function, based on a rejection method and needing two uniform deviates.

Given two uniform deviates $x_1 \in (-1, 1)$ and $x_2 \in (-1, 1)$ then the point (x_1, x_2) lies in the unit circle $x_1^2 + x_2^2 \leq 1$. Reject values outside the unit circle. Since uniform deviates (x_1, x_2) are uniformly distributed over the area of the unit circle, the probability of a value lying in the sector $(\theta, \mathrm{d}\theta)$ is $\mathrm{d}\theta/2\pi$. Hence, if $\theta \in (0, 2\pi)$ is a uniform deviate, $\tan\theta = y = x_1/x_2$ and $p(y) = 1/(1+y^2)$.

$$y = x_1/x_2 \tag{9.A.7}$$

Similarly, $\sin\theta = x_1/\sqrt{x_2 + x_2^2}$ and $\cos\theta = x_2/\sqrt{x_1^2 + x_2^2}$.

An alternative method for *delete* sine and *delete* cosine terms not requiring the surds is: given two uniform deviates $x_1 \in (-1, 1)$ and $x_2 \in (-1, 1)$ lying in the unit circle. Applying the double angle trigonometric formulae $\tan\theta = 2x_1x_2/(x_1^2 - x_2^2)$ and $\sin\theta = 2x_1x_2/(x_1^2 + x_2^2)$, $\cos\theta = (x_1^2 - x_2^2)/(x_1^2 + x_2^2)$.

Normal (Gaussian) distribution with a probability density distribution

$$p(y) = \frac{1}{\sqrt{2\pi}} \exp\left(-\frac{y^2}{2}\right) \tag{9.A.8}$$

has a zero mean and unit variance. In view of its importance, there is a wide range of methods to determine random variates.

9.A.3.1 Box–Muller Method

This is a rejection method based on the circle deviates discussed above. The probability of two variates (y_1, dy_1) and (y_2, dy_2) is

$$p(y_1, y_2)dy_1dy_2 = \frac{1}{2\pi}\exp\left\{-\frac{y_1^2 + y_2^2}{2}\right\} dy_1\, dy_2 \tag{9.A.9}$$

Transforming to polar co-ordinates $r = \sqrt{(y_1^2 + y_2^2)}$ and $\theta = \arctan(y_1/y_2)$, the probability may be written in terms of (r, θ),

$$p(r,\theta)drd\theta = \frac{1}{2\pi}\exp\left(\frac{r^2}{2}\right) dr\, d\theta \tag{9.A.10}$$

Therefore, the cumulative probability of values $(0 \le r)$, $(0 \le \theta)$ is $P(r, \theta) = \exp\left(r^2/2\right)\theta$ provided $r \le 1$. Applying the transformation method to yield the required deviates y_1, y_2. From two uniform deviates x_1, x_2, reject any when $x_1^2 + x_2^2 > 1$; two independent uniform deviates are given by

$$y_1 = \sqrt{-2\ln(x_1)}\cos(2\pi x_2) \qquad y_2 = \sqrt{-2\ln(x_1)}\sin(2\pi x_2) \tag{9.A.11}$$

9.A.4 Gamma Distribution

The gamma distribution is the generalisation of the 'waiting time' picture associated with the exponential deviate. The gamma distribution represents, for example, the delay before the ath event occurs. The probability density

$$p_a(y)\,dy = y^{a-1}e^{-y}/\Gamma(a) \quad dy \tag{9.A.12}$$

is the probability of the nth event occurring in the interval $(y,\ dy)$, where $\Gamma(n)$ is the gamma function.

This result is most easily proved by induction from the value for $p_1(y) = e^y$ (Eq. (9.A.4)) for a single event. Assume $p_{a-1}(y')\,dy'$ is the probability of the $(a-1)$th event taking place in (y', dy') and the following event (the ath) occurring in $(y - y')dy$ with probability $p_1(y - y')\ \exp(y - y')\,dy$ then

$$p_a(y)\,dy = \int dy'\, p_{(a-1)}(y')\, p_1(y - y')\,dy = e^{-y}\,dy\,/\,\Gamma(a) \tag{9.A.13}$$

establishes the result.

The simplest method of obtaining a deviate follows immediately. Generate a set of n uniform deviates each giving an event value. From these, we can construct the sum of the event times

$$y = \sum_{i=1}^{n} y_i = -\ln\left(\prod_{i=1}^{n} x_i\right) \tag{9.A.14}$$

where the set $x_i \in (0,1)$ is a set of uniform deviates.

An alternative method, preferable if the number of events is not small, uses the rejection method. Since the gamma distribution has a 'bell shaped' profile, the Lorentz distribution offers an appropriate comparison curve $\phi = c_0/[1 + (y - y_0)/a_0]^2$. The values of c_0, y_0 and a_0 are chosen to minimise the product (area) $a_0\,c_0$ whilst ensuring that $\phi(y) \ge p(y)$. Therefore, $y_0 = (a-1)$ is the peak of the distribution, $a_0 = \sqrt{2a-1}$ and

$c_0 = [(a-1)^{(a-1)}\, e^{(a-1)}]/\Gamma(a)$. The deviate z from the Lorentz distribution is shifted to mean y_0 and width a_0, namely $y = a_0 z + y_0$ and the comparison function written as

$$\frac{p(y)}{\phi(y)} = \left[1 + z^2\right] \left(\frac{y_0}{y}\right)^{(a-1)} \exp - \left(y - y_0\right) \tag{9.A.15}$$

The value y is accepted as the required deviate if the value of $p(y)/\phi(y)$ is less than a uniform deviate x.

9.A.5 Poisson Distribution

Suppose an event occurs on average λ times in the interval, what is the probability of n events in the interval?

Let us divide the interval into a large number m of small sub-intervals in each of which the probability of an event in m is $p = \lambda/m$ and the probability of no event correspondingly $q = 1 - p$. To obtain the *binomial or Bernoulli distribution*, the probability of n events in the full interval of width 1 unit is therefore given by the binomial distribution by noting that the n events may be arranged in any order amongst the m sub-intervals. Making use of the limit $\lim\limits_{n\to\infty} (1 + x/n)x^n = \exp(x)$ and that $\lambda = mp$, the expected number of events is

$$\begin{aligned}\binom{m}{(m-n)} p^n q^{m-n} &= \lim_{n\to\infty} \frac{m(m-1)(m-2)\cdots(m-n+1)}{n!} p^n \left(1-p)^{(m-n)}\right) \\ &= \frac{(\lambda)^n}{n!} \exp(-\lambda)\end{aligned} \tag{9.A.16}$$

Making use of the limit $\lim\limits_{n\to\infty} (1 + x/n)x^n = \exp(x)$ and that $\lambda = mp$, the expected number of events in the interval, we obtain the required probability for n counts, known as the *Poisson distribution*: namely, the probability of n events in an interval when the average is λ:

$$\text{Probability of } n \text{ counts:} \qquad P(n, \lambda) = p^n e^{-\lambda}/n! \tag{9.A.17a}$$

The mean and the variance are both equal to λ.

Alternatively, we may identify the length of an interval with unit average, i.e. one count per interval on average; and count the probability of n counts in m intervals

$$\text{Probability of } n, \text{ counts:} \qquad P(n, m) = p^n e^{-m}/n! \tag{9.A.17b}$$

The Poisson distribution is closely related to the gamma distribution but considers integer rather continuous values. We may identify two different random deviates associated with this probability:

i) The random deviate is the number of counts in m intervals.
ii) The random deviate is the number of intervals needed before n counts are achieved.

Two simple methods are available for easy calculation of Poisson deviates

- **Cumulative method**: In the context of the exponential distribution, where on average one event occurs per interval, we may regard this situation as one in which a series of m exponential events takes place over λ intervals. Since each exponential deviate defines an event associated with one interval, the Poisson derivative is just the number of exponential events occurring before the end of the total interval, namely λ. Thus, if we generate a series of independent exponential deviates $x_1, x_2, \ldots, x_n$ until their sum (namely the total interval) exceeds the event average (namely λ unit intervals), the number of deviates m is the required Poisson deviate.

- **Rejection method**: We may imagine as a stepped function of a continuous variable $p(y) = P(n, m)$ where $m = \lfloor y \rfloor$, $\lfloor y \rfloor$ being the largest integer less than y, $\lfloor y \rfloor$ is then the required integer deviate.
  In this case, m is similar to the gamma deviate and may be dealt with by similar means. Treating m as a stepped continuous variable, we may again use a Lorentz distribution (deviate z) as comparison to give a trail value $m = \lfloor y \rfloor$ where $y = a_0 z + y_0$. The values in the Lorentzian probability function $\phi(y) = c/[1 + (y - y_0)/a_0]^2$ should be taken as shift $y_0 = n$, width $a_0 = \sqrt{2n}$ and c_0 chosen so that the comparison function
  $$\frac{p_m}{\phi(y)} = 0.9\left[1 + z^2\right] n^{(m-n)} \frac{n!}{m!} \tag{9.A.18}$$
  Note that we do not need to know an exact value of c_0. The value of 0.9 ensures that the comparison function always exceeds the required distribution in the neighbourhood of $m = n$.

9.A.6 Planck Distribution

The frequency distribution of photons emitted by a black body satisfies Planck's law for which the normalised density distribution function is

$$p(y) = \frac{15}{\pi^4} \frac{y^3}{(\exp(y) - 1)} = \frac{15}{\pi^4} \sum_{n=1} y^3 e^{-ny} \tag{9.A.19}$$

The mean and variance are easily calculated using the following integral identity:

$$\int \frac{y^m}{\exp(y) - 1}\, dy = \Gamma(m)\, \zeta(m) \tag{9.A.20}$$

where $\Gamma(y)$ is the gamma function and $\zeta(m) = \sum_{n=1} \frac{1}{n^m}$ is Riemann's zeta function.

$$p(y) dy = \sum_{n=1} y^{(m+3)} e^{-ny} dy = \sum_{n=1} \frac{1}{n^{m+4}} = \zeta(m + 4) \tag{9.A.21}$$

The maximum of the Planck distribution curve is at $y_{\max} = 2.821439$. The mean value is 3.83223 and the variance 4.11326.

We may calculate deviates by two methods:

a) **Composition method:** Writing the probability as a sum

$$p(y) dy = \sum_{n=1} f_n\, g_n(y)\, dy \tag{9.A.22}$$

where

$$f_n = \frac{90}{\pi^4 n^4} \qquad g_n(z) = \frac{1}{6} z^3 e^{-z|} \tag{9.A.23}$$

where $z = ny$.

We need first to calculate a deviate to the normalised probability $f_n = 90/\pi^4 n^4$ where $\sum f_n = 1$. This is done by calculating a uniform deviate x and finding the smallest value of n such that $\frac{90}{\pi^4}\sum_{\ell=1}^{n+1}\frac{1}{\ell} > x$.

Once n is known, the function $g_n(z)$ is simply a set of four gamma distributions of mean n in $z = ny$, and the required deviate is obtained from a further four uniform deviates $x_1 \cdots x_4$ as $y = \frac{z}{n} = -\frac{1}{n}\ln\left\{\prod_{i=1}^{4} x_i\right\}$.

b) **Rejection method:** Since the Planck distribution function is a 'bell-shaped' curve with values only for $y \geq 0$, we may use a truncated Lorentzian distribution for comparison with $\phi(y) = c_0/(1 + [(y - y_0)/a_0])^2$, where $y_0 = 2.82144, a_0 = 2.770253$ and $c_0 = 0.218865$ as comparator with 71% efficiency.
We require only values in the range $-y_0 \leq y \leq \infty$ for the corresponding uniform deviate, from which the truncated Lorentz deviate is generated, which is in the range $[\pi/2 - \arctan(x_0/a_0)]/\pi \leq x < 1 \equiv 0.2470862 < x < 1$.

Since the overall efficiency of the rejection method is only about 71%, it is usually better to use the composition method provided a fast and efficient uniform deviate generator is available.

9.A.7 Watt Distribution

Following the fission of a U^{235} nucleus,[9] neutrons are released with a range of energies according to the distribution.[10]

$$f(\epsilon) = \sqrt{\frac{2}{\pi\epsilon}}\sinh\left[\sqrt{2\epsilon}\right]\mathrm{e}^{-\epsilon} \tag{9.A.27}$$

where $f(\epsilon)\mathrm{d}\epsilon$ is the probability that a neutron has energy $\epsilon \to \epsilon + \mathrm{d}\epsilon$ MeV, where $f(\epsilon_{\max}) = 0.35557$ MeV. The mean energy is 2 MeV and the variance $\frac{1}{2}$ MeV^2. Although the distribution can be analytically integrated, the resulting cumulative probability is not easily inverted.

We therefore seek to use the rejection method, seeking as efficient a method as possible.

9 We detail this method to show how the rejection method may be easily improved using simply inverted functions.

10 The normalisation integral is evaluated as follows:

$$\sinh\left[\sqrt{2\epsilon}\,\mathrm{e}^{\epsilon}\right] = \frac{1}{2}\left\{\mathrm{e}^{\sqrt{2\epsilon}} - \mathrm{e}^{-\sqrt{2\epsilon}}\right\}\mathrm{e}^{-\epsilon} = \frac{1}{2}\mathrm{e}^{1/2}\left\{\exp\left[\left(1/\sqrt{2}+\epsilon\right)^2\right] - \exp\left[-\left(1/\sqrt{2}-\epsilon\right)^2\right]\right\} \tag{9.A.24}$$

$$= \frac{\sqrt{\mathrm{e}}}{2}\left\{\exp(x_+^2) - \exp(x_-^2)\right\} \tag{9.A.25}$$

where $x_\pm = 1/\sqrt{2} \pm \epsilon$. The integral

$$\int_0^\infty \sinh\left[\sqrt{2\epsilon}\right]\mathrm{e}^{-\epsilon}\mathrm{d}\epsilon = \sqrt{\frac{\mathrm{e}}{2}}\left\{\int_{\sqrt{1/2}}^{\infty}\exp(-x_+^2)\mathrm{d}x_+ + \int_{-\infty}^{\sqrt{1/2}}\exp(-x_-^2)\mathrm{d}x_-\right\} = \sqrt{\frac{\pi\mathrm{e}}{2}} \tag{9.A.26}$$

Since the curve has a 'bell-shaped profile', we try as comparison function a Lorentz distribution

$$\phi(\epsilon) = \alpha / \left[1 + (\epsilon - \epsilon_{max})^2 / a^2\right] \tag{9.A.28}$$

which passes through the peak ($\alpha = f(\epsilon_{max})$) and matches the actual distribution as closely as possible whilst $\phi(\epsilon \geq f(\epsilon))$. In general, one would seek the match at the peak by setting $\mathrm{d}^2 f/\mathrm{d}\epsilon^2|_{\epsilon_{max}} = \mathrm{d}^2\phi/\mathrm{d}\epsilon^2|_{\epsilon_{max}}$, but if this is done, there is a second intersection at $\epsilon \sim 2$. We therefore seek the value of a such that $\phi(\epsilon)$ has a second tangent point with $f(\epsilon)$, namely when $a = 1.8408$. The efficiency of this case is 48.63%.

The first improvement we can make is to note that values of the Lorentz deviate need only be calculated when $\epsilon \geq -\epsilon_{max}$, i.e. In the form $\tan\theta$ if $\theta \in (-0.37265, \pi/2)$ the efficiency is increased to 78.61%.

We could alternatively take an exponential as comparison

$$\phi(\epsilon) = \frac{1}{L}\exp\left(-\frac{\epsilon}{L}\right) \tag{9.A.29}$$

which has an optimum value of L when the ratio of $\phi(\epsilon)$ to $f(\epsilon)$ is a minimum when the curves are tangent to each other. This occurs when $L = 2{,}04787$ and $\beta = f(\epsilon)/\phi(\epsilon) = 1.29196765$. The comparison function is therefore $\beta'/L\exp(-\epsilon/L)$. Efficiency is 77%.

A major problem with using a Lorentzian as comparator is its weak approach to zero as the argument becomes large, as evidenced in its inability to form a variance. Combining the exponential with the Lorentzian enables a good representation at both the maximum, $\epsilon \sim \epsilon_{max}$, and in the limit $\epsilon \to \infty$. Thus, the comparator becomes

$$\phi(\epsilon) = \begin{cases} \alpha/(1 + (\epsilon - \epsilon_{max})^2/a^2) & \text{if } \epsilon < \epsilon' \\ \beta/L\exp(-\epsilon/L) & \text{otherwise} \end{cases} \tag{9.A.30}$$

where ϵ' is the point at which the Lorentzian and exponential curves intersect. There are three such intersections

$$\epsilon' = \begin{cases} \epsilon'_1 & 1.57249 & \text{Efficiency } 88.7\% \\ \epsilon'_2 & 2.24751 & \text{Efficiency } 88.6\% \\ \epsilon'_3 & 4.88026 & \text{Efficiency } 89.4\% \end{cases} \tag{9.A.31}$$

A further improvement can be made at values $\epsilon < \epsilon_{max}$ avoiding the Lorentz term by introducing the small order term

$$\phi(\epsilon) = \frac{2}{\sqrt{\pi e^1}}\sqrt{\epsilon} \qquad \epsilon < \epsilon'_0 \tag{9.A.32}$$

where $\epsilon'_0 = 0.2362$. The efficiency is either 90.6% or 91.3% for ϵ'_1 or ϵ'_3, respectively.

The cumulative term is easily evaluated

$$s(\epsilon) = \int_0^{\epsilon} \phi(\epsilon)\mathrm{d}\epsilon = \frac{2}{\sqrt{\pi e}} \begin{cases} \frac{2}{3}\epsilon^{3/2} & \epsilon < \epsilon_0 \\ s(\epsilon_0) + \alpha\, a(\arctan[(\epsilon - \epsilon_{max})/a] - \\ \quad \arctan[(\epsilon_0 - \epsilon_{max})/a]) & \epsilon_0 \leq \epsilon \leq \epsilon' \\ s(\epsilon') + \beta\left[\exp(-\epsilon'/L) - \exp(-\epsilon/L)\right] & \epsilon > \epsilon' \end{cases} \tag{9.A.33}$$

for the appropriate value of ϵ' from Eq. (9.A.31).

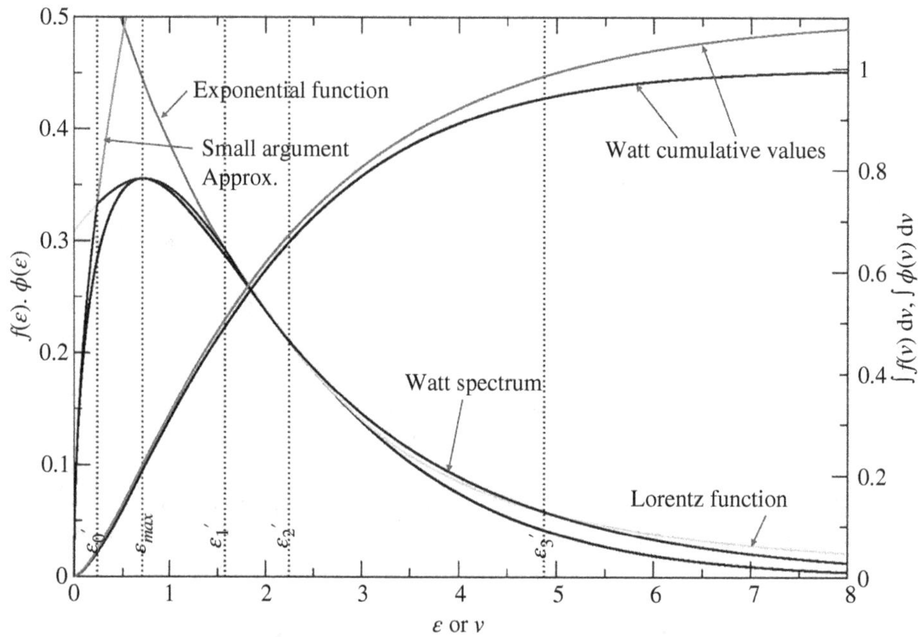

Figure 9.A.1 Plot of the Watt spectrum and its composite comparison function as labelled. The transition points are shown. The cumulative integrals of the probability functions are plotted for transitions at ϵ_0' and ϵ_1'. Exponential and Lorentz functions are also plotted along with the small argument approximation for the Watt function.

However, since the *log* term is more rapidly calculated than the *tan*, any efficiency gain may be outweighed by an increase in CPU time.

Figure 9.A.1 illustrates this method comparing the incorporation of the different components and gives the cumulative value for the Watt profile and for the comparator with ϵ_1'.

9.A.8 Maxwell–Boltzmann Distribution

This distribution gives the probability of finding gas particles with a given speed in equilibrium. As it is one of the most useful deviates, it is important to make it as efficient as possible. The distribution function is

$$f(v) = \sqrt{\frac{2}{\pi}} v^2 \exp\left\{-\frac{v^2}{2}\right\} \tag{9.A.34}$$

The mean is $\sqrt{8/\pi}$ and the variance $(3 - 8/\pi)$. The deviate may be calculated in several ways.

a) **Cumulative method:** The distribution is equivalent to Gaussian distributions in each of the three component directions. Thus, if x_1, x_2 and x_3 are three normal derivatives, $y = \sqrt{x_1^2 + x_2^2 + x_3^2}$ is a Maxwell deviate.[11]

11 Note this fits in well with the Box–Muller method for computing normal deviates, which generates two values for each call.

b) **Rejection method 1:** The substitution $v^2 = 2[u/(1-u)]^{2/3}$ transforms the velocity distribution to the range $v \in 0, 1$ for which a rejection method may be used.

$$v^2\frac{dy}{du} = \frac{1}{3}\frac{dy^3}{du} = \frac{2\sqrt{2}}{3}\frac{d}{du}\left\{\frac{u}{1-u}\right\} = \frac{2\sqrt{2}}{3}\frac{d}{du}\left\{\frac{u}{(1-u)}\right\}$$
$$= \frac{2\sqrt{2}}{3}\left\{\frac{1}{(1-u)} + \frac{u}{(1-u)^2}\right\} = \frac{2\sqrt{2}}{3}\frac{1}{(1-u)^2}$$

A truncated Lorentz distribution for the comparison term is appropriate with $\phi(u) = c_0/\{1 + [(u-u_0)/a_0]^2\}$ with the peak at $u_0 = 0.78240$ and scaling parameters $a_0 = 0.77465$ and $c_0 = 1.519695$. Lorentz deviates are required in the range $0 \le u \le 1$ only, i.e. $[\pi/2 - \arctan(u_0/a_0)]/\pi \le x \le [\pi/2 + \arctan(\{1-u_0\}/a_0)]/\pi$.[12] The efficiency is 80%.

Figure 9.A.2 shows plots of the comparison function $\phi'(u)$ against the transformed Maxwellian $\phi(u)$ and plots of a comparison function against a Maxwellian. It can be seen that equivalence is bad for values $u > u_0 \approx 0.78240$ corresponding to $v \approx 2$, i.e. in the tail of the distribution, where the method will be very inefficient.

c) **Rejection method 2:** Calculate deviates against a comparison function with a three-way split. For small uniform deviates $0 < x < 0.71343$ a quadratic distribution based on the small order Taylor series is used,[13] for medium $0.71343 < v < 2.29870$ a

$$\phi(y) = \sqrt{2/\pi}\begin{cases} y^2 & \text{if } y < y_1 \\ \left(\sqrt(2) - gg(y - \sqrt{2})^2\right) & y_1 < y < y_2 \\ (a\exp\{-by\}) & \text{otherwise}\end{cases} \tag{9.A.35}$$

12 The range of uniform deviates required is $0.248416 \le x \le 0.587167$. This is easily obtained from a normal uniform deviate ($x \in (0,1)$) as $x' = 0.248416 + 0.271279x$.

13 The second derivative at the turning point is given by

$$\frac{d\phi}{dv^2} = \sqrt{2/\pi}\cdot(1 - \frac{1}{2}v^2)e^{-v^2/2} \qquad \frac{d\phi}{dv} = \frac{dv^2}{dv}\frac{d\phi}{dv^2} = \sqrt{2/\pi}\cdot 2v(1 - v^2/2)\exp - v^2/2 \tag{9.A.36}$$

If $d\phi/dv = 0$ then $v = \sqrt{2}$ and

$$\frac{d^2\phi}{dv^2} = \sqrt{2/\pi}\cdot\left\{(2 - 3v^2) - 2v^2(1 - v^2/2)\right\}e^{-v^2/2} \tag{9.A.37}$$

Therefore, at the maximum

$$\frac{d^2\phi}{dv^2} = -\sqrt{2/\pi}\cdot 4/e \lessapprox \sqrt{2/\pi}\cdot 2gg \tag{9.A.38}$$

where $gg = 1.25/e = 0.459844$, $a_2 = 19.7971$ and $b = 1.711162$ and $y_1 = \sqrt{2/e}$ and $y_2 = 1.924065$

$$\int\phi(y)dy = \sqrt{2/\pi}\begin{cases} \frac{1}{3}y^3 & \text{if } y < y_1 \\ \left.\begin{array}{l}\frac{1}{3}y_1^3 - \sqrt{2}(y_1 - y) \\ -\frac{gg}{3}\left[(y-\sqrt{2})^3 - (y_1 - \sqrt{2})^3\right]\end{array}\right\} & \text{if } y_1 < y < y_2 \\ \left.\begin{array}{l}\frac{1}{3}y_1^3 - \sqrt{2}(y_1 - y_2) \\ -\frac{gg}{3}\left[(y_2-\sqrt{2})^3 - (y_1 - \sqrt{2})^3\right] \\ +\frac{a}{b}\left(e^{-by} - e^{-by_2}\right)\end{array}\right\} & \text{otherwise}\end{cases} \tag{9.A.39}$$

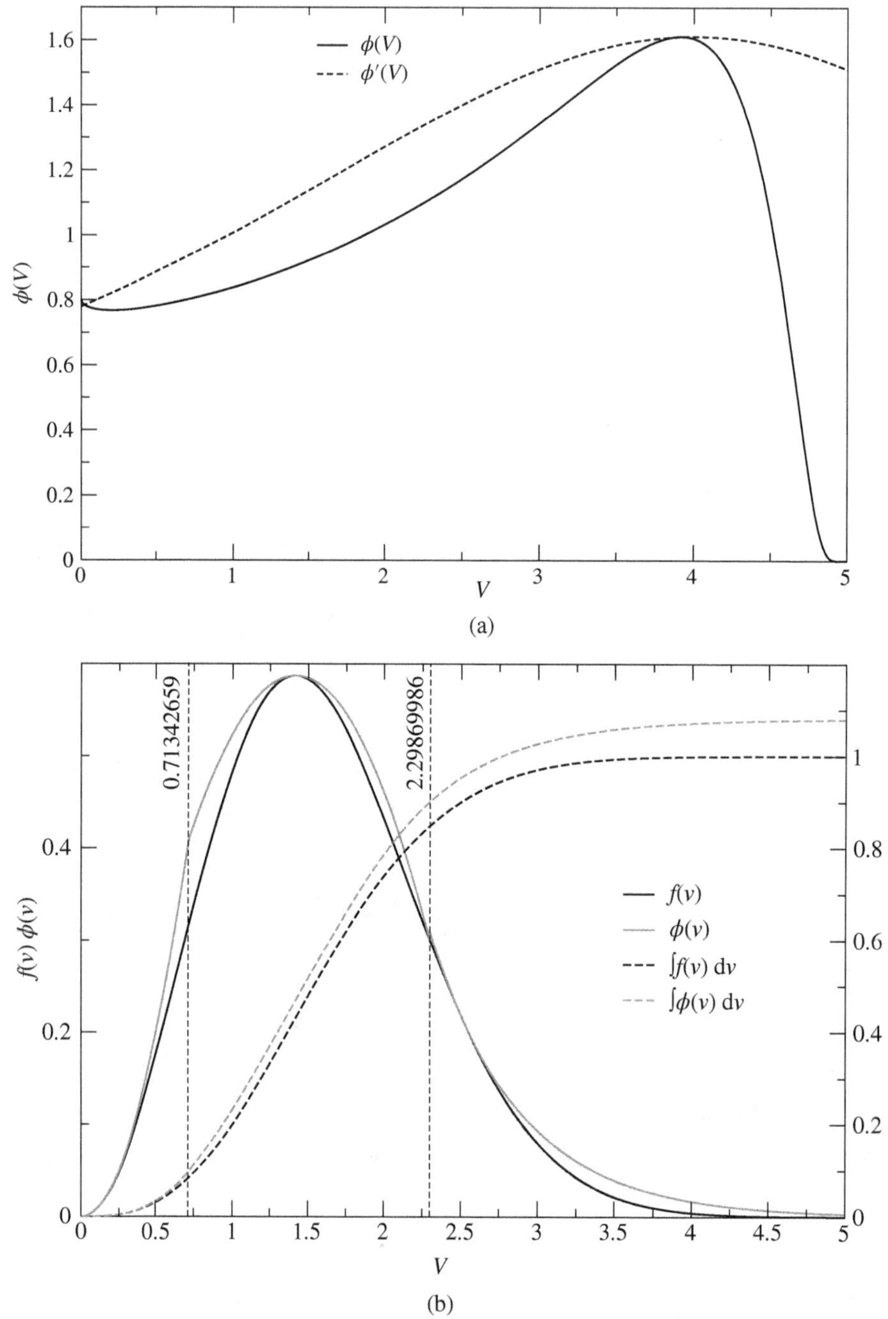

Figure 9.A.2 Comparisons of rejection method Maxwellian distribution functions and comparator functions. (a) Plots of the transformed Maxwellian $\phi(u)$ and the comparison function $\phi'(u)$ against the transformed velocity u and (b) plots of the Maxwellian $f(v)$ and the comparison function $\phi(v)$ against the velocity v. The cumulative (integrated) values are shown as dashed lines.

parabolic distribution touching at the peak, and for large $2.29870 < v < 1$ a touching exponential

$$\phi(y) = \sqrt{2/\pi}\begin{cases} y^2 & \text{if } y < \sqrt{2/e} \\ \left(\sqrt{(2)} - 0.459844(v - \sqrt{2})^2\right) & \sqrt{2/e} < y < 1.924065 \\ (19.7971\exp\{-1.711162y\}) & \text{otherwise} \end{cases} \quad (9.A.40)$$

This rather complex form gives a good fit between the distribution and the comparison, the efficiency is about 92%. The further advantage is that the integrated form is simple, but unfortunately involves cubic terms. Since the derivative has a turning point, there are three roots; the guess for the random number can be obtained by solution of the cubic equation using the standard method (Press et al. 1986) at the expense of arccos and cos evaluations. As a result, the cumulative method using the Box–Müller form is probably to be preferred.

9.A.9 Holtsmark Distribution

The microfield-field variations at a point induced by uniformly distributed charges determines the spectral profile due to stationary ions (known as quasi-static ion broadening). The distribution also finds application in astrophysics and the theory of liquids. The profile is given by the probability function

$$p(y) = \frac{\pi}{y}\int_0^{\infty} t\sin(t)\exp\left\{-\left(\frac{t}{y}\right)\right\}^{3/2} dt \quad (9.A.41)$$

for $0 \leq y < \infty$. The mean is 3.413, but since $p(y \to y_{5/2}$ as $y \to infty)$ the variance does not exist. The probability function is easiest calculated from a tabulated data table (Chandrasekhar, 1943; Griem, 1964). Deviates are most easily calculated by the **Markov method**, provided correlations are not important. Alternatively, deviates may be calculated by the **rejection method** using a displaced Lorentz function as comparator with $\phi(x) = c_0/(1 + [x - x_0]/a_0^2)$ with displacement $x_0 = 1.6$ and with scaling parameter $a_0 = 1.357802$ and $c_0 = 0.36633$. Lorentz deviates are needed only in the $-x_0, \infty$, i.e. uniform deviates are required only in the range $\left[\pi/2 - \arctan(x_0/a_0)\right]$. The efficiency is 82%.

Appendix 9.B Kinematics of Elastic Scattering

Consider two particles with mass m_1 and m_2, and velocities $\mathbf{v}_1$ and $\mathbf{v}_2$. The centre-of-mass velocity $\mathbf{V} = (m_1\mathbf{v}_1 + m_2\mathbf{v}_2)/(m_1 + m_2)$. The particles have velocities $\mathbf{u}_1 = \mathbf{v}_1 - \mathbf{V} = m_2/(m_1 + m_2)\mathbf{q}$ and $\mathbf{u}_2 = \mathbf{v}_2 - \mathbf{V} = m_1/(m_1 + m_2)(-\mathbf{q})$ where $\mathbf{q} = \mathbf{v}_1 - \mathbf{v}_2$ is the velocity of approach.

On elastic scattering, the magnitude of the velocity of approach is unchanged but its vector is rotated through an angle θ. Thus, after scattering the velocity of approach of the particles is $\mathbf{q}' = (q\cos\theta, q\sin\theta)$. The velocities in the lab frame are

$$\mathbf{v}_1' = \mathbf{V} + m_2/(m_1 + m_2)(q\cos\theta,\ q\sin\theta) \quad (9.B.1a)$$

$$\mathbf{v}_2' = \mathbf{V} - m_1/(m_1 + m_2)(q\cos\theta,\ q\sin\theta) \quad (9.B.1b)$$

The speed of particle ①

$$v_1'^2 = \left[\frac{(m_1 v_1 + m_2 v_2)}{(m_1 + m_2)} + \frac{m_2}{(m_1 + m_2)}(v_1 - v_2)\cos\theta\right]^2 + \left[\frac{m_2}{(m_1 + m_2)}(v_1 - v_2)\sin\theta\right]^2 \tag{9.B.2}$$

$$= \left[\frac{(m_1 v_1 + m_2 v_2)}{(m_1 + m_2)}\right]^2 + 2\left[\frac{m_2}{(m_1 + m_2)^2}(m_1 v_1 + m_2 v_2)\right](v_1 - v_2)\cos\theta + (v_1 - v_2)^2 \tag{9.B.3}$$

If particle ② is at rest, $v_2 = 0$ and the resulting speed of particle ① is

$$v_1'^2 = v_1^2 \frac{[(m_1 v_1 + m_2 v_2)^2 + 2m_1(m_1 + m_2)\cos\theta + 1]}{(m_1 + m_2)^2} \tag{9.B.4}$$

The fractional energy loss in an elastic scattering with a nucleus of mass number $A = m_2/m_1$ at rest $v_2 = 0$ is therefore

$$\frac{v_1'^2}{v_1^2} = \frac{A^2 + 2A\cos\theta + 1}{(A+1)^2} \tag{9.B.5}$$

Thus far, we have considered the scattering in two dimensions only with respect to the incoming and outgoing velocities at angle θ. However, there is a preferred direction, namely the radius vector $\mathbf{r}$ from the origin to the scattering point. The outgoing vector $\mathbf{v}'$ can, therefore, be rotated about the line $\mathbf{v}$ at angle ϕ to the plane of $\mathbf{r}$ and $\mathbf{v}$ at a random angle ϕ. If this is done, the angle between $\mathbf{r}$ and the outgoing velocity $\mathbf{v}'$ is changed from that of the incident particle ψ.

Consider a set of polar co-ordinates with axis $\mathbf{v}$ and the zero of azimuth set on the plane normal to the axis at $\phi = 0$ at the plane of $\mathbf{r}, \mathbf{v}$. The scattered vector in the centre-of-mass frame in polar co-ordinates is (v', θ, ϕ).

Transforming to Cartesian co-ordinates in the lab frame with the x-direction taken along the line $\mathbf{r}, \mathbf{v}$, and adding the centre-of-mass velocity from Eq. (9.B.1) the scattered velocity is

$$\mathbf{v}' = \frac{v'}{(m_1 + m_2)}\left\{(m_1 + m_2\cos\theta), (m_2\sin\theta)\cos\phi, m_2\sin\theta\sin\phi\right\} \tag{9.B.6}$$

Taking the scalar product of $\mathbf{v}' \cdot \mathbf{r}$ and utilising Eq. (9.B.5), we obtain the new value of the direction of travel, namely

$$\cos\psi' = \frac{\mathbf{r}\cdot\mathbf{v}'}{|\mathbf{r}|\,|\mathbf{v}'|} = \frac{(1 + A\cos\theta)\cos\psi_0 + A\sin\theta\sin\psi_0\cos\phi}{\sqrt{1 + 2A\cos\theta + A^2}} \tag{9.B.7}$$

References

Chandrasekhar, S. 1943. “Stochastic problems in physics and astronomy.” *Rev Mod Phys* 15:1–89.

Cooper, N G, ed. 1989. *From Cardinals to Chaos.* Cambridge University Press. Originally published as: Stanislaw Ulam 1909–1984 in *Los Alamos Science No.15 1987.*

Davis, D H. 1963. Critical size calculations for neutron systems by Monte-Carlo methods. In *Statistical Physics*, ed. B Alder S Fernbach and M Rotenberg. Number 1 *in* "Methods in Computational Physics" New York: Academic Press pp. 1–45.

Griem, H R. 1964. *Plasma Spectroscopy*. New York: McGraw-Hill.

Hill, J F. 1957. Nuclear Reactor Physics. Technical report Research reactor Dept., A.E.R.E., Harwell. Originally published as monthly articles in the magazine Nuclear Power between May 1956 and May 1957.

Knuth, D E. 1998. *The Art of Computer Programming*. Vol. 2 Seminumerical Algorithms Third ed. Reading, MA: Addison-Wesley.

MacKeown, P K. 1997. *Stochastic Simulation in Physics*. Berlin: Springer-Verlag.

Margalia, G. 2013. "Xorshift RNGs." *J Stat Softw* 8:1–6.

Pert, G J. 2021. *Introduction to Plasma Physics*. London: Wiley.

Press, W H, B P Flannery, S A Teukolsky and W T Vetterling. 1986. *Numerical Recipes - The Art of Scientific Computing*. Cambridge, U.K.: Cambridge University Press.

Press, W H, B P Flannery, S A Teukolsky and W T Vetterling. 2007. *Numerical Recipes - The Art of Scientific Computing*. Third ed. Cambridge, U.K.: Cambridge University Press. The third edition of this series of texts uses C^{++}as the language for the listed routines.

Spanier, J and E M Gelbard. 2008. *Monte Carlo Principles and Neutron Transport Problems*. Mineola, New York: Dover books. Originally published by Addison-Wesley, Reading, MA, 1968.

Woolfson, M M and G J Pert. 1999. *An Introduction to Computer Simulation*. Oxford: Oxford University Press.

10

Numerical Diffusion Schemes

10.1 Introduction

Diffusion is due to the drift of particles when their distribution is weakly perturbed spatially by a directed 'force'. It is found under many conditions: particles (particle diffusion), momentum (viscosity), heat transfer (thermal conduction), etc. The kinetic theory description is well known since the classic work of Maxwell (1867). A full discussion of the process and its mathematical description can be found in our earlier work (Pert, 2021). For our purposes it suffices to know that it can be generally described by the flux $\mathbf{f}$ of the quantity X, which is determined by its spatial gradient. Since the process is conservative, the divergence of the flux $\mathbf{f}$ defines the rate of increase of the quantity of X per unit volume. Considering a volume V with outwards bounding surface S

$$\frac{\partial}{\partial t}\int_V X dV = -\int_S \mathbf{f} \cdot \mathrm{d}S \tag{10.1}$$

If the perturbation is weak, so that the particle distribution is only weakly perturbed, the flux $\mathbf{f}$ is linearly dependent on the gradient of the driving perturbation, which itself is proportional to the quantity X. Therefore, in finite-difference form

$$f_i = -D_{i,j}\frac{\partial X_j}{\partial x_i} \tag{10.2}$$

where $D_{i,j}$ is the diffusivity tensor.

In a uniform medium with Cartesian geometry equations (10.1) and (10.2)

$$\frac{\partial X}{\partial t} + \sum_i \frac{\partial}{\partial x_i}\left(D\frac{\partial X}{\partial x_i}\right) = 0 \tag{10.3}$$

which is a parabolic partial differential equation.

The inclusion of diffusion processes into the body of large hydro-dynamic codes has stimulated the use of implicit solutions to the diffusion equations which converge in some well-behaved fashion onto the equilibrium (uniform) solution. Such solutions are clearly of value, since it frequently happens that regions in which diffusion processes are extremely rapid are of limited physical interest. One is thus faced with the partial differential equation equivalent of the stiff equation problem in ordinary differential equations. In a similar fashion, one seeks solutions that are stable, and convergent for the fast processes of little interest, yet accurate for the slower ones. The restrictions of computer store and CPU time imposed

Modelling and Simulation in Plasma Physics for Physicists and Mathematicians, First Edition. Geoffrey J. Pert.

by the mesh of a large fluid code usually necessitate that no more than two time levels of data be stored simultaneously. In consequence, one is forced to consider two-step integration schemes in which the data is advanced from one time level to the next. In the course of this time-step, the characteristic rate coefficients are assumed to take constant values varying in space and from time-step to time-step. As a result of its convergence property for large time-steps, Laasonen, or fully implicit temporal differencing (Richtmyer and Morton, 1967), is often preferred to the split-time-step or Crank–Nicolson scheme, despite the loss in small order accuracy introduced. Fully implicit temporal diffencing also avoids the oscillations inherent in the split time-step method.

The spatial differencing of the diffusion equation is usually required to satisfy various constraints inherent within the governing differential equation. These are usually consistency (Richtmyer and Morton, 1967) and conservation. More recently, Kershaw (1978) has drawn attention to the need for the matrix representing this difference – the diffusion matrix – to be non-positive definite. We shall show that this condition essentially expresses the second law of thermodynamics, and in consequence, the mathematical stability of diffusion (Friedman, 2008). A hitherto neglected (in this context) physical property of diffusion is contained in the maximum principle (Morse and Feshbach, 1953), namely extremal decay in time. We examine the form of spatial and temporal differencing necessary for the numerical solution to satisfy this condition, i.e. to be extremal. Laasonen (Richtmyer and Morton, 1967) has drawn attention to this property for fully implicit differencing in an isotropic medium on an orthogonal mesh, but in the more general case, the properties of an extremal form have not been considered.

There are two dissimilar methods for solving the diffusion equation in more than one dimension. The first is based on multiple single-step stable integrations and has been used for many years in different guises: alternate direction integration (ADI) and time splitting; these use the simple Thomas algorithm to solve the matrix equation associated with the implicit stages of the calculation.

The second method is newer and uses sparse matrix solvers (ICCG) to solve the implicit problems. The recent major advance in tackling implicit problems of this type, namely the development of the ICCG sparse matrix solving routine by Meijerink and van der Vorst (1977) (see also Stoer and Bulirsch, 1991, S8.7.10) and its implementation by Kershaw (1978) for the solution of the diffusion matrix, generated a need for improvements in the earlier finite-difference representation in generalised geometry. In particular, two problems exist:

a) Convergence for large rates requires fully implicit temporal differencing, with relatively poor accuracy. On the other hand, the improved accuracy given by split-time-step schemes necessitates the use of limited time-steps.
b) The spurious generation of occasional negative values, which violate essential physical constraints, is an unavoidable occurrence in calculations with existing methods. These must be removed by an ad hoc procedure (which normally violates conservation) such as 'reset to zero'. The application of extremal principles is shown to generate algorithms to overcome both these difficulties.

Our approach to the general problem, when the diffusion matrix is not symmetric, will be similar to that of Kershaw (1981), who derived conditions for the simpler symmetric case.

The close relationship of this approach with the theory of stiff equations is emphasised by considering the spatial differencing (thereby yielding the diffusion matrix), independently of the temporal, and allowing the conditions on the diffusion matrix necessary to re-produce those of the exact equations to be determined. Implicit temporal differencing, and its relation to the extremal principle, is examined separately.

Two examples of extremal schemes are constructed. In each case, the simplest finite-difference representation is not extremal. In one case, resource to the known physical behaviour allows the diffusion matrix to be made extremal. In the other case, no general consistent extremal representation of the diffusion matrix exists: under restricted conditions, it is therefore necessary to introduce an extremum-limited flux term to treat this problem. It is believed that the two methods outlined in these examples are sufficient to cover all such problem.

The use of extremum-limited solutions to generate weights for the temporal integration is shown to allow split-time-step differencing for small rates and fully implicit for large rates. This approach is very similar to that devised for integrating positivity maintaining rate equations (Section 12.2.1).

10.2 Split-Time-Step and ADI Methods for Solving Diffusion Problems in Orthogonal Cartesian Grid Systems

The solution of the diffusion equation in one-dimensional orthogonal geometries using finite differences is particularly easy to programme stably and exemplifies the problems associated with this equation. We can then extend these solutions initially to multidimensional problems and then to the situation in a general spatial configuration.

10.2.1 One-Dimensional Orthogonal Grid

Diffusion is a conservative phenomenon in which a quantity ϵ, such as heat (thermal energy per unit mass), is transported conservatively through an isotropic medium. The energy diffusing in a one-dimensional volume is treated by writing the diffusion equation (10.3) in the form with the thermal diffusivity $K = \chi/c$, where χ is the thermal conductivity and c the specific heat.[1] We may identify the total flux $\sum_{\text{All faces}} \nabla\epsilon| \cdot \mathbf{S}$ with the thermal fluxes out of the computational cell faces $\mathbf{S}$ as shown in Figure 10.1. Thus, in a one-dimensional isotropic system with no magnetic field, the total diffusion out of the cell

$$\sum_i D_i = \left[(SK)_{i+1/2}\left(\epsilon_{i+1} - \epsilon_i\right) - (SK)_{i-1/2}\left(\epsilon_i - \epsilon_{i-1}\right)\right]/\Delta x \tag{10.4}$$

where $S_{i+1/2}$ is the cross-sectional area of the face $i + 1/2$ (see Figure 10.1)

$$\begin{aligned}\epsilon_i^{n+1} = \epsilon_i^n &- \left\{(SK)_{i+1/2}\left[\theta\left(\epsilon_{i+1}^{n+1} - \epsilon_i^{n+1}\right) + (1-\theta)\left(\epsilon_{i+1}^n - \epsilon_i^n\right)\right]/\Delta x_{i+1/2}\right\} Dt/V_i \\ &+ \left\{(SK)_{i-1/2}\left[\theta\left(\epsilon_i^{n+1} - \epsilon_{i-1}^{n+1}\right) + (1-\theta)\left(\epsilon_i^n - \epsilon_{i-1}^n\right)\right]/\Delta x_{i-1/2}\right\} Dt/V_i\end{aligned} \tag{10.5}$$

1 The diffusivity is assumed constant over the time step, but varying spatially.

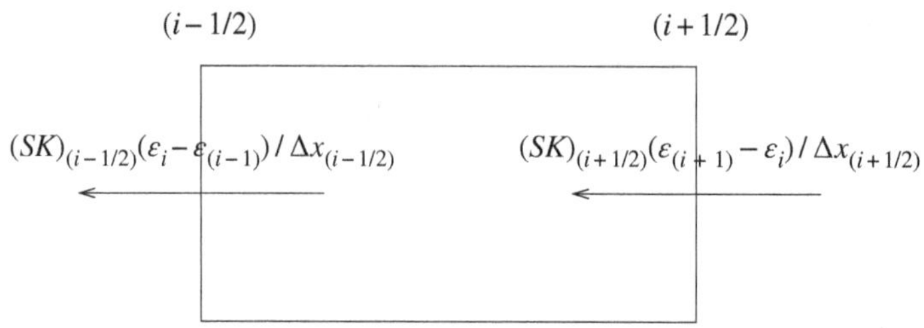

Figure 10.1 Flux across cell boundaries in one-dimensional diffusion on a one-dimensional non-uniform mesh.

The value of ϵ at the end of the time step is given by the solution of the tri-diagonal matrix equation

$$-A_i\,\epsilon_{i+1}^{n+1} + B_i\,\epsilon_i^{n+1} - C_i\,\epsilon_{i-1}^{n+1} = D_i \tag{10.6}$$

where

$$A_i = \theta\,(SK)_{i+1/2}\,Dt/\left(\Delta x_{i+1/2}\,V_i\right) \qquad C_i = \theta\,(SK)_{i-1/2}\,Dt/\left(\Delta x_{i-1/2}\,V_i\right)$$
$$B_i = \theta\,\left[(SK)_{i+1/2}/\Delta x_{i+1/2} + (SK)_{i-1/2}/\Delta x_{i-1/2}\right]\,Dt/V_i \tag{10.7}$$

and

$$D_i = -A_i\,\epsilon_{i+1}^{n} + B_i\,\epsilon_i^{n} - C_i\,\epsilon_{i-1}^{n} \tag{10.8}$$

On a regular (uniform) mesh, the volume V_i and the cell surface $S_{i+1/2}$ are related by $S/\Delta x\,V = 1/\Delta x^2$ and we obtain a more familiar equation

$$\left[\epsilon_i^{(n+1)} - \epsilon_i^{n}\right]/Dt = -\left[K_{(i+1/2)}\left(\epsilon_{(i+1)} - \epsilon_i\right) - K_{(i-1/2)}\left(\epsilon_i - \epsilon_{(i-1)}\right)\right]/\Delta x^2 \tag{10.9}$$

which is the simple finite-difference form of Eq. (10.3) in one dimension. This equation is particularly easy to solve by computer using the so-called *Thomas algorithm* (Richtmyer and Morton, 1967, p 213) as detailed at the end of this Chapter.

10.2.1.1 Stability Constraints

On a mesh with constant coefficients, let us represent the variable set ϵ_i by a Fourier series, so that the finite-difference relationship[2] becomes

$$\epsilon_i^{n+1} + \frac{\theta\,K\,Dt}{\Delta x^2}\left(\epsilon_{i+1}^{n+1} - 2\epsilon_i^{n+1} + \epsilon_{i-1}^{n+1}\right) = \epsilon_i^{n} - \frac{(1-\theta)\,K\,Dt}{\Delta x^2}\left(\epsilon_{i+1}^{n} - 2\epsilon_i^{n} + \epsilon_{i-1}^{n}\right) \tag{10.10}$$

The amplification factor

$$G = \frac{1 - 2\,(1-\theta)\,\left[1 - \cos(k\,\Delta x)\right]/\Delta x^2}{1 + 2\theta\,\left[1 - \cos(k\,\Delta x)\right]/\Delta x^2} \tag{10.11}$$

so that if $\sigma = 2K\,Dt/\Delta x^2\left[1 - \cos(k\Delta x)\right]$, von Neumann's stability criterion yields

$$-1 \le \frac{1-(1-\theta)\,\sigma}{1+\theta\sigma} \le 1 \qquad \text{or} \qquad [\sigma \ge 0 \text{ and } (1-2\theta)\,\sigma \le 2] \tag{10.12}$$

2 There a number of different finite-difference forms, whose form and properties are described in detail by Richtmyer and Morton (1967, §8.2).

Since $0 \leq 1 - \cos(k\Delta x) \leq 2$, we require that max $\left(\sigma = 4K\,Dt/\Delta x^2\right) \leq 2/(1-\theta)$. Thus, an explicit solution $\theta = 0$ is only stable if $2K\,Dt/\Delta x^2 \leq 1$, which limits the time-step for integration to the conditional stability condition $Dt \leq \Delta x^2\ /2K$.

Implicit solutions which have non-zero values of θ are unconditionally stable (i.e. stable for all values of Dt) iff $\theta \geq 1/2$.

- *Centred difference*, $\theta = 1/2$, or Crank–Nicolson solutions, which are unconditionally stable, i.e. stable for all values of K and therefore of the time step Δt of the implicitness parameter. The solution is of second order and are therefore more accurate than that generated by other values of the implicitness parameter θ.
- *Fully implicit*, $\theta = 1$, or Laasonen, is only first-order accurate. This solution has the merit that for large time steps it converges to a solution, namely the equilibrium solution.

Apart from the fully explicit solution, all these methods with $\theta \neq 0$ require the solution of a set of simulataneous equations. In one dimension this leads to the tri-diagonal set described earlier, which are easily solved by Thomas algorithm. In two dimensions the Peaceman–Rachford method described later gives a second-order solution whose solution requires no more effort than the one-dimensional problem. It is not difficult to show that all stable solutions are unsatisfactory, in that for large steps the solutions become oscillatory. For example the Crank–Nicolson scheme oscillates unless

$$Dt \leq \alpha\epsilon_i/\left[K|\epsilon_{i+1} - \epsilon_i|/\Delta x^2\right] \tag{10.13}$$

where α is a constant usually set to $\sim$0.1. This is clearly less restrictive condition than the time step-limit required by the explicit solution.

Fully implicit solutions although of only first order have the merit of satisfying the extremal condition, for if ϵ_i is a maximum then $\epsilon_{i+1} - 2\epsilon_i + \epsilon_{i-1} \leq 0$ and

$$\epsilon_i^{n+1} = \epsilon_i^n - \frac{K\,Dt}{\Delta x}\left(\epsilon_{i+1}^n - 2\epsilon_i^n + \epsilon_{i-1}^n\right) = \epsilon_i^n\left(1 - \frac{K\,Dt}{\Delta x^2}\right) \leq \epsilon_i^n \tag{10.14}$$

As a result, no new extremum can be created by the finite-difference form of the diffusion equation. Hence, if at a minimum is positive $\epsilon_i^n > 0$, it follows that $\epsilon_i^{n+1} \geq 0$ must also be non-negative.

It can be shown under extremely general conditions (Pert, 1981b) the finite-difference operator may be cast into a finite-difference form such that the fully implicit scheme obeys the extremal condition.

Implicit solutions although, are an improvement on the explicit approach unless the time-step is small, but their use is limited if required for programmes where the time-step is required to be limited only by the hydrodynamics. As a result of the favourable behaviour for arbitrary time-steps, the fully implicit method is frequently incorporated in laser-scale computer modelling codes to treat thermal conduction.

10.2.2 Multidimensional – Orthogonal Grid

A number of viable efficient multi-step schemes satisfying stability, convergence and accuracy requirements have been devised. However, requiring multiple scans through the mesh limits their suitability for rapid calculation in a multi-physics application such as our interest here. The underlying idea behind these schemes is to use the one-dimensional stable Crank–Nicolson method, which when extended to multiple dimensions over several integration steps remains stable and replaces a complex set of simultaneous equations by a series of simple ones solved by the efficient Thomas algorithm (Appendix 10.A).

10.2.2.1 Alternating Direction (ADI)

These methods work by constructing two or more sweeps through the mesh in the different directions. For two dimensions there is one implicit and one explicit pass with half a time step each. There are several different forms of these algorithms (Richtmyer and Morton, 1967, §8.8). We will consider only the first and most straightforward, namely that due to Peaceman and Rachford (1955) and Douglas (1955) [alternating direction integration (ADI)], where an implicit x sweep and explicit y sweep are alternate with an implicit y sweep and explicit x sweep; by alternating the order of the sweeps, any bias (x versus y) is avoided in a uniform medium with Cartesian geometry. For simplicity we will assume the diffusivity is a scalar constant and that the cell widths Δx and Δy are equal; the generalisation to variable diffusivity and different scale lengths in x and y is straight forward. Thus

$$\epsilon_{i,j}^{n+1/2} = \epsilon_{i,j}^{n} - \left\{\mathcal{K}_x \left[\epsilon_{(i+1),j}^{n+1/2} - 2\epsilon_{i,j}^{n+1/2} + \epsilon_{(i-1),j}^{n+1/2}\right] + \mathcal{K}_y \left[\epsilon_{i,(j+1)}^{n} - 2\epsilon_{i,j}^{n} - \epsilon_{i,(j-1)}^{n}\right]\right\} \tag{10.15a}$$

and with the explicit and implicit steps interchanged

$$\epsilon_{i,j}^{n+1} = \epsilon_{i,j}^{n+1/2} - \left\{\mathcal{K}_x \left[\epsilon_{(i+1),j}^{n+1/2} - 2\epsilon_{i,j}^{n+1/2} + \epsilon_{(i-1),j}^{n+1/2}\right] + \mathcal{K}_y \left[\epsilon_{i,(j+1)}^{n+1} - 2\epsilon_{i,j}^{n+1} - \epsilon_{i,(j-1)}^{n+1}\right]\right\} \tag{10.15b}$$

where $\mathcal{K}_x = K\,Dt/2\Delta x^2$ and $\mathcal{K}_y = K\,Dt/2\Delta y^2$.

To investigate the von Neumann stability of this scheme, we consider $\exp(\imath(k_x x + k_y y))\epsilon^n$ from which by substitution in Eq. (10.15) we obtain the equivalent of Eq. (10.11) which enables the components of the amplification to be calculated

$$G(k_x, k_y) = G_1(k_x) \cdot G_1(k_y) = \frac{1 - (1 - \cos\beta_x)}{1 + (1 - \cos\beta_x)} \cdot \frac{1 - (1 - \cos\beta_y)}{1 + (1 - \cos\beta_y)} \tag{10.16}$$

where $\beta_x = k_x \Delta x$; $\beta_y = k_y \Delta y$.

Since $\left(1 - \cos\beta_{x/y}\right) \geq 0$, it follows that $0 \geq G \geq 1$, and hence the scheme is unconditionally stable.

However, as in the 1D case, the solution oscillates if the time step is large.

The ADI concept can be extended to more than two dimensions relatively simply, but at the obvious expense of further repeated passes through the mesh. Unfortunately, the obvious extension to three dimensions is unstable. (Douglas, 1962) showed how this could be done by re-writing the two-dimensional equations (10.15) in terms of new variables as

$$\epsilon_{i,j}^{*n+1} = \epsilon_{i,j}^{n} - \mathcal{K}_x \left\{\left[\epsilon_{(i+1),j}^{*n+1} - 2\epsilon_{i,j}^{*n+1} + \epsilon_{(i-1),j}^{*n+1}\right] + \mathcal{K}_y \left[\epsilon_{i,(j+1)}^{n} - 2\epsilon_{i,j}^{n} - \epsilon_{i,(j-1)}^{n}\right]\right\} \tag{10.17a}$$

$$\epsilon_{i,j}^{n+1} = \epsilon_{i,j}^{n} - \mathcal{K}_x \left\{\left[\epsilon_{i,j}^{*n+1} - 2\epsilon_{i,j}^{*n+1} + \epsilon_{(i-1),j}^{*n+1}\right] + \mathcal{K}_y \left[\epsilon_{i,(j+1)}^{n+1} - 2\epsilon_{i,j}^{n+1} - \epsilon_{i,(j-1)}^{n+1}\right]\right\} \tag{10.17b}$$

which are easily seen to be the set of Eqs. (10.15) with the notational change $\epsilon_{i,j}^{n+1/2} \equiv \epsilon_{i,j}^{*n}$.

This procedure introduces the next-dimensional set of variables $\epsilon^{n}_{i,j,k}$, $\epsilon^{*n}_{i,j,k}$, $\epsilon^{**n}_{i,j,k}$, $(\epsilon^{***n}_{i,j,k} = \epsilon^{n+1}_{i,j,k})$

$$\begin{aligned}\epsilon^{*n+1}_{i,j,k} &= \epsilon^{n}_{i,j,k} - \mathcal{K}_x\left\{\left[\epsilon^{*n+1}_{(i+1),j,k} - 2\epsilon^{*n+1}_{i,j,k} + \epsilon^{*n+1}_{(i-1),j,k}\right] + \left[\epsilon^{n}_{(i+1),j,k} - 2\epsilon^{n}_{i,j,k} + \epsilon^{n}_{(i-1),j,k}\right]\right\}\\ &\quad - 2\,\mathcal{K}_y\left[\epsilon^{n}_{i,(j+1),k} - 2\epsilon^{n}_{i,j,k} + \epsilon^{n}_{i,(j-1),k}\right] - 2\,\mathcal{K}_z\left[\epsilon^{n}_{i,j,(k+1)} - 2\epsilon^{n}_{i,j,k} + \epsilon^{n}_{i,j,(k-1)}\right]\end{aligned} \tag{10.18a}$$

$$\begin{aligned}\epsilon^{**n+1}_{i,j,k} &= \epsilon^{n}_{i,j,k} - \mathcal{K}_x\left\{\left[\epsilon^{*n+1}_{(i+1),j,k} - 2\epsilon^{*n+1}_{i,j,k} + \epsilon^{*n+1}_{(i-1),j,k}\right] + \left[\epsilon^{n}_{(i+1),j,k} - 2\epsilon^{n}_{i,j,k} + \epsilon^{n}_{(i-1),j,k}\right]\right\}\\ &\quad - \mathcal{K}_y\left\{\left[\epsilon^{**n+1}_{(i+1),j,k} - 2\epsilon^{**n+1}_{i,j,k} + \epsilon^{**n+1}_{(i-1),j,k}\right] + \left[\epsilon^{n}_{(i+1),j,k} - 2\epsilon^{n}_{i,j,k} + \epsilon^{n}_{(i-1),j,k}\right]\right\}\\ &\quad - 2\,\mathcal{K}_z\left[\epsilon^{n}_{i,j,(k+1)} - 2\epsilon^{n}_{i,j,k} + \epsilon^{n}_{i,j,(k-1)}\right]\end{aligned} \tag{10.18b}$$

$$\begin{aligned}\epsilon^{***n+1}_{i,j,k} &= \epsilon^{n}_{i,j,k} - \mathcal{K}_x\left\{\left[\epsilon^{**n+1}_{(i+1),j,k} - 2\epsilon^{**n+1}_{i,j,k} + \epsilon^{**n+1}_{(i-1),j,k}\right] + \left[\epsilon^{n}_{(i+1),j,k} - 2\epsilon^{n}_{i,j,k} + \epsilon^{n}_{(i-1),j,k}\right]\right\}\\ &\quad - \mathcal{K}_y\left\{\left[\epsilon^{**n+1}_{i,(j+1),k} - 2\epsilon^{**n+1}_{i,j,k} + \epsilon^{**n+1}_{i,(j-1),k}\right] + \left[\epsilon^{n}_{i,(j+1),k} - 2\epsilon^{n}_{i,j,k} + \epsilon^{n}_{(i-1),j,k}\right]\right\}\\ &\quad - \mathcal{K}_z\left\{\left[\epsilon^{***n+1}_{i,j,(k+1)} - 2\epsilon^{***n+1}_{i,j,k} + \epsilon^{***n+1}_{i,j,(k-1)}\right] + \left[\epsilon^{n}_{i,j,(k+1)} - 2\epsilon^{n}_{i,j,k} + \epsilon^{n}_{i,j,(k-1)}\right]\right\}\end{aligned} \tag{10.18c}$$

The algorithm is based around the Crank–Nicolson scheme and is therefore second order and unconditionally stable (Richtmyer and Morton, 1967, §8.9).

10.2.2.2 Time Splitting

As an alternative to ADI, we may use the equivalent method of *split-time-step*, which was developed in the supplementary material (M.4.i) as a general method for tackling multidimensional problems. Using the Crank–Nicolson, centred difference integration algorithm (10.10), with the implicitness parameter $\theta = \frac{1}{2}$ as the key step, will guarantee consistency and stability (Yanenko, 1971, §2.4). For two dimensions Eq. (M.4.i) takes the simple form where $m = 2$.

First pass

$$\epsilon^{n+1/2}_{i,j} = \epsilon^{n}_{i,j} - \left\{\mathcal{K}_x\left[\epsilon^{n+1/2}_{(i+1),j} - 2\epsilon^{n+1/2}_{i,j} + \epsilon^{n+1/2}_{(i-1),j}\right] + \mathcal{K}_y\left[\epsilon^{n}_{i,(j+1)} - 2\epsilon^{n}_{i,j} - \epsilon^{n}_{i,(j-1)}\right]\right\} \tag{10.19a}$$

Second pass

$$\epsilon^{n+1/2}_{i,j} = \epsilon^{n+1/2}_{i,j} - \left\{\mathcal{K}_y\left[\epsilon^{n+1/2}_{(i+1),j} - 2\epsilon^{n+1/2}_{i,j} + \epsilon^{n+1/2}_{(i-1),j}\right] + \mathcal{K}_y\left[\epsilon^{n+1}_{i,(j+1)} - 2\epsilon^{n+1}_{i,j} - \epsilon^{n+1}_{i,(j-1)}\right]\right\} \tag{10.19b}$$

Moving on to three-dimensional diffusion we have $m = 3$, so that $Dt' = Dt/3$, and therefore $\mathcal{K}'_x = \frac{1}{3}\mathcal{K}_x$, $\mathcal{K}'_y = \frac{1}{3}\mathcal{K}_y$ and $\mathcal{K}'_z = \frac{1}{3}\mathcal{K}_z$.

We first treat the implicit x sweep

$$\begin{aligned}\epsilon^{n+1/3}_{i,j,k} &= \epsilon^{n}_{i,j,k} - \mathcal{K}_x\left\{\left[\epsilon^{n+1/3}_{(i+1),j,k} - 2\epsilon^{n+1/3}_{i,j,k} + \epsilon^{n+1/3}_{(i-1),j,k}\right] + \left[\epsilon^{n}_{(i+1),j,k} - 2\epsilon^{n}_{i,j,k} + \epsilon^{n}_{(i-1),j,k}\right]\right\}\\ &\quad - 2\,\mathcal{K}_z\left[\epsilon^{n}_{i,j,(k+1)} - 2\epsilon^{n}_{i,(j-1),k} - \epsilon^{n}_{i,j,(k-1)}\right] - 2\,\mathcal{K}_y\left[\epsilon^{n}_{i,(j+1),k} - 2\epsilon^{n}_{i,j,k} - \epsilon^{n}_{i,(j-1),k}\right]\end{aligned} \tag{10.20a}$$

then, the second sweep tackles the implicit y derivatives

$$\epsilon_{i,j,k}^{n+2/3} = \epsilon_{i,j,k}^{n+1/3} - \mathcal{K}_y \left\{ \left[\epsilon_{i,(j+1),k}^{n+2/3} - 2\epsilon_{i,j,k}^{n+2/3} + \epsilon_{i,(j-1),k}^{n+2/3} \right] + \left[\epsilon_{i,(j+1),k}^{n+1/3} - 2\epsilon_{i,j,k}^{n+1/3} + \epsilon_{i,(j-1),k}^{n-1/3} \right] \right\}$$
$$- 2\,\mathcal{K}_x \left[\epsilon_{(i+1),j,k}^{n+1/3} - 2\epsilon_{i,j,k}^{n+1/3} + \epsilon_{(i-1),j,k}^{n+1/3} \right] - 2\,\mathcal{K}_z \left[\epsilon_{i,j,(k+1)}^{n+1/3} - 2\epsilon_{i,j,k}^{n+1/3} + \epsilon_{i,j,(k-1)}^{n+1/3} \right] \quad (10.20b)$$

and finally the implicit z sweep

$$\epsilon_{i,j,k}^{n+1} = \epsilon_{i,j,k}^{n+2/3} - \mathcal{K}_z \left\{ \left[\epsilon_{i,j,(k+1)}^{n+2/3} - 2\epsilon_{i,j,k}^{n+2/3} + \epsilon_{i,j,(k-1)}^{n+2/3} \right] + \left[\epsilon_{i,j,(k+1)}^{n+1} - 2\epsilon_{i,j,k}^{n+1} - \epsilon_{i,j,(k-1)}^{n+1} \right] \right\}$$
$$- 2\,\mathcal{K}_x \left[\epsilon_{i+1,j,k}^{n+2/3} - 2\epsilon_{i,j,k}^{n+2/3} - \epsilon_{i-1,j,k}^{n+2/3} \right] - 2\,\mathcal{K}_z \left[\epsilon_{i,(j+1),k}^{n+2/3} - 2\epsilon_{i,j,k}^{n+2/3} - \epsilon_{i,(j-1),k}^{n+2/3} \right] \quad (10.20c)$$

The algorithm is based on the Crank–Nicolson scheme and is therefore second-order and unconditionally stable (Richtmyer and Morton, 1967, §8.9).

The close relationship with the equivalent equation for ADI (Eq. (10.18)) should be noted, although the two algorithmic forms slightly differ in lower order (see Richtmyer and Morton, 1967, p. 217). The two methods ADI and time splitting are therefore entirely equivalent. Which is normally selected for use is a matter of individual choice, governed by ease of programming.

Both time splitting and ADI are much more versatile than simply suitable for solving multidimensional diffusion problems. Indeed, both can be used to integrate a wide variety of problems involving parabolic and hyperbolic equations (Douglas and Dunn, 1964; Yanenko, 1971).

10.3 The Diffusion Matrix

The simple formulation of finite-difference equations for the diffusion matrix on an orthogonal grid, as demonstrated in Sections 10.2.2.1 and 10.2.2.2, has limitations in its scope. A more general method of establishing finite differencing in conservative systems, namely the *control volume method*, was investigated earlier in Section 1.11 and will be used here.

The diffusion of some intrinsic (real) quantity, ϵ, for example specific energy, is governed by an equation of the form

$$\frac{\partial \epsilon}{\partial t} = \nabla \cdot q \quad (10.21)$$

$$\frac{\partial \epsilon}{\partial t} = -\nabla .\, (K \nabla \epsilon) \quad (10.22)$$

and $\mathbf{q}$ is the diffusive flux, given by

$$\mathbf{q} = \boldsymbol{\kappa} \cdot \nabla \epsilon \quad (10.23)$$

and $\boldsymbol{\kappa}$ is the diffusivity tensor. We consider the behaviour of the above equation within an irreducible isolated enclosure, such that there is no flux through the walls:

$$\boldsymbol{\kappa} \cdot \nabla \epsilon|_n = 0 \quad (10.24)$$

where n is normal to the bounding surface S.

In general, the diffusion operator may be considered as the sum of symmetric, $\mathcal{D}_S$, and anti-symmetric, $\mathcal{D}_a$, components, such that $\mathcal{D}_S$ is self-adjoint, and $\mathcal{D}_A$ equal to minus its adjoint. For example, in the case of diffusion in a magnetic field:

$$\boldsymbol{\kappa} \cdot \nabla \epsilon = \kappa_{\|} \nabla_{\|} \epsilon + \kappa_{\perp} \nabla_{\perp} \epsilon + \boldsymbol{\kappa}_{\wedge} \wedge \nabla \epsilon \tag{10.25}$$

where the suffixes $\|$, $\perp$ and $\wedge$, respectively, have their usual meaning, parallel, perpendicular and cross product to the magnetic field. The first two terms are clearly symmetric and the third anti-symmetric. The diffusion operator is differential, i.e. if ϵ is constant throughout space, of value ϵ_0, then

$$\mathcal{D}\epsilon_0 = 0 \tag{10.26}$$

and ϵ is constant in time. The diffusion equation is conservative:

$$\frac{\partial}{\partial t}\int_V \epsilon \, \mathrm{d}V = -\int_S \mathrm{ds} \cdot \boldsymbol{\kappa} \cdot \nabla \epsilon = 0 \tag{10.27}$$

The properties of the diffusion operator are discussed in detail in ((Pert, 1981b) to which reference may be made for further details. It is a consequence of the second law of thermodynamics that the operator must be non-positive definite in the sense:

$$\int_V \epsilon \mathcal{D}(\epsilon) \mathrm{d}V < 0 \tag{10.28}$$

Furthermore, the familiar picture of diffusion in terms of a decay of extrema is contained in the *extremal condition*: namely that if a is independent of time

$$\max(\epsilon(t_0), \epsilon_0) \geq \max(\epsilon(t)) > \min(\epsilon(t)) > \min(\epsilon(t_0), \epsilon_0) \tag{10.29}$$

where $t > t_0$, and max and min are the largest and smallest values of the set denoted, respectively, and ϵ_0 are the boundary values of ϵ over the interval t_0 to t. This condition ensures the essential positivity of the physical quantity c during diffusion.

The extremal condition is stronger than the previous one, for an extremal operator has eigenvalues whose real part must be non-positive, and therefore if not defective must be non-positive definite. We may remark that one eigenvalue is necessarily zero since the operator is differential and corresponds to the steady state.

Thus far, our discussion of the properties of diffusion has been quite general in that arbitrary forms of the system parameters a and K are allowed. In a finite-difference scheme, the integration of Eq. (10.21) follows a step-by-step procedure, integrating from one time-step to the next, the system parameters a and $\boldsymbol{\kappa}$ being held constant during this time interval. In the following, we assume that such a procedure is to be adopted, and therefore investigate the integration properties under the assumption that a and $\boldsymbol{\kappa}$ are temporal constants, i.e. we are considering integration over one time-step. The constants a and $\boldsymbol{\kappa}$ will of course change their values from one time-step to next.

10.3.1 Properties of the Diffusion Matrix

If the diffusivity tensor κ is constant in time, then Eq. (10.21) may be represented by an algebraic equation.

Presently, we will assume that diffusion is a linear process, so that the diffusion operator $\mathcal{D}(\epsilon)$ is linear, which we may represent by a linear matrix $\mathcal{D} \to D_{ij}$.

$$\frac{\partial \epsilon_i}{\partial t} = \sum_j D_{ij} \epsilon_j \tag{10.30}$$

where ϵ_i is the value of ϵ at some point $\mathbf{r}_i$. To be a consistent representation the diffusion matrix must possess the same essential physical properties as the original operator. The spatial finite difference of the diffusion equation therefore takes a linear (matrix) form. D is the real diffusion matrix, a consistent representation of the diffusion operator $\mathcal{D}$. Since $\mathcal{D}$ is defined within an irreducible space, D is itself irreducible. The matrix D is found by applying the control volume method to a cell surrounding the point $\mathbf{r}$ and can be split into two components corresponding to the symmetry of parts of $\mathcal{D}$. Thus, if V is a real positive diagonal matrix whose component V_{ii} is the volume of the cell surrounding the point $\mathbf{r}_i$ and the matrix:

$$B = VD \tag{10.31}$$

The representation B of the symmetric parts of $\mathcal{D}$ must in the limits as the mesh is refined, satisfy

$$\int \delta \mathcal{D}_S \epsilon \mathrm{d}V = \sum_{i,j} \delta_i B_{i,j} \epsilon_j = \int \epsilon \mathcal{D}_s \delta \mathrm{d}V = \sum_{i,j} \epsilon_i B_{i,j} \delta_j \tag{10.32}$$

for all vectors δ and ϵ. Hence, if the representation is consistent, B must be symmetric to terms of lowest order of the mesh spacing. In practice, this implies that B must be symmetric. Similarly, the anti-symmetric term $\mathcal{D}_A$ must have an anti-symmetric representation B_A. The representation must be differential, i.e. if $\epsilon_i = \epsilon_j = \epsilon_0$ for all i and j then

$$\sum_j D_{ij}\, \epsilon_j = 0 \tag{10.33a}$$

and

$$\sum_j D_{ij} = \sum_j B_{ij} = 0 \tag{10.33b}$$

which ensures the existence of a steady-state solution, and that D is singular (Pert, 1981b). We shall call any matrix satisfying condition (10.33) a *differential matrix*. The matrix representation is conservative if

$$\frac{\mathrm{d}}{\mathrm{d}t} \sum_i A_{ii} \epsilon_i V_{ii} = O \tag{10.34}$$

for all ϵ_{ij}. Therefore,

$$\sum_i B_{ij} = 0 \tag{10.35}$$

This result is clearly equivalent to (13) if the matrix is purely symmetric (Kershaw, 1978). Any matrix satisfying Eq. (10.35) is called a conservative matrix.

The matrix equation (10.30) is clearly stable if the eigenvalues of $(A^{-1}D)$, namely λ, have real parts satisfying $\Re(A) \leq 0$. This condition is ensured if the matrix B is non- positive definite,[3] a condition equivalent to that for the operator $\mathcal{D}$. Let $\Lambda = \Phi + \imath\Pi$ be an eigenvector of $(A^{-1}D)$ with eigenvalue $\lambda = \phi + \imath\psi$. Consider:

$$\begin{aligned}(\Lambda^*, B\Lambda) =&(\Phi, B\Phi) + (\Psi, B, \Psi) + \imath\,[(\Phi, (B\Psi)) - (\Psi, (B\Phi))]\\ =&\lambda\,(\Lambda^*, (V\Lambda)\Lambda)\\ =&(\phi + \imath\psi)\,[(\Phi, (VA)\Phi) + (\Psi, (VA)\Psi)]\end{aligned} \tag{10.36}$$

But (VA) is a positive real diagonal matrix and is therefore positive definite. Hence,

$$\Re(\lambda) = \phi = [(\Phi, B\Phi) + (\Psi, B\Psi)]\,/\,[(\Phi, A\Phi) + (\Psi, A\Psi)] \tag{10.37}$$

and is non-positive if B is non-positive definite. If $(A^{-1}D)$ is not defective this is also a necessary condition.

We defer a discussion of the extremal properties until later, since its form is more appropriately derived in connection with temporal differencing. We may, however, remark that it may be proved by a similar analysis that Eq. (10.30) is extremal if and only if the matrix D is differential and non-negativity maintaining (i.e. if $\epsilon_0 \geq 0$, then $\epsilon(t) \geq 0$ for all $t \geq 0$), i.e. D is the negative of an **M**-matrix form.[4]

10.3.2 The Temporal Finite Difference

The set of linear equations (10.30) forms a stiff system. If the matrix $\mathbf{D}$ is defined physically we have seen that the eigenvalues of this matrix $\left([A]^{-1}\mathbf{D}\right)$ have non-positive real parts, so that the solutions are decaying. A general two-step form of Eq. (10.30) is

$$\epsilon = [\mathbf{A} - \theta\,\mathbf{D}\,\Delta t]^{-1}\,[\mathbf{A} + (1-\theta)\,\mathbf{D}\,\Delta t]\,\epsilon^0 \tag{10.38}$$

where ϵ^0 is the value of ϵ at time $(t - \Delta t)$, and θ is an implicitness parameter. Assuming the matrix $\left(\mathbf{A}^{-1}\mathbf{D}\right)$ is not defective, we may solve this equation formally in terms of the projections C_ℓ of the vector ϵ onto the eigenvectors Λ_ℓ of $\mathbf{A}^{-1}\mathbf{D}$

$$\mathbf{C}_\ell = \left(1 - \theta\,\lambda_\ell\,\Delta t\right)^{-1}\left[1 + (1-\theta)\,\lambda_\ell\,\Delta t\right]\mathbf{C}^0_\ell \tag{10.39}$$

For stability we require that the projections C_ℓ be bounded as the number of repetitions of Eq. (10.38) tends to infinity. Thus, if $\mathbf{D}$ is time independent, then

$$\theta \geq \frac{1}{2} + \frac{\phi_\ell}{\left(\phi_\ell^2 + \psi_\ell^2\right)\Delta t} \tag{10.40}$$

where ϕ_ℓ and ψ_ℓ are the real and imaginary parts of the eigenvalues λ_ℓ, respectively. This may be recognised as a generalisation of the usual stability condition for diffusion. Noting that the above inequality cannot be satisfied as $\Delta t \to 0$ if $\Re\lambda > 0$ (anti-diffusion) we conclude that the two-step system is unconditionally stable if and only if $\mathbf{A}^{-1}\mathbf{D}$ has only non-positive real part eigenvalues and $\frac{1}{2} \leq \theta \leq 1$.

It is clear that unless $\theta = 1$, the solution C_ℓ does not decay uniformly, as required by the exact solution, but may oscillate for large values of Δt. Such oscillations will be inhibited if the extremal condition can be applied.

---

3 Note these definitions of 'definiteness' involve a generalisation of the usual form to asymmetric matrices.
4 An **M**-matrix is by definition non-singular (Varga, 1962) but D is singular. We therefore use the above term to describe such matrices.

10.3.3 Extremal Forms

We define an extremal operator, $\mathcal{G}$, as one whose operation on a vector set ϵ^0 yields a set ϵ:

$$\epsilon = \mathcal{G}\epsilon^0 \tag{10.41}$$

whose values are extremal with respect to the set ϵ^0.

Theorem 10.1 A linear operator is extremal if and only if it is both differential (i.e. if ϵ^0 is a uniform set), then $\epsilon = \epsilon^0$ and non-negativity maintaining (i.e. if $\epsilon^0 \geq 0$ then $\epsilon \geq 0$).

Consider two vector sets ϵ^0 and δ^0 such that $\epsilon^0 \geq \delta^0$, since $\mathcal{G}$ is non-negativity maintaining, we have that

$$\epsilon = \mathcal{G}\epsilon^0 \geq \mathcal{G}\delta^0 = \delta$$

Let δ^0 be a uniform vector whose components are all equal to the smallest component of ϵ^0, $\delta^0 = \min\ \epsilon^0$, and a uniform vector equal to the largest value of ϵ^0, $\gamma^0 = \max\ \epsilon^0$. If $\mathcal{G}$ is a differential operator:

$$\gamma = \gamma^0 = \max\ \epsilon^0$$
$$\delta = \delta^0 = \min\ \epsilon^0$$

Therefore, since $\gamma^0 \geq \epsilon^0 \geq \delta^0$

$$\gamma \geq= \max\ \epsilon^0 \geq \epsilon > \min\ \epsilon^0 = \delta \tag{10.42}$$

i.e. the solution is extremal. It is clear that an operator, which is extremal, must be differential and non-negativity maintaining, and the above condition is proven.

For a two-step finite-difference representation (10.38), $\mathcal{G}$ has a matrix form

$$\mathbf{G} = \mathbf{F}^{-1}\mathbf{E} \tag{10.43}$$

when $\mathbf{E}$ is an explicit operation:

$$\mathbf{E} = \mathbf{I} + \left[(1-\theta)\,\mathbf{A}^{-1}\mathbf{D}\Delta t\right] \tag{10.44}$$

and $\mathbf{F}$ an implicit one:

$$\mathbf{F} = \mathbf{I} - \left[\theta\mathbf{A}^{-1}\mathbf{D}\Delta t\right] \tag{10.45}$$

where $\mathbf{I}$ is the identity matrix. It is readily shown that $\mathcal{G}$ is differential if $(\mathbf{E} - \mathbf{F}) = \mathbf{A}^{-1}\Delta t$ is a differential representation, i.e. if $\mathbf{D}$ satisfies Eq. (10.33). If $\mathcal{G}$ is non-negativity preserving and $\mathbf{E}$ is non-negative (i.e. if $\epsilon^0 \geq 0$), then $\mathbf{F}$ is monotone (Collatz, 1960) (i.e. if $\epsilon^0 \geq 0$ then $\epsilon = \mathbf{F}^{-1}\epsilon^0 \geq 0 = \mathbf{F}^{-1}\epsilon_0 \geq 0$); similarly, since $\mathbf{E}$ and $\mathbf{F}$ commute, if $\mathbf{F}$ is monotone then $\mathbf{E}$ must be non-negative. This reciprocity occurs naturally within the matrices $\mathbf{E}$ and $\mathbf{F}$, for if $\mathbf{E}$ is non-negative, all its components, $E_{ij} \geq 0$ and

$$D_{ij} \geq 0 \qquad i \neq j$$
$$D_{ii} = -\sum_{i \neq j} \geq -A_{ii}/\left[(1-\theta)\,\Delta t\right] \tag{10.46}$$

D is therefore the negative of an **M**-matrix form. When **D** has this form, **F** is an M- matrix and therefore monotone (Varga, 1962).

In general, **D** is a local matrix such that it has non-zero components between only restricted set of mesh points, which we call *neighbours*. If **D** is a local **M**-matrix form, then the operation **G** is locally extremal. Thus, if the set of values ϵ^0 and ϵ at the neighbours of i is ξ_i, then

$$\min\left(\xi_i, \epsilon_i^0\right) \leq \epsilon_i \leq \max\left(\xi_i, \epsilon_i^0\right) \tag{10.47}$$

a condition closely related to the more general maximum principle of the governing differential equation. This result follows since

$$F_{ii}\epsilon_i = \left(E_{ii}\epsilon_i^0 + \sum_{j\neq i} E_{ij} + |F_{ij}|\right) \qquad \leq \left\{E_{ii}\epsilon_i^0 + \sum_{j\neq i} E_{ij} + |F_{ij}|\right\} \max\left(\xi_i, \epsilon_i^0\right)$$

and

$$F_{ii} = E_{ii} + \sum_{i\neq j}\left(E_{ij} + |F_{ij}|\right) \tag{10.48}$$

An extremal operator is never divergent, and therefore the associated matrix $\left(\mathbf{A}^{-1}\mathbf{D}\right)$ has non-positive real part eigenvalues, i.e. it is derived from a Lyapunov semi-stable matrix. We may illustrate these results by the simple example of one-dimensional diffusion on a uniform Cartesian mesh with constant diffusivity, $\boldsymbol{K}$, for which the only non-zero elements are

$$\left(\mathbf{A}^{-1}\mathbf{D}\right)_{i,i+1} = \left(\mathbf{A}^{-1}\mathbf{D}\right)_{i,i-1} = \boldsymbol{K} \qquad \left(\mathbf{A}_{-1}\mathbf{D}\right)_{i,i} = -2\boldsymbol{K}. \tag{10.49}$$

An explicit calculation $\theta = 0$ is extremal only if $\boldsymbol{K}\Delta t \leq \frac{1}{2}$, which is also the well- known stability condition (Richtmyer and Morton, 1967). For a split-time-step calculation ($\theta = \frac{1}{2}$), the operation is unconditionally stable, Eq. (10.40), but extremal only if $\boldsymbol{K}\Delta t \leq 1$, reflecting the well-known 'overshoot' of this algorithm. Fully implicit schemes ($\theta = 1$) are unconditionally extremal and, of course, stable.

The **M**-matrix form of **D** is clearly a sufficient condition that **G** be extremal for a restricted range of the implicitness parameter θ. It is not, however, necessary, for example a monotone matrix is not necessarily an **M**-matrix (Collatz, 1960); indeed we shall consider later a case where the diffusion matrix, **D**, cannot be cast into **M**-matrix form but yields a monotone operation, **F**. In this case, the explicit form, **E**, is not positively maintaining, and the extremal form must be fully implicit ($\theta = 1$). The **M**-matrix form of **D** is however a very useful 'necessary' condition in a general sense providing a least restrictive condition on the form of **D** suitable for rapid test during repetitive calculation, whilst still remaining extremal.

Some general properties of the implicit operator, **F**, may be derived from the theorems in see (Pert, 1981b). In particular, it follows from Theorem 1 that if **D** is non-positive definite **F** is non-singular and a solution of Eq. (10.41) exists. Furthermore from (Pert, 1981b) it follows that as $\Delta t \to \infty$, the matrix **F** is monotone and converges to the steady state, provided **D** is differential and conservative. Thus, the fully implicit form is always extremal provided the time-step, Δt, is sufficiently large. The matrices **E** and **F** may be defined in a number of ways equivalent to Eqs. (10.44) and (10.45), of which the following is particularly useful:

$$\mathbf{E} = M + (1 - \theta)\,\mathbf{B}\Delta t \tag{10.44a}$$

and

$$\mathbf{F} = \mathbf{M} - (\theta \mathbf{B}\Delta t) \tag{10.45a}$$

where **B** is given by (10.31) and $M = AV$ is the diagonal cell 'mass' matrix.

10.3.4 Extremal Finite-Difference Representations in Orthogonal Two-Dimensional Geometries

We consider the finite-difference representation of the diffusion operation with both symmetric and anti-ti-symmetric terms in Cartesian (r, z) space with either planar or cylindrical symmetry. The diffusion flux may be written as

$$\mathbf{q} - \kappa_0 \nabla \epsilon - \kappa_1 \hat{\mathbf{h}} \wedge \nabla \epsilon \tag{10.50}$$

where $\hat{\mathbf{h}}$ is a unit vector perpendicular to the plane of the co-ordinates (in practice, usually taken to be along the magnetic field). The differential equation (10.21) takes the form

$$\begin{aligned} a\frac{\partial \epsilon}{\partial t} =& \frac{\partial}{\partial z}\left(\kappa_0 \frac{\partial \epsilon}{\partial z}\right) + \frac{1}{\tilde{r}}\frac{\partial}{\partial r}\left(\tilde{r}\kappa_0 \frac{\partial \epsilon}{\partial r}\right) \\ &+ \frac{1}{\tilde{r}}\left\{\frac{\partial}{\partial r}\left(\tilde{r}\kappa_1\right)\frac{\partial \epsilon}{\partial z} - \frac{\partial}{\partial z}\left(\tilde{r}\kappa_1\right)\frac{\partial \epsilon}{\partial r}\right\} \end{aligned} \tag{10.51}$$

where $\tilde{r} = \left\{\begin{matrix} 1 & \text{planar} \\ r & \text{cylindrical}\end{matrix}\right\}$ geometry.

The difference form may be found by a standard centred difference about the centre $(r_{ij}\ z_{ij})$ of the cell (i, j). Thus, for example

$$\frac{\partial}{\partial z}\left(\kappa_0 \frac{\partial \epsilon}{\partial z}\right)_{ij} = \left[\left(\kappa_0 \frac{\partial \epsilon}{\partial z}\right)_{i,(j-1/2)} - \left(\kappa_0 \frac{\partial \epsilon}{\partial z}\right)_{i,(j+1/2)}\right] \Big/ \left[z_{i,j+1/2} - z_{i,j-1/2}\right] \tag{10.52}$$

where ϵ_{ij} is the value of ϵ at the mesh point $(r_i j,\ z_{ij})$; the values at intermediate points being determined by a suitable interpolation.

Similarly,

$$\frac{\partial \epsilon}{\partial z} = \frac{\epsilon_{i,j+1/2} - \epsilon_{i,j-1/2}}{z_{i,j+1/2} - z_{i,j-1/2}} = \frac{1}{2}\,\frac{\epsilon_{i,j+1/2} - \epsilon_{i,j-1/2}}{z_{i,j+1} - z_{i,j-1}} \tag{10.53}$$

Since the volume of the cell $V_{i,j} = \tilde{r}_{i,j}\left(r_{i+1/2,j} - r_{i-1/2,j}\right)\left(z_{i,j+1/2} - z_{i,j-1/2}\right)$ we obtain the matrix **B** whose only non-zero components are:

$$\begin{aligned} B_{i,j,(i+1),j} =& \left(K_0\right)_{i+1/2,j}\tilde{r}_{i+1,j}\left(z_{i,j+1/2} - z_{i,j-1/2}\right) / \left(r_{i+1,j} - r_{i,j}\right) \\ &- \frac{1}{4}\left[\left(\tilde{r}\kappa_1\right)_{i,j+1} - \left(\tilde{r}\kappa_1\right)_{i,j-1}\right] \\ B_{(i,j),(i,j+1)} =& \left(K_0\right)_{i,j+1/2}\tilde{r}_{i,j}\left(r_{i+1/2,j} - r_{i-1/2,j}\right) / \left(r_{i,j} - r_{i,j-1}\right) \\ &+ \frac{1}{4}\left[\left(\tilde{r}\kappa_1\right)_{i+1,j} - \left(\tilde{r}\kappa_1\right)_{i-1,j}\right] \end{aligned}$$

$$
\begin{aligned}
B_{(i,j),(i-1,j)} &= \left(K_0\right)_{i-1/2,j} \tilde{r}_{i,j} \left(z_{(i,j+1/2)} - z_{(i,j-1/2)}\right) / \left(r_{i,j} - r_{i-1,j}\right) \\
&\quad + \frac{1}{4}\left[\left(\tilde{r}\kappa_1\right)_{i,j+1} - \left(\tilde{r}\kappa_1\right)_{i,j-1}\right] \\
B_{(i,j),(i,j+1)} &= \left(K_0\right)_{i,j1/2} \tilde{r}_{i,j} \left(r_{(i+1/2,j)} - r_{(i-1/2,j)}\right) / \left(z_{i,j} - z_{i-1,j}\right) \\
&\quad + \frac{1}{4}\left[\left(\tilde{r}\kappa_1\right)_{i,j+1} \left(\tilde{r}\kappa_1\right)_{i,j-1}\right] \\
B_{(i,j),(i,j)} &= -\left(z_{i,(j+1/2)} - z_{i,(j-1/2)}\right)\left[\tilde{r}_{(i+1),j}\left(K_0\right)_{(i+1/2),j} / \left(r_{(i+1),j} - r_{i,j}\right)\right. \\
&\quad \left. + \tilde{r}_{i,j}\left(K_0\right)_{(i-1/2),j} / \left(r_{i,j} - r_{(i-1),j}\right)\right] - \tilde{r}_{i,j}\left(r_{(i+1/2),j} - r_{(i-1/2),j}\right) \\
&\quad \times \left[\left(K_0\right)_{i,(j+1/2)} / \left(z_{i,(j+1)} - z_{i,j-1}\right) + \left(K_0\right)_{i,(j+1/2)} / \left(z_{i,j} - z_{i,(j-1)}\right)\right]
\end{aligned} \tag{10.54}
$$

The matrix **B** is clearly both differential (10.33) and conservative (10.35), and from the derivation also consistent. It is not, however, an **M**-matrix, and therefore not necessarily extremal. This defect is readily identified as associated with the anti-symmetric terms in Eq. (10.50).

Examination of these terms shows that they have a form similar to those describing advection, and suggests that they may be conveniently differenced by forward or backward differencing depending on the direction of the flux: thus,

$$
\frac{\partial \epsilon}{\partial z} = \begin{cases} \dfrac{\epsilon_{i,j+1} - \epsilon_{i,j}}{z_{i,j+1/2} - z_{i,j-1/2}} & \text{if } \dfrac{\partial}{\partial r}\left(\tilde{r}\kappa_1\right) > 0 \\[2ex] \dfrac{\epsilon_{i,j} - \epsilon_{i,j-1}}{z_{i,j+1/2} - z_{i,j-1/2}} & \text{otherwise} \end{cases} \tag{10.55}
$$

As before, this term is differential and conservative, and can also be shown to be consistent. The error terms are of lower order than those of the set (10.54). Against this, however, the modified set converges properly to a uniform state for large time-steps. It is therefore suggested that the set (10.54) be used unless one of the non-diagonal elements of **B** is negative, in which case the appropriate forward or backward differencing for that term be used (with care to ensure that conservation is maintained). This algorithm, in a slightly modified correctly symmetrised form, has been extensively used by the author to treat diffusion in magnetic field (Pert, 1981a). The advantages of the **M**-matrix form have been clearly demonstrated by the removal (with no ill effects) of a fix to zero check, which was necessary to avoid occasional negative values arising in the calculation using finite-difference form Eq. (10.54).

10.3.5 Weighted Temporal Integration

The solution to problems involving diffusion is given by the integral of Eq. (10.30). If we require unconditional stability in a simple two-step algorithm, this demands an implicit solution with parameter $\theta \geq \frac{1}{2}$ as demonstrated in Section 10.2. Unfortunately, as we saw the highest-order terms $\theta = \frac{1}{2}$ generate oscillatory solutions if the time step is large. The simple solution is only well behaved for large time-steps if fully implicit $\theta = 1$ (Section 10.2.1.1).

The question arises: Is it possible to improve the accuracy of the solution for arbitrary time-steps by suitably adjusting the implicitness parameter $\theta = 1$. Referring to Section 12.2.1 it is easily seen that the diffusion matrix **B** is identical in form to the collisional-radiative matrix **X** defined in Eq. (12.2).

For small time steps, Δt, the most accurate integration is given by a centred time difference, $\theta = \frac{1}{2}$, which is second-order accurate. On the other hand, for large time-steps, centred-difference integration with $\theta = 1$, converges uniformly to the final steady state and provides a more accurate representation for large time-steps. For intermediate time-steps, there exists a set of weights W_{ij} such that

$$\epsilon = \left[\mathbf{A} = \mathbf{D}_1 \Delta t\right]^{-1} \left[\mathbf{A} + \mathbf{D}_2 \Delta t\right] \epsilon^0 \tag{10.56}$$

where $\mathbf{D}_1$ and $\mathbf{D}_2$ are matrices such that

$$\begin{aligned} D_{1(ij)} &= W_{ij} D_{ij} \quad \text{for} \quad i \neq j \\ \mathbf{D}_1 + \mathbf{D}_2 &= \mathbf{D} \end{aligned} \tag{10.57}$$

where the weight W_{ij} satisfies $\frac{1}{2} \leq W_{ij} \leq 1$. In principle, the weights could be chosen so as to coerce the finite-difference form to the exact integration. Such an approach is extremely difficult, and involves an a priori knowledge of the solution before the weights are determined. We may, however, note that the term

$$\left[\left(\mathbf{VDV}^{-1}\mathbf{A}^{-1}\right)_{ij} + \left(\mathbf{VDV}^{-1}\mathbf{A}^{-1}\right)_{ji}\right]$$

essentially determines the relaxation rate between the elements ϵ_i and ϵ_j. The weight **W** determines which time level of ϵ_j to use in calculating ϵ_i. Clearly, if ϵ_i and ϵ_j rapidly relax, a near steady-state distribution will be maintained between them, i.e. $W_{ij} \approx 1$. On the other hand, if the relaxation is weak, the appropriate weight is $W_{ij} \approx \frac{1}{2}$, giving highest-order accuracy. Thus, the weights are assumed to take the form

$$W_{ij} = F(\lambda_{ij})$$

where

$$\begin{aligned} \lambda_{ij} &= \left\{ \left(\left[VDV^{-1}A^{-1}\right]_{ij} + \left[VDV^{-1}A^{-1}\right]_{ji} \right) \Delta t \right\} \\ &= \left\{ \left(\left[DA^{-1}\right]_{ij} + \left[DA^{-1}\right]_{ji} \right) \Delta t \right\} \ \text{if } (\mathbf{VD}) = \mathbf{B} \text{ is symmetric} \end{aligned} \tag{10.58}$$

and

$$\begin{aligned} F(\lambda) &= \frac{1}{2} \qquad \text{if } \lambda = 0 \\ &= 1 \qquad \lambda \to \infty \end{aligned} \tag{10.59}$$

such a form is obtained for a two element system for which

$$F(\lambda) = l/[1 - \exp(-\lambda)] - 1/\lambda \tag{10.60}$$

coerces the finite-difference form to the exact analytic solution.

In the more general case, we may require $F(\lambda)$ to be of a form which maintains the physical properties of the solution. For example, if **D** can be differenced in an extremal form, we may require the solution of Eq. (10.55) to be extremal. Thus, in the case both $\mathbf{D}_1$ and $\mathbf{D}_2$ take an **M**-matrix form

$$\left|\left(A^{-1}D_2\right)_{ij}\right| \Delta t \leq 1 \tag{10.61}$$

The element D_{1ij} is evaluated from the condition that Eq. (10.55) be conservative, if the weights W are independent, i.e.

$$\left(VD_2\right)_{ii} = -\sum_{j \neq i} \left(VD_2\right)_{ij} \tag{10.62}$$

Examination of the exact two element weights (10.58) shows that as $\lambda \to \infty$, the weight $W \to 1 - l/\lambda$ and suggests that a similar form

$$W_{ij} = \max\left\{\frac{1}{2}, \left|1 - \frac{1}{\alpha_{ij}\lambda_{ij}}\right|\right\} \tag{10.63}$$

The temporal integration of diffusion on a two-dimensional regular grid is accomplished by evaluating the diffusion equation in the form of Eq. (10.51). Since the diffusion matrix D is sparse, this is reasonably easily accomplished using a sparse matrix solver such as ICCG (Kershaw, 1978). The resulting routine is comparable in speed with ADI or split-time-step.

10.3.6 Multidimensional, Non-orthogonal Grid

Turning now to the general case where the values are assigned on a non-orthogonal grid, as in the case of a 2d Lagrangian fluid code (Section 4.3). The establishment of the finite-difference form is most easily accomplished with the equations for the transformation onto a regular mesh used in the Jacobian method as in Eq. (4.39). When this approach is carried out the identification of the finite-difference forms follow the same methodology using the control volume technique as in the orthogonal case (Section 10.3.4). The resulting equations are differential and conservative, and may be shown to be extremal. The inclusion of weighted implicitness follows the same line as before. The only change is the need to treat the additional corner differences. Since this involves much messy algebra, but no new methodology, it is convenient to refer to the paper (Pert, 1981b) where these derivations are given.

Appendix 10.A Thomas Algorithm – Tri-diagonal Matrix Equation Solver

We wish to solve the tri-diagonal equation in n unknowns X_i

$$A_i X_{i+1} + B_i X_i + C_i X_{i-1} = D_i \tag{10.A.1}$$

with boundary conditions $A_1 =$ and $C_n = 0$.

We use Gaussian elimination to eliminate X_{i+1} and obtain on the first sweep through the mesh

$$X_i = \begin{cases} F_n & i = 1 \\ F(i) - E_i F_{i+1} & i = 2,3,4,\ldots,n-1 \end{cases} \tag{10.A.2}$$

$$E_i = \begin{cases} C_i \,/ B_i & i = 1 \\ C_i \,/\, \left(B_i - C_i E_{i-1}\right) & i = 2,3,4,\ldots,n-1 \end{cases} \tag{10.A.3}$$

and the second sweep

$$F_i = \begin{cases} D_i \,/ B_i & i = 1 \\ \left(D_i - A_i E_{i1}\right) \,/\, \left(A_i - C_i E_{i-1}\right) & i = 2,3,4,\ldots,n-1 \end{cases} \tag{10.A.4}$$

References

Collatz, L. 1960. *The Numerical Treatment of Differential Equations*. Berlin: Springer-Verlag.

Douglas, Jr, J. 1955. "On the numerical integration of $\frac{\partial^2 u}{\partial x^2} + \frac{\partial^2 u}{\partial y^2} = \frac{\partial u}{\partial t}$ by implicit means." *J Soc Ind Appl Math* 3:42–65.

Douglas, Jr, J. 1962. "Alternating direction methods for three space variables." *Numer Math* 4:41–63.

Douglas, Jr, J and J Dunn. 1964. "A general formulation of alternating direction methods: Part I. Parabolic and hyperbolic problems." *Numer Math* 6:428–453.

Friedman, A. 2008. *Partial Differential Equations of the Parabolic Type*. New York: Dover Books.

Kershaw, D S. 1978. "The incomplete Cholesky-conjugate gradient method for the iterative solution of systems of linear equations." *J Comput Phys* 26:43–65.

Kershaw, D S. 1981. "Differencing of the diffusion equation in Lagrangian hydrodynamics." *J Comput Phys* 39:375–395.

Meijerink, J A and H A van der Vorst. 1977. "An iterative solution method for linear systems of which the coefficient matrix is a symmetric M-matrix." *Math Comput* 31:148–162.

Morse, P M and H Feshbach. 1953. *Methods of Theoretical Physics*. New York: McGraw-Hill.

Peaceman, D W and H H Rachford, Jr. 1955. "The numerical solution of parabolic and elliptic differential equations." *J Soc Ind Appl Math* 3:28–41.

Pert, G J. 1981a. "Algorithms for the self-consistent generation of magnetic fields in plasmas." *J Comput Phys* 43:111–163.

Pert, G J. 1981b. "Physical constraints in numerical calculations of diffusion." *J Comput Phys* 39:20–52.

Pert, G J. 2021. *Introduction to Plasma Physics*. London: Wiley.

Richtmyer, R D and K W Morton. 1967. *Difference Methods for Initial Value Problems*. New York: Interscience.

Stoer, J and R Bulirsch. 1991. *introduction to Numerical Analysis*. Second ed. New York: Springer.

Varga, R S. 1962. *Matrix Iterative Analysis*. Englewood Cliffs, NJ: Prentice-Hall.

Yanenko, N N. 1971. *The Method of Fractional Steps: The Solution of Problems of Mathematical Physics in Several Variables*. Berlin: Springer-Verlag.

11

Ion–Electron Equilibration

The collisional exchange of energy between ions and electrons is an exponential process leading to the eventual temperature equalisation of the two groups of particles with a rate determined by the plasma thermo-dynamic state. As such within the context of plasma dynamics, it is determined by a stiff pair of equations, and best treated by a quasi-analytic method. Ion–electron collisional equilibration is described by the familiar Spitzer (1956) formula for energy exchange between the particles

$$\frac{d\epsilon_i}{dt} = -\frac{d\epsilon_e}{dt} = -c_{vi}\frac{(T_i - T_e)}{\tau_{eq}} = -\frac{c_{vi}{}^{-1}\left(\epsilon_i - \left[c_{vi}/c_{ve}\right]\epsilon_e\right)}{\tau_{eq}} \tag{11.1}$$

where c_{vi} is the ion-specific heat per unit plasma mass and c_{ve} the electron-specific heat per unit plasma mass; neglecting ionisation, etc. the ratio of specific heats $c_{ve}/c_{vi} = Z$. Introducing the effective temperature difference $\Delta = T_i - T_e = (1/c_{vi})\left[\epsilon_i - \left(c_{vi}/c_{ve}\right)\epsilon_e\right]$, and substituting for ϵ_1 and ϵ_e we obtain

$$\frac{d\epsilon_i}{dt} = -\frac{d\epsilon_e}{dt} \qquad\qquad \frac{d\Delta}{dt} = -\frac{\Delta}{\tau_{eq}} \tag{11.2}$$

which have the elementary integrals

$$\epsilon_i + \epsilon_e = \epsilon_{i0} + \epsilon_{e0} \qquad\qquad \Delta = -\Delta_0 \exp\left\{-t/\tau'_{eq}\right\} \tag{11.3}$$

where $\tau'_{eq} = \tau_{eq}/\left(1 + c_{ve}/c_{vi}\right)$ and ϵ_{i0}, ϵ_{e0} and Δ_0 are the initial values at the time $t = 0$. Since equilibration is collisional, it is an irreversible process. Thus, if the direction of time is reversed the solution becomes unstable.

Let us consider a direct finite difference of Eq. (11.2), namely

$$\frac{\epsilon_i^{n+1} - \epsilon_i^n}{Dt} = -\frac{\epsilon_e^{n+1} - \epsilon_e^n}{Dt} = \frac{\theta\epsilon_i^{n+1} - (c_{ve}/c_{vi})(1-\theta)\left[\epsilon_i^n - (c_{ve}/c_{vi})\right]\epsilon_i^n - (c_{vi}/c_{ve})\epsilon_e^n}{\tau_{eq}} \tag{11.4}$$

or alternatively in terms of the temperature difference

$$\epsilon_i^{n+1} + \epsilon_e^{n+1} = \epsilon_i^n + \epsilon_e^n \qquad \text{and} \qquad \frac{\Delta^{n+1} - \Delta^n}{Dt} = -\frac{\left(\theta\Delta^{(n+1)} + (1-\theta)\Delta^n\right)}{\tau'_{eq}} \tag{11.5}$$

Modelling and Simulation in Plasma Physics for Physicists and Mathematicians, First Edition. Geoffrey J. Pert.

which guarantees energy conservation and the correct limit $\Delta \to 0$ as $Dt \to \infty$. The last of these equations is not satisfactory as it may be re-written as

$$\Delta^{n+1} = \Delta^n \frac{\left[1 - (1-\theta)Dt/\tau'_{eq}\right]}{\left[1 + \theta Dt/\tau'_{eq}\right]} \tag{11.6}$$

- Thus, the fully explicit case $\theta = 0$ is unstable if $t/\tau'_{eq} \geq 2$ and is ill-behaved (oscillates) if $t/\tau'_{eq} \geq 1$, which is not satisfactory for a general purpose routine suitable for use with arbitrary time steps.
- Similarly, the solution for a centred time-step ($\theta = \frac{1}{2}$) is stable but oscillates if $t/\tau'_{eq} \geq 2$ and is therefore also unsatisfactory.
- Only the fully implicit form $\theta = 1$ fulfils the necessary conditions unconditionally, but is inaccurate for large $Dt \to \infty$, although it does achieve equilibrium in this limit.

This behaviour is typical of the integration of a pair of stiff equations, such as coupled rate equations, by a simple two-step scheme.

However, the analytic solution Eq. (11.3) allows their integration directly, provided any variation of the rate parameter, in this case τ_{eq}, can be accurately taken into account. A satisfactory way of simply accomplishing this is to use a predictor–corrector scheme, i.e. perform a predictor step with the initial values τ_{eq0} and to obtain a predicted value at the end of the step to form Δ_P, which is then used to form $\Delta_C = \frac{1}{2}(\Delta_0 + \Delta_P)$ to give τ_{eqC} in the corrector.

Reference

Spitzer, Jr, L. 1956. *Physics of Fully Ionized Gases*. New York: Interscience.

12

Ionisation–Recombination Models

12.1 Introduction

The interest in generating laser action in the soft X-ray region of the electromagnetic spectrum in the 1970s led to the need to develop simple computer models which could be used to model the history of the plasma state self-consistently with the plasma dynamics. As computer power has increased, these codes, which were designed to use the most powerful mainframes of that period, can now run comfortably on today's personal computers. However, the efficient methods developed then can still be advantageous. The underlying philosophy of their design was to obtain the maximum speed with a minimum memory requirement, yet generate results of sufficient accuracy to guide the design of experiments. To that aim, use is made whenever possible of possible shortcuts allowed by the underlying physical processes.

We need to investigate the time development of ionisation in plasma as the system is heated and undergoes hydrodynamic change. Under most circumstances, this is a result of collisional ionisation induced by the high temperature of the heated gases, as electrons are stripped off or added to the core of the atomic system. In addition, the bound electrons are excited or de-excited by collisions with hot electrons. The total number of electronic states is very large and, in principle, transitions between a large number of them may be permitted. Calculations involving all possible systems, which may be allowed in the steady state (*collisional-radiative model*), may become intractable over many time time-steps in a multi-celled code.

To avoid the problems associated with convective transport of particles of different ionisation, Lagrangian fluid models are preferred with the same particles remaining within a particular cell; thereby avoiding the complications associated with the transport of different ions between calculational elements. Because of its simplicity, the hybrid model is especially suited to tasks requiring both minimum store and giving good calculational speed.

Fortunately, many plasmas are sufficiently tenuous to be optically thin; the nature of the ionisation–recombination process allows us to reduce this to a combination of two basic processes in many situations: ionisation and excitation. If the specific ion density is sufficiently high that the optical depth of the transition is not small, the effect is an effective reduction of the transition probability by re-absorption. This effective reduction may be accounted for by a *trapping factor*. This quasi-analytic approximation avoids the need for a

Modelling and Simulation in Plasma Physics for Physicists and Mathematicians, First Edition. Geoffrey J. Pert.

treatment of the radiation transfer in the plasma. Ionisation is a much more probable transition than excitation to a bound state. In consequence, ionisation tends to take the 'atom' from one ionisation stage to the next in a single jump, the population of the excited states being small, and taking account of the principle of detailed balance, the same must apply to recombination. The ionisation–recombination balance between the different atomic ionisation stages using averaged ionisation and recombination rates; the intermediate excited states being treated implicitly with the averaged ionisation–recombination rates; the validity of this simplification was investigated in some detail for hydrogenic ions, see Pert (1990).

The ion populations may be described by a single value conveniently thought of as the 'ground state' value, but actually it represents the total population in levels below the 'collision limit' or 'bottleneck' (Pert, 1990, see). We may therefore treat transitions between one ion stage and the next higher as distinct processes due to excitation/de-excitation between the bound states and the ionisation continuum of neighbouring stages. This will limit the available energy transfer per ion to some value of available electron energy determined by the heat deposited per time step Δ_e' and a fraction of the thermal energy kT_e. The parameter $\alpha \sim 1$ is set by the user. We can write

$$\Delta E \sim Z \left(\Delta_e' + \alpha \frac{3}{2} kT_e\right) \tag{12.1}$$

12.2 Collisional-Radiative Model

Interest in the feasibility of generating population inversions in highly ionised species led to the need to study ionisation processes in developing plasma. To this end, simple algorithms will be described to allow the inclusion of time-dependent ionisation and recombination within a multi-celled fluid code. In this chapter, we further develop this work by considering algorithms for the more general collisional-radiative equations.

The collisional-radiative equations (12.2) form a general group of rate equations with transitions between arbitrarily designated levels. Although their physical properties, for example conservatism, equilibration and positivism, are well known (Bates, Kingston and McWhirter, 1962; McWhirter and Hearn, 1963), these have not been generally proven in time-varying plasmas. We therefore, in this chapter, initially establish these properties by considering the behaviour of the equations with a set of arbitrary rates. These results are used to define classes of numerical operators, with properties which mimic the exact equations.

The differential equations associated with the time-dependent collisional-radiative equations are by their nature stiff (see Supplement M.2) and involve the calculation of $n(n-1)$ different rate coefficients dependent in a complicated fashion on the state variables (principally temperature) of the gas. These should clearly be calculated as infrequently as possible, i.e. on the characteristic time scale of the hydrodynamics of the system. However, in general, the characteristic time scale associated with the rates may be much shorter than that of the hydrodynamics. The algorithm should therefore be capable of stable operation in a fully stiff mode. Accurate multi-step methods for treating this problem are available (Gear, 1971, see), but these involve the storage of data from several time steps, typically $8n$ elements for each fluid cell. As this rapidly becomes prohibitively large, we have concentrated on two-step schemes, which have the advantage of A-stability (Dahlquist, 1963).

Two methods are considered: an exact method involving the eigenvector decomposition and a weighted finite-difference approach. Since the latter involves approximately one-sixth as many multiplications as the former, it is to be preferred unless accuracy is at a premium. In this context, it should be remembered that the rates themselves are only inaccurately known, so it is unlikely that a method of high accuracy can be justified in practice.

12.2.1 The Collisional-Radiative Equations

In general, the fractional population of a state, i, of some stage of ionisation determined by an equation of the form (Bates, Kingston and McWhirter, 1962; McWhirter and Hearn, 1963) is

$$\mathrm{d}q_i/\mathrm{d}t = \sum_{j \neq i} \left(X_{i,j}\, q_j - X_{j,i}\, q_i\right) \tag{12.2}$$

where $\mathbf{Q}$ is the population vector and $\mathbf{X}$ the excitation rate matrix. The sum is taken over all states of that particular 'atomic' species with direct transitions between the states i and j. In principle, this sum is extended to include all the states of all the ionic stages of that 'atom' by defining the excitation matrix, $X_{i,j}$, which gives the transition probability per unit time from state j to state i. The relative population of state q_i is such that the density of particles in state i is $n_i = q_i\, n_a$, where n_a is the total atomic density of 'atoms' per unit volume. Since the atom must exist in some, but only one, state, we conclude that $\sum_i q_i = 1$, a condition easily seen to be satisfied by Eq. (12.2). Ionisation and recombination between ionisation stages Z and $Z + 1$ is included by allowing j to be a state of stage Z and i one in the next stage $Z + 1$, or vice versa. This defines an appropriate sparsity pattern of the matrix $\mathbf{X}$. The matrix, X, for a given element must be irreducible, for by the principle of detailed balance, every transition $X_{i,j}$ must be accompanied by its inverse, $X_{j,i}$, and furthermore, we may assume that there exists at least one transition into every state, albeit with zero probability.

The rate equation may be written in terms of the transition rate as

$$\frac{\mathrm{d}q_i}{\mathrm{d}t} = \sum_{j=1}^{n} R_{i,j}\, q_j \tag{12.3}$$

where the transition matrix $R_{i,j}$ is given by

$$R_{i,i} = -\sum_{j=1} X_{j,i} \qquad \text{and} \qquad R_{i,j} = X_{i,j} \quad j \neq i \tag{12.4}$$

Clearly, the negative of the diagonal terms of $\mathbf{R}$ represents the decay loss of the population of the state i by transitions to other states.

The sparsity pattern imposed on the transition matrix, $\mathbf{X}$, will depend, in practice, on the energy level structure associated with each ionic stage. This may range from a tri-diagonal form (ground states only) to a completely dense matrix (one ionisation stage alone). The rate coefficients, $\mathbf{X}$, are by definition real, non-negative and bounded, so $\mathbf{R}$ is an M-matrix form (Collatz, 1960; Varga, 1962). Such matrices have some important properties; in particular, the eigenvalues have non-negative real parts.

In the presence of chemical interactions, the transition matrix for different atomic species is separable, and we may consider the behaviour of each element alone. We shall, therefore, henceforward consider one element alone.

The set of Eqs. (12.2) has some important properties, which are necessary for their physical interpretation. In the past, these have been assumed on physical grounds; however, since the rates, $X_{i,j}$, are not exactly known, it is important to show that these properties follow as a direct consequence of the above form of the equations and do not depend on the exact values of the rates.

a) $\sum q_i$ is conserved, for clearly

$$\sum_i \mathrm{d}q_i/\mathrm{d}t = 0 \tag{12.5}$$

In conformity with the usual definition we put

$$\sum_i q_i = 1 \tag{12.6}$$

and consistent with the equation of conservation there is therefore always one eigenvalue of the rate matrix **R** with zero eigenvalues and $q_i = 1$ eigenfunction, which is therefore singular. This state must not be confused with those corresponding to the final equilibrium state.

b) The equations are positivity preserving, i.e. if the vector set $q^0 \geq 0$ at an initial time, t_0, then the set $q \geq 0$, at a later time, t. The formal solution of Eq. (12.2) maybe obtained by integration by parts (Appendix 12.A)

$$q_i(t) = \underbrace{\left\{ q_i^0 \exp\left[-\int_{t_0}^{t} \sum_{j\neq i} X_{j,i}(t')\,\mathrm{d}t' \right] \right\}}_{\text{Decayed initial population of } i} + \left\{ \underbrace{\sum_{j\neq i} \int_{t_0}^{t} X_{i,j}(t')\,q_j(t')\,\mathrm{d}t}_{\text{From } j \text{ to } i \text{ from } t_0 \text{ to } t} \; \underbrace{\exp\left[-\int_{t'}^{t} \sum_{j\neq i} X_{j,i}(t'')\,\mathrm{d}t'' \right]}_{\text{Decay of } i \text{ from } t' \text{ to } t} \mathrm{d}t' \right\} \tag{12.7}$$

This equation may be readily understood. The first term represents the decay of the initial population of state i as a result of direct transitions from state i to state j. The second term is the population growth of state i from arbitrary state j and its subsequent decay into other arbitrary state k. The coefficients $X_{i,j}$ are all non-negative and bounded. Therefore, if $q_j(t') > 0$ for all $j \neq i$ and $t_0 < t' < t$, then $q_i(t) \geq 0$. Hence, by induction, since $q_0 \geq 0$ and for real arguments $\exp(\alpha) \geq 0$, it follows that $q_i > 0$ for all i. Since the matrix X is irreducible we may extend this result to conclude that $q(t) > 0$ for all finite time intervals $t > t_0$.

c) The equations are convergent. It follows from the exponential decay of the terms in Eq. (12.7) that as $t \to \infty$ the excitation and decay terms in Eq. (12.2) come into balance, where $X_{i,j}\,q_j = X_{j,i}\,q_i$ for all i, j are in equilibrium. The equations are therefore stable. The exact solution of the set of differential equations can be obtained by reducing **R** to its Jordan normal form.

The resultant equations are stable if and only if the non-zero eigenvalues of **R** have positive real parts. The nature of the eigenvalues of **R** is readily established by consideration of its transpose for

$$R^T_{i,i} = -\sum_{j\neq i} R^T_{j,i} \tag{12.8}$$

which has the same eigenvalues as **R**. The matrix $\mathbf{R}^T$ has non-zero positive real parts (Taussky, 1949, Theorem VII). The equations are therefore stable and, if the rates are constant, and converge to an equilibrium state.

The matrix $\mathbf{R}^T$ is singular, **R** therefore has a zero determinant det $\mathbf{R} = 0$. Consequently, due to the singularity, one eigenvalue is zero and corresponds to a constant total population q_i

$$\sum_{j\neq i} q_i = \text{const} = \sum_{j\neq i}^{n} q_i^0 \tag{12.9}$$

where q_i^0 is the initial population at $t = 0$. The corresponding eigenfunction is somewhat arbitrarily determined by the value of this sum, for example

$$\begin{pmatrix} 1 \\ 1 \\ \vdots \\ 1 \end{pmatrix}$$

usually taken as unity, namely the steady state, the result of conservation.

We may remark on the relation of this condition to those of a) and b), for in the set, q, the term q_i is bounded by $\sum_i q_i$, which is constant. This does not ensure convergence to a steady state since purely oscillatory solutions may occur. These are not permitted by the convergence condition, which ensures that any oscillatory term decays to the final steady state.

d) The equilibrium state is unique. The equilibrium state is described by

$$\mathrm{d}\mathbf{Q}/\mathrm{d}t = \mathbf{R} \cdot \mathbf{Q} = 0 \tag{12.10}$$

i.e. by condition $\mathbf{R} = 0$. However, since $\sum q_i = \text{const}$, one row of the array may be removed from Eq. (12.2) to remove the singularity before any direct numerical calculation of the equilibrium state is attempted. In principle, the equilibrium state may not be unique, and the final equilibrium condition may depend on the path by which it is approached. Such phenomena are found in phase equilibria, for example. In this case, it follows from Eq. (12.7) that the path to equilibrium is unique for single atomic species.

The populations of the equilibrium state q_i^s satisfy the condition established by the alternative form of the collisional-radiative equations (12.2) by the principle of detailed balance

$$q_i^s/q_j^s = X_{i,j}/X_{j,i} \qquad \text{and} \qquad \sum_i q_i = 1 \tag{12.11}$$

Finally, $R_{i,i} \leq 0$ and $R_{i,j} \geq 0$

Let $\mathbf{Q}'$ be an eigenvector of **R** with non-zero eigenvalue, and let Δ_i be the departure of the state i from this equilibrium:

$$\Delta_i = q_i - q_i^s : \qquad \sum_{i=1}^{n} \Delta_i = 0 \tag{12.12}$$

and

$$\mathrm{d}\Delta_i/\mathrm{d}t = \sum_j R_{i,j}\,\Delta_j \tag{12.13}$$

We may reduce the order of the equation from n to $(n-1)$ by eliminating one component (say q_k) to give

$$\mathbf{D} = \mathrm{d}\Delta_i/\mathrm{d}t = \sum_{j=}^{n} \mathbf{S}_{i,j}\,\Delta_j \tag{12.14}$$

with $\Delta_i = q_i + q_t - \sum_{j=1}^{k-1} + \sum_{j=k+1}^{n-1} q_j$. Clearly, S and R have the same eigenvalues and eigenvectors, except that the eigenvector due to q_s, the singularity resulting from conservation, is removed from the matrix S. The components of $\mathbf{S}$ are

$$S_{i,j} = X_{i,n} - X_{i,j} \quad i \neq j \tag{12.15a}$$

$$S_{i,i} = X_{i,n} + \sum_{j\neq i} X_{j,i} \tag{12.15b}$$

Since at least one pair $(X_{i,n}, X_{n,i})$ must be non-zero, the matrix $\mathbf{S}$ is non-singular (Eq. 12.C). The matrix, $\mathbf{R}$, therefore has only one zero eigenvalue and the equilibrium state is unique. Since the matrix $\mathbf{X}$ is irreducible and non-negative, it follows that no component of the equilibrium vector q_s is zero.

e) The equilibrium state is physical, for if one component of q^s (say q_n^s) is positive, then since R is an irreducible M-matrix and therefore monotone (Collatz, 1960; Varga, 1962), all the remaining $(n-1)$ components are solutions of the $(n-1)$ equations:

$$\sum_{j=1}^{n-1} R_{i,j}\, q_i/q_i^s = R_{in}\, q_n^s/q_n \tag{12.16}$$

and therefore positive.

$$\mathbf{X}\cdot\boldsymbol{\Delta} = 0 = \mathbf{X}\cdot\boldsymbol{\Delta}^s \tag{12.17}$$

and hence

$$\frac{\mathrm{d}\boldsymbol{\Delta}^T}{\mathrm{d}t} = \mathbf{X}\cdot\boldsymbol{\Delta}^s \tag{12.18}$$

The solution of Eq. (12.2) converges uniformly to the steady state when the rates, X, are constant, i.e. the maximum norm of the fractional deviation vector (Q_i/q_i^s) at a time, t, satisfies

$$\|Q_i\, q_i^s\| < \|Q_i^0\, q_i^s\| \tag{12.19}$$

where Q_0 is the value of Q at an earlier time, $t_0 < t$, and is assumed to be non-zero. We define the vector

$$Q_i = q_i/q_i^s \tag{12.20}$$

and Eq. (12.2) becomes

$$\mathrm{d}Q_i/\mathrm{d}t = \sum_{j=1}^{n} R'_{i,j} q_j \tag{12.21}$$

where $R'_{i,j}$, $= (q_i^s/q_j)R_{i,j}$. If L is a uniform vector, i.e. one whose components are equal, then clearly

$$\sum R'_{i,j} L_j = 0 \tag{12.22}$$

for the case $Q_1 = Q_2 = \cdots = Q_n = L$, corresponds to the steady state. Let L be the smallest component of Q^0 at time, t_0, then

$$(Q - L) > 0, \qquad (Q_0 - L) \geq 0 : \qquad t > t_0 \tag{12.23}$$

since Eq. (12.21) must be positivity preserving. Hence, since L is constant in time,

$$\min(Q) > \min\, Q_0 \tag{12.24}$$

Applying this result to the set $(-Q)$ we obtain

$$\max\, Q < \max\, Q_0 \tag{12.25}$$

Since $\sum_i q_i \sum_1 q_i^0$, it follows that $\max(Q) \geq 1$ and $\min(Q) \leq 1$. Since $Q_i = 1 + Q_i/q_i^0$; the result, Eq. (12.19), follows.

That this solution converges onto the equilibrium state follows from the exact solution (Eq. (12.7)) in terms of the eigenvector projection of q, which decays exponentially.

f) The excitation/de-excitation rates between levels may vary over large ranges in magnitude, and therefore the eigenvalues of **X**, the set of collisional-radiative equations, are stiff M.2. This will give rise to serious problems in repetitive calculations.

Since this method involves calculating the full set of eigenvalues and eigenfunctions at every time, it is lengthy and expensive, and consequently falls outside the remit of our survey.

Finite-difference methods although limited by the stiff nature of the system provide a more realistic approach.

12.2.2 Numerical Solutions

In general, we wish to evaluate the vector set q at some time, t, given the values of the set, q_0, at an earlier time, t_0, by means of a linear operation,[1]

$$q(t) = G(q^0) \tag{12.26}$$

In view of the nature of the collisional-radiative equation, the operation $G(q^0)$ is likely to be one of matrix multiplication. We have identified several important properties which we may require preserved by the operator G, namely:

a) G is conservative if

$$\sum_i q_i = \sum_i q_i^0 \tag{12.27}$$

b) G is positivity maintaining if for any non-negative initial stage,

$$q \geq 0 \quad \text{if } q_i^0 \geq 0 \tag{12.28}$$

1 If G is linear, then $G(aq_1, aq_2, \ldots) = aG(q_1, q_2, \ldots)$, where a is a constant multiplier of the complete set, q. We note that since the average ionisation $\overline{Z} = \sum_i Z_i q_i / \sum_i q_i$, where Z_i is the ionic charge of the state i, we may include variation of the rates due to ionisation-induced changes of electron density, G, remaining linear.

Furthermore, by analogy with the exact solution, we require that if the operation G is repeated a sufficient number of times, then $q > 0$.

c) G is equilibrating if and only if

$$q = q^0 \quad \text{if and only if} \;\; q^0 = q^s \tag{12.29}$$

where q^s is the unique set of steady-state values of q corresponding to G. Clearly, if G is consistent with Eq. (12.2), such a set must exist and be unique.

Two important theorems govern the repeated application of the operation G.

Theorem 12.1 If G is conservative and positivity preserving, then G is stable.

Proof 12.1 Since the sum, $\sum_i q_i$, of a set of positive values q_i is constant, each value q_i is bounded, and the theorem follows.

Theorem 12.2 If, and only if, G is equilibrating and positivity maintaining, then G is convergent; i.e. repeated application of G converges uniformly to the steady state, q^sq.

Proof: The proof of this result follows an approach similar to that of the analogous theorem for the differential equation. Thus, we consider the behaviour of the vector Q by means of the operation

$$Q_i = q_i^s G(q_1^s, Q_1^0 \ldots) = G'(Q_0^s \ldots) \tag{12.30}$$

Since G is equilibrating, G' is differential, i.e. $Q_i = Q_i^0$ if and only if Q_i^0 is uniform. Furthermore, since G is consistent with Eq. (12.2), G' must be linear in Q. Thus, $\mathcal{L}$ is the uniform vector whose values equal the smallest component of Q^0. Finally, since the operator G is linear in q, we conclude that the solutions, q, must converge to the equilibrium state, q^s.

A convergent operator is clearly stable.

12.2.3 Two-Stage Schemes Part 1

As an example of the application of these theorems, we consider a general two-stage scheme for which there is a simple linear form. In these methods, the rate coefficient matrix is assumed to be tri-diagonal in form. The ionisation change across a pair of isolated levels q_r and q_{r+1} is then calculated, Δ_r. The net ionisation is obtained by recursively evaluating

$$q_i = q_i^0 + \Delta_{i-1} - \Delta_i = 0 \tag{12.31}$$

subject to $\Delta_0 = 0$. The scheme is clearly conservative. It is not, however, positivity maintaining. This condition is readily achieved by a simple limit on Δ, such that

$$\Delta_i < (q_i^0 + \Delta_{i-1}) \qquad \text{and} \qquad \Delta_i > -q_i^0 \tag{12.32}$$

if the recursion in (12.32) is performed with increasing i. This check, which is readily performed in the recursion, has been routinely included in codes using this method. In our experience, the clipping is extremely rarely required. The two-stage scheme in this form is stable. If the form of Δ_i is chosen so that

$$\Delta_i = 0 \qquad \text{if} \qquad q_i/q_i^s = q_{i+1}^s/q_{i+1} \tag{12.33}$$

as in the method outlined in Section 12.3, then the scheme is equilibrating. The positivity-maintaining form is therefore convergent. We note that these general proofs are applicable to all two-stage methods.

12.2.3.1 The Two-Level System

If we consider a system with just two levels between which upward and downward transitions take place from lower level, q_1, to upper level, q_2, with rates S and R, respectively, the population equation takes the form

$$\frac{dq_1}{dt} = [-Sq_1 + Rq_2] \tag{12.34}$$

$$\frac{dq_2}{dt} = [Sq_1 - Rq_2] \tag{12.35}$$

Introducing the total rate and the departure of levels ① and ②

$$\Pi_1 = q_i - q_i^s = q_i - \frac{R}{(R+S)}(q_1 + q_2) = \frac{Sq_1 - Rq_2}{R+S} \tag{12.36}$$

$$\Pi_2 = q_2 - q_2^s = q_i - \frac{S}{(R+S)}(q_1 + q_2) = \frac{Rq_2 - Sq_1}{R+S} \tag{12.37}$$

from equilibrium

$$\Pi_1 = -\Pi_2 \qquad \text{and} \qquad \frac{d\Pi_1}{dt} = -\lambda\Pi_1 \tag{12.38}$$

where $\lambda = (R + S)$, the equations become

$$\frac{d\Pi_1}{dt} = -\lambda\Pi_1 \quad \text{and} \quad \frac{d\Pi_2}{dt} = -\lambda\Pi_2; \quad \text{and} \quad \frac{d}{dt}\{\Pi_1 + \Pi_2\} = 0 \tag{12.39}$$

Integrating from t_0

$$\Pi = \Pi_0 \exp(-\lambda t) \qquad \text{and} \qquad \Pi_1 + \Pi_2 = \Pi_1^0 + \Pi_2^0 \tag{12.40}$$

The two levels therefore approach their equilibrium at a rate $\lambda = S + R$ determined by the sum of their ionisation and recombination rates. The change over one time step is therefore

$$\Delta\Pi_1 = -\Delta\Pi_2 = \Pi_{10}\left[1 - \exp(-\lambda\Delta t)\right] \tag{12.41}$$

and thus we obtain the value of $q_{1/2}$ at the end of the time-step Δt from $\Delta_{1/2} = \Delta_{1/2}^0 + \Delta\Delta q_{1/2}$.

12.2.4 Eigenfunction Decomposition and Integration

In principle, the exact solution of Eq. (12.2) can be obtained by a similarity transformation of $\boldsymbol{X}$ into its Jordan normal form. In practice, this method, although feasible, is complicated and involves excessive computer operations when compared to the more straightforward finite-difference approximations to be described later. However, it should be borne in mind that this method does, in principle, provide exact solutions, whereas those of the more practicable finite difference are only approximate. Having found the eigenvalues and eigenfunctions of the collisional-radiative matrix, **X**, which we assume are non-degenerate, we may apply the similarity transformation

$$\mathbf{X} = \mathbf{P}^{-1}\mathbf{S}\mathbf{P} \tag{12.42}$$

where **P** is an orthogonal matrix and **S** is a diagonal matrix whose values are the eigenvectors.[2]

$$\mathbf{X} = \mathbf{P}^{-1}\mathbf{S}\mathbf{P} \tag{12.43}$$

where **P** is an orthogonal matrix and **S** is a diagonal matrix whose values are the eigenvectors. Substituting for **X**, the collisional-radiative equation becomes

$$\frac{\mathrm{d}}{\mathrm{d}t}\mathbf{Q} = \mathbf{P}^{-1}\mathbf{SPQ} \Rightarrow \frac{\mathrm{d}}{\mathrm{d}t}\mathbf{P}\cdot\mathbf{Q} = \mathbf{SPQ} \tag{12.44}$$

Thus, assuming that the rates are constant, we obtain the solution

$$(\boldsymbol{PQ})(t) = (\boldsymbol{PQ})_i^0\exp(\lambda t) \Rightarrow \mathbf{Q}(t) = \boldsymbol{P}^{-1}\mathbf{exp}(\boldsymbol{\lambda_i t})(\boldsymbol{PQ})_i^0 \tag{12.45}$$

Since the matrix is singular we avoid the difficulty by evaluating the steady state solution q_s

$$\frac{\mathrm{d}}{\mathrm{d}t}\mathbf{Q} = \mathbf{P}^{-1}\mathbf{SPQ} \Rightarrow \frac{\mathrm{d}}{\mathrm{d}t}\mathbf{P}\cdot\mathbf{Q} = \mathbf{SPQ} \tag{12.46}$$

Thus, assuming that the rates are constant, we obtain the solution

$$(\boldsymbol{PQ})(t) = (\boldsymbol{PQ})_i^0\exp(\lambda t) \Rightarrow \mathbf{Q}(t) = \boldsymbol{P}^{-1}\mathbf{exp}(\boldsymbol{\lambda_i t})(\boldsymbol{PQ})_i^0 \tag{12.47}$$

Since the matrix, **X**, is singular, we may remove the singularity to evaluate the steady-state solution q_s. We may reduce the order of the matrix by one using **S**, and the set $Q = q - q^s$, as in Eq. (12.9). The projection of the set $\boldsymbol{Q}$ onto the set of eigenvectors of **S** is straightforward if **S** is not defective or nearly defective. However, the problems introduced by defectiveness into a general algorithm, which together with the large number of operations required make this approach unattractive for large repetitive calculations, are discussed in Appendix 12.B.

Example 12.1 Two-level system. We consider the two-level system described in Section 12.2.3.1 whose rate matrix

$$X = \begin{pmatrix} -S & R \\ S & -R \end{pmatrix} \tag{12.48}$$

We identify the population deviation vector $\mathbf{\Pi} = \begin{pmatrix} \Pi_1 \\ \Pi_2 \end{pmatrix}$. The two eigenvalues and eigenvectors are easily obtained, since rate matrix is singular, therefore the product of the

2 The similarity transformation consists of a transition from the original basis to one in which the matrix is diagonal and back to the original basis.

eigenvalues is zero; and the trace their sum $-(R+S)$

$$\lambda = 0 \qquad \lambda = -(S+R) \tag{12.49}$$

$$\lambda = -(R+S) \tag{12.50}$$

$$-S\Lambda_1 + R\Lambda_2 = S\Lambda_1 - R\Lambda_2 \qquad R\Lambda_1 - S(S+R)\Lambda_1 = -(S+R)\Lambda_2 \tag{12.51}$$

$$\Lambda_1 = -\Lambda_2 \tag{12.52}$$

and

$$\lambda = 0 \tag{12.53}$$

$$-S\Lambda_1 + R\Lambda_2 = 0 \qquad S\Lambda_1 = R\Lambda_2 \tag{12.54}$$

and

$$\lambda_1 = S + R; \qquad \mathbf{\Pi}_1 = \begin{pmatrix} -1 \\ 1 \end{pmatrix} \tag{12.55}$$

and

$$\lambda_2 = 0; \qquad \mathbf{\Pi}_2 = \begin{pmatrix} \Pi_1^s \\ \Pi_2^s \end{pmatrix} = \mathbf{\Pi}^s = 0 \tag{12.56}$$

so that the solution at time t is

$$\mathbf{\Pi} = \mathbf{\Pi}_0 \exp(\lambda_1 t) \tag{12.57}$$

The second eigenfunction clearly corresponds to the steady-state equilibrium state, and the first to a progressive approach to the steady state.

Since this method involves calculating the full set of eigenvalues and eigenfunctions at every time it is lengthy and expensive, and consequently falls outside the remit of our survey.

Finite-difference methods although limited by the stiff nature of the system provide a more realistic approach.

12.2.5 Finite-Difference Algorithms

A more satisfactory approach is to use a finite-difference scheme. If we require such a scheme to be unconditionally stable, i.e. for all step lengths, Δt, it is necessary that the finite-difference form of Eq. (12.2) be stable over the complete positive complex half-plane of $\lambda\Delta t$, the routine must therefore be A-stable $\lambda\Delta t$ (Dahlquist, 1963). There is a fundamental theorem (Dahlquist, 1963) which states that multi-step schemes cannot be A-stable if their order is greater than two. Therefore, if we require an algorithm which is unconditionally stable for all forms of the rate Eq. (12.2), we must restrict ourselves to two-level multi-step schemes. Such a restriction is not, of course, fully compatible with our storage requirement discussed earlier.

We may write Eq. (12.2) in a finite-difference form, using a weighted mean for the terms on the right-hand side,

$$\begin{aligned}\left(q_i^{N+1} - q_i^N\right)/\Delta t = \sum_{j \neq i} X_{i,j} &\left[W_{i,j} q_j^{n+1} + (1 - Wi, j) q_j^N\right] \\ &- \left[W_{j,i} q_i^{n+1} + (1 - Wj, i) q_j^n\right]\end{aligned} \tag{12.58}$$

yielding n equations, which we may solve by matrix inversion. We may reduce the dimensions of the system (and hence the work) by making use of the conservation

$$\sum_i q_i^{n+1} = q_i^n \tag{12.59}$$

to eliminate q_i from the set of equations. Hence, we obtain the set

$$\sum_{j=1}^{n-l} C_{i,j} q_j^{n+1} = B_i \tag{12.60}$$

where

$$C_{i,i} = 1 + \left[X_{i,n} W_{i,n} + \sum_{\substack{j \neq 1 \\ j=1}}^{n-1} X_{j,i} W_{j,i} \right] \Delta t;$$

$$C_{i,j} = \left[X_{i,n} W_{i,n} - X_{i,j} Wi, j \right] \Delta t \tag{12.61}$$

and

$$B_i = q_i^n \left[X_{i,n} W_{i,n} - \sum_{\substack{j = \neq i \\ j=1}}^{n-1} X_{j,i} \left(1 - W_{j,i}\right) \right] \Delta t$$
$$+ \sum_{\substack{j = \neq i \\ j=1}}^{n-1} q_j^N X_{i,n} + X_{i,j} \left(1 - W_{i,j}\right) \Delta t \tag{12.62}$$

An alternative form of Eq. (12.60), suitable for use with an energy limit, considering the change in q_i, is obtained by

$$\Delta_i = q_i^{n+1} - q_i^n; \qquad \Delta_n = -\sum_{j=1}^{n-1} \Delta_j \tag{12.63}$$

to give

$$\sum^{n-1} C_{i,j} \Delta_j = B_i' \tag{12.64}$$

where

$$B_i' = \sum_{\substack{j \neq 1 \\ j=1}}^{n} \left\{ X_{i,j} q_j^n - X_{j,i} q_i^N \right\} \Delta t \tag{12.65}$$

The matrix C is, in general, dense and Eq. (12.60) or (12.64) must be solved by either an elimination or an iterative method. In view of the fact that one usually only works with a restricted set of states n, the demands of a compact elimination method are not severe. We note that since C is the form, particularly as the change of Δ from one time iteration N to the next $N + 1$ should not be large, so that Δ^N provides a good set of initial values. We note that C is a matrix of the type considered in Appendix 12.A, and is therefore nonsingular. The set of Eqs. (12.58) is clearly conservative. The derived set (12.60) is obtained using the conservation law (12.59) and must therefore be conservative.

In equilibrium if

$$\sum_{j\neq i} X_{i,j}\, q_j^s = \sum_{j\neq i} X_{j,i}\, q_i^s \tag{12.66}$$

Hence, if $q^N = q^s$, $B' = 0$, then, since C is non-singular, $\Delta = 0$. Equation (12.64) and hence Eqs. (12.60) and (12.58) are equilibrating, the steady state being that of the true solution.

This general algorithm is not in general positivity maintaining. To incorporate this requirement we could include a post-evaluation check, as in the two-stage scheme. A more satisfactory approach, however, is to assign the weights so as to ensure this condition is upheld.

We write Eq. (12.58) in the form

$$q = M^{-1}\, P\, q_0 \tag{12.67}$$

which is positivity maintaining if and only if M is monotone (Collatz, 1960) and P a non-negative matrix. In fact,

$$\begin{aligned} P_{i,i} &= 1 - \sum_{j\neq i}(1 - W,j,i) X_{j,i}\, \Delta t; \qquad & P_{i,j} &= (1 - W_{i,j}) X_{i,j}\, \Delta t \\ M_{i,i} &= 1 + \sum_{j\neq i} W_{j,i} X_{j,i}\, \Delta t; \qquad & M_{i,j} &= W_{i,j} X_{i,j}\, \Delta t \end{aligned} \tag{12.68}$$

Since $X_{i,i}$ is non-negative, M is clearly an M-matrix, and therefore monotone, and P is non-negative if and only if $P_{i,i} \leq 0$. If this condition is upheld, the algorithm, and those derived from it, are both stable and convergent.

12.2.5.1 The Choice of Weights

The weights, $W_{i,j}$, must be chosen in such a way as to enable the solution of the finite-difference equations to approach the exact one for arbitrary values of the step Δt. Clearly, this requires that in the limits

$$\begin{aligned} W_{i,j} &\to \frac{1}{2} \qquad \text{as} \qquad \Delta t \to 0 \\ W_{i,j} &\to 1 \qquad \text{as} \qquad \Delta t \to \infty \end{aligned} \tag{12.69}$$

The weights are used in an attempt to represent the degree of equilibration of a pair of states i and j during the time step. Thus, if the rates are large, $(X_{i,j} + X_{j,i})\Delta t \gg 1$, the states will be instantaneously equilibrated throughout the time step, and $W_{i,j} = 1$ is appropriate. On the other hand, if $(X_{i,j} + X_{j,i})\Delta t \ll 1$, the states will not equilibrate and $W_{i,j} = 1/2$ is suitable. It is clear that each weight is associated with a transition and therefore

$$W_{i,j} = W_{j,i} = f(\lambda_{i,j}) \tag{12.70}$$

where $\lambda_{i,j} = (X_{i,j} + X_{j,i})\, \Delta t$, and f is a function with the limits (12.69). We may solve Eq. (12.2) for the case of a two-level system to give Eq. (12.41)

$$\Delta_i = \left(X_{12}\, q_2^0 - X_{01}\, q_1^0\right)\, \Delta t\, \left\{1 - \exp\left(-\lambda_{12}\right)\right\} / \lambda_{12} \tag{12.71}$$

in contrast to the finite-difference result,

$$\Delta_i = \left(X_{12}\, q_2^0 - X_{21}\, q_1^0\right)\, \Delta t / \left(1 + W_{12}\lambda_{12}\right) \tag{12.72}$$

Thus, the finite-difference form is exact if the weights

$$W_{12} = 1/\left\{1 - \exp\left(-\lambda_{12}\right)\right\} - 1/\lambda_{12} \tag{12.73}$$

We note that this function has the limits (12.69) and is therefore a suitable weight function.

In view of the nature of the weights, to allow for rapid equilibration between two strongly coupled states, we may expect that it is an appropriate form.

Alternative functions, simpler but of similar form, are

$$W = \max\ \{1/2,\ (1 - 1/\lambda)\} \tag{12.74}$$

and

$$= \max\ \{1/2,\ 1 - [1/\lambda - \exp(-\lambda)]\} \tag{12.75}$$

Numerous tests are presented by Pert (1981). The accurate solution was calculated using Gear's routine (Gear, 1971). A comparison with the values obtained by a split-time-step $W = 1/2$ and with fully implicit $W = 1$ calculations are shown. There is a marked improvement obtained by using the weights, which can be can be clearly seen. Of the three weight functions the two-level form, Eq. (12.73), is marginally superior, a result consistent with other tests of this set.

These choices of weights do not in general ensure positivity. This requires that

$$\sum_{j\neq i} \left(1 - w_{j,i}\right)\ |X_{j,i}\ \Delta t \leq 1 \tag{12.76}$$

for all i. Let us suppose that level i has N_i non-zero transition connecting it to other states; then the above condition is satisfied if

$$W_{j,i} \geq 1 - 1/N_i X_{j,i}\ \Delta t \tag{12.77}$$

This may be put into a form consistent with Eq. (12.69):

$$W_{i,j} = W_{j,i} \max \begin{cases} \dfrac{1}{2} \\ 1 - 1/\max(N_i, N_j)\lambda_{i,j} \end{cases} \tag{12.78}$$

or by analogy with Eq. (12.73)

$$W_{i,j} = W_{j,i} = \max \begin{cases} l/[1 - \exp(-\lambda_{i,j})] - 1/\lambda_{i,j} \\ 1 - 1/\max(N_i\ ,\ N_j)\ \lambda_{i,j} \end{cases} \tag{12.79}$$

Of these, the first, which is computationally simpler, is preferred. It was found that the results using weighting are considerably more accurate than those obtained without the positivity limit. Calculations show that the simpler form (Eq. (12.78)) is marginally more accurate, and since it is simpler it is to be preferred.

12.2.5.2 The Energy Limit

It was shown in Pert (1978) that if the system is near equilibrium, the energy exchange between ionisation and the electron temperature resulting from collisional ionisation and three-body equilibration can lead to instability. In a similar fashion, the inclusion of general collisional excitation and de-excitation energy exchange may also result in an unstable scheme. Thus, near equilibrium

$$\Delta_i q_i \approx \left(dq_i/dT_e\right)\Delta T_e \tag{12.80}$$

and the electron energy change due to a change Δ_i in the population of state i

$$= -V_i' \Delta q_i \tag{12.81}$$

per ion, where $V_i' = V_I + \frac{3}{2} Z_I k T_e$, and V_i is the total excitation and ionisation energy (from the ground state of the neutral atom). Thus, as before, considering an iteration for the electron temperature, we obtain on the next iteration[3]

$$\frac{3}{2}\overline{Z} k \Delta T_e \approx -\sum V_i \cdot \left(\mathrm{d}q_i/\mathrm{d}T_e\right) \tag{12.82}$$

and is unstable if $\sum V_i \cdot \left(\mathrm{d}q_i/\mathrm{d}T_e\right) > 1$. This difficulty may be overcome by specifying a limiting energy transfer specifying a limiting energy transfer from the electron thermal energy to the ionisation ΔE by replacing Δq_i by

$$\Delta q_i \, \Delta_1' \Big/ \left(1 + \sum V_i \Delta_i' / \Delta E\right) \tag{12.83}$$

where Δ_1' is the value given by Eq. (12.63). A suitable value for ΔE is given later (Eq. (12.109)).

12.2.6 Summary

The finite-difference algorithm described here has been widely used by the author in several studies. In an extensive series of tests, the maximum error of the optimally weighted scheme did not exceed 0.07 and convergence to the equilibrium value was observed to be uniform. The results of two representative tests are shown in Pert (1981). The maximum error is about 0.06 and is less than 0.07 over the range in which all states simultaneously equilibrate. By comparison, the maximum error of the fully implicit scheme was 0.18 and that of centred difference 0.8; the latter method, in common with all non-positivity-maintaining weights (12.73)–(12.75), yielding negative values.

The most expensive computational element in the algorithm is the solution of the set of linear equations (12.63), which involves $\frac{1}{3}(n-1)^3$ multiplications. A significant reduction in the computational work required can be achieved by the reduction of the algorithm to a two-stage form. In its most extreme form, this involves the treatment of each ionisation separately by either a collisional-radiative form or a single level (bottleneck/collision-limit) form approximation. In this form, the sparse areas of the matrix, R, which have the form of a set of diagonal overlapping blocks are separated. The algorithm involves an independent solution for each block, R. The number of multiplications involved is then $\frac{1}{3}\prod_{i=1}^{N}(n_i-1)^3$ when there are N blocks each of n_i levels. Clearly,

$$n = 1 + \sum_{i=1}^{n} \left(n_i - 1\right) \tag{12.84}$$

This method is particularly convenient if the element contains a large number of ionisation stages, many of which are only briefly involved in the calculation. It is a straightforward generalisation of the two-stage method (Section 12.3) which is, of course, stable and convergent.

3 This correction term is the first iteration using either the secant or *regula falsi* methods of

$$W \sum [V_i' \Delta q_i + \Delta(Z k T_e) \Delta T_e = 0]$$

In principle, the number of discrete levels of each ionisation stage is limited only by the depression of the ionisation level. In practice, one must introduce some appropriate limit to the number of levels considered. This can only be determined by consideration of the individual characteristics of the system, for example its L.T.E. limit, and time variation. The levels above this limit are included by some appropriate averaging such as is used in the collision-limit approximation.

12.3 Two-Stage Model

12.3.1 Introduction

The number of energy levels required for a complete calculation of the ionic state in a complex non-hydrogenic atom is extremely large, typically in the upper hundreds. Since a complete calculation involves the inversion of large matrices without any clear sparsity pattern, this conflicts with our aims that the resultant computer program be both reasonably fast and compact. It therefore becomes necessary to identify methods to reduce the size or the code and improve its speed.

Fortunately, the physics of ionisation–recombination comes to our aid in two ways:

i) Firstly, there is generally a clear demarcation between the lower levels linked with the ground state and the upper tied to the ionisation continuum of the next ionisation stage. This was initially defined by the 'bottle-neck' state in collision dominated plasma by Hinnov and Hirschberg (1962) and Pitaevskii (1962) or in radiatively dominated plasma as the collision limit by (Griem, 1964); the relationship between the two being identified in Pert (1990): the appropriate limit being at the upper level. Working in this approximation, the total population of the lower levels is assigned to the ground state of the ion Z and the upper states to the ground state of the next higher ion $Z+1$. By this means, the calculation of the population of the excited states in the ion stage Z are subsumed into a quasi-analytic form with averaged ionisation and recombination rates for that ion stage. In this way, the large number of states in the full collisional radiative equations can be greatly reduced by a semi-analytic calculation for stage Z using analytic forms for S and R

$$\frac{dq_Z}{dt} = \left\{S_{Z-1}\, q_{Z-1} - R_Z\, q_Z\right\} - \left\{S_Z\, q_Z - R_{Z+1}\, q_{Z+1}\right\} \tag{12.85}$$

the brackets {} indicate the detailed balance relation and equilibrium condition $q^s_Z/q^s_{Z-1} = R_{Z-1}/S_Z$.

ii) The general calculation of a specific ionisation stage requires the determination of the occupation probability of the atomic states of that stage q_i as discussed earlier. In principle, as we have seen in Section 12.2, this requires the full determination of the entire atomic system, involving a large number of states, thereby treating n electrons in both free and bound orbitals, and thus includes both the parent atom and daughter ions. We shall restrict our models to cases where the density is sufficiently low that the interaction between atomic systems is small and can be treated as a perturbation rather than requiring a generalisation to include nuclei (i.e. molecular) states. The general problem of atomic dynamics is simply stated in terms of a probability q_i that a given state is occupied by the rate equation

$$\frac{dq_i}{dt} = \sum_j [X_{ij} q_j - X_{ji} q_i] \tag{12.86}$$

where X_{ij} is the probability per unit time of a transition from state j to i. The transition may be spontaneous, or radiative or maybe collision induced.

iii) For a system involving only two ionisation stages, there exists a simple analytic form to describe the ionisation process which can be easily modified to allow for variations in the ionisation–recombination rates as the electron density changes, consistent with ionisation. Furthermore, a simple modification allows for the energy change during ionisation as thermal energy is converted into ionisation energy – correction necessary to avoid instability due to this exchange.

In general, the solution of Eq. (12.86) is intractable due to the large number of states and lack of detailed information about the full set of rates X_{ij}. As a result, it is necessary to reduce the system to manageable limits by judicious selection of integration methods, bearing in mind the criteria identified in the Section (12.3.1). Firstly, the free electron orbitals are compressed to a single group with the electrons assumed to have a Maxwell–Boltzmann distribution amongst them. Secondly, the bound orbitals are treated by one or two simplified models: the collisional-radiative or average atom models. Thirdly, only transitions involving one or two electron jumps are considered. Finally, the ionisation–recombination of a single electron ionisation stage is treated. When this is done, simplifications allow the calculations to be brought to acceptable levels.

We reduce the number of equations to be integrated by introducing the population change across an ionisation stage during a time step,

$$\Delta_i = q_i - q_i{}^0 + \Delta_{i-1} \qquad \Delta_0 = 0 \tag{12.87}$$

where q_i is the value at some arbitrary time and $q_i{}^0$ the initial value. Substituting

$$\begin{aligned}\frac{d\Delta_i}{dt} &= \mathcal{R}_{i+1} q^0_{i+1} - \mathcal{S}_i q_i{}^0 \\ &\quad + \mathcal{R}_{i+1}\,\Delta_{i+1} - (\mathcal{R}_i + \mathcal{S}_i)\Delta_i + \mathcal{S}_i \Delta_{i-1}\end{aligned} \tag{12.88}$$

where the ionisation and recombination rates vary with electron density

$$\mathcal{S}_i = S_i\, n_e \qquad \mathcal{R}_i = R_1\, n_e + R_2\, n_e{}^2 \tag{12.89}$$

The fractional ionisation after the $(n+1)$th time step

$$Z^{n+1} = Z^n + \sum_i \Delta_i \tag{12.90}$$

where the electron density $n_e = Z\, n_i$ and n_i the total ion density.

It is useful to note that Eq. (12.88) has the form of the finite-difference diffusion equation with spatially varying diffusivity. The solutions therefore have the properties exhibited by diffusion.

12.3.2 Integration of the Rate Equations

Equation (12.86) is representative of the general family of stiff equations for which standard methods of solution are available (Gear, 1971). As can be seen, Eq. (12.88) has the form of a diffusion equation and for stability must be integrated implicitly. As noted earlier, computer CPU limits restrict us to two-step schemes, i.e. trapezoidal integration, although the tri-diagonal nature of the resulting equations allows an efficient matrix solution. In the simple second-order form, the integration step is limited by the fastest rate to ensure positivity.

$$\Delta_i - (1 - W_i)\,\mathscr{R}_{i+1}\Delta_{i+1} - \left(\mathscr{R}_{i+1} + \mathscr{S}_i\right)\Delta_i + W_i\mathscr{S}_i\Delta_{i-1} = \left(\mathscr{R}_{i+1}q_{i+1}{}^0 - \mathscr{S}_i q_i{}^0\right)\Delta t \quad (12.91)$$

To ensure positivity, we may introduce a stage weight, which guarantees stability and ensures positivity by limiting the changes, forcing the ionisation across the stage into equilibrium when the rates are large. Referring to Eq. (12.94) when the correction for varying electron density is neglected ($D = 0$), we obtain the weight

$$W_i = \text{Min}\left\{\begin{array}{l} 0.5 \\ 1/C_i - \exp(-C_i) \end{array}\right. \quad (12.92)$$

where $C_i = \left(\mathscr{R}_{i+1} + \mathscr{S}_i\right)$. Although the use of this simple weighting function allows considerable accuracy improvement at large rates ($C_i \gg 1$), it implies the use of time splitting. The limitations of the simple implicit schemes were studied in Pert (1978) and their performance without and with iteration noted when used in a code where the time-step control is limited by the hydrodynamic rather than the ionisation rates, which may be very large if the density is high Pert (1978) (Figure 12.1). Substantial improvement is obtained when iterated by writing algorithm in terms of the set Δ_i and linearising directly into the implicit scheme. These iterated schemes may be viewed in a number of ways from Eq. (12.91): as a Newton–Raphson solution of the non-linear equations in Δ, as obtained by a single predictor–corrector integration step, as an iteration of a series of integrations across a time-step in keeping with time-step splitting.

12.3.3 Two-Stage Scheme

Let us consider the split-time-step approximation for the solution treating the integration across each step independently. The problem considering the population change Δ_i across a single ionisation stage i was used to derive the weights W_i (Eq. (12.92)). Across the stage boundary level, $i + 1$ is populated by ionisation at a rate $\mathscr{S}_i$ from its parent stage i and de-populated by recombination at rate $\mathscr{R}_{i+1}$ from its daughter. Hence, the rate equation takes the form

$$\frac{dq_i}{dt} = \mathscr{R}_{i+1}\,q_{i+1} - (\mathscr{R}_i + \mathscr{S}_i)\,q_i + \mathscr{S}_{i-1}\,q_{i-1} \quad (12.93)$$

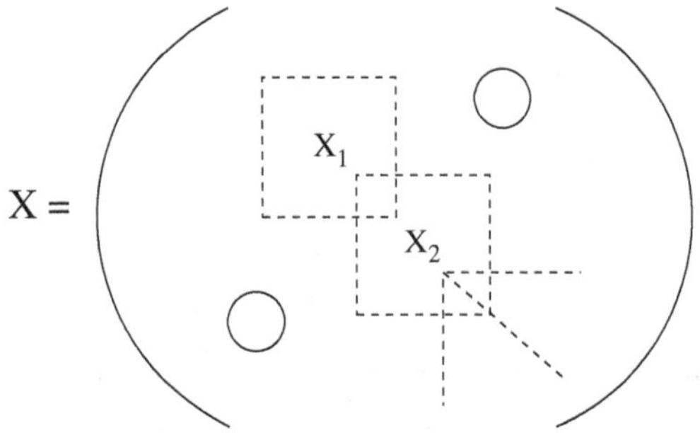

Figure 12.1 Diagram to show how the total set X is the union over all the ion stages subsets $\mathbf{X} = \mathbf{X}_1 \cup \mathbf{X}_2 \ldots \cup \mathbf{X}_n$, each with an intersection at the boundary with its neighbours ($\mathbf{X}_{i-1} \cap \mathbf{X}_i \cap \mathbf{X}_{i+1}$) representing ionisation–recombination with the lower and upper neighbouring ion stage. Notice the simplifying sparsity introduced by this approximation.

The rates depend on the electron density

$$S_i = S_i n_e \qquad \mathscr{R}_i = R_{1\,i} n_e + R_{2\,i} n_e^{\,2} \tag{12.94}$$

where S_i is the ionisation coefficient and $R_{1\,i}$ and $R_{2\,i}$ respectively, the two-body and three-body recombination coefficients depend on the electron temperature but only weakly on the other plasma parameters, such as electron density.

Since the ionisation and recombination rate coefficients vary linearly with electron density, we may include this density variation across the stage boundary; substituting for n_e and treating S_i and $R_{i+1} = R_{1(i+1)} + R_{2(i+1)}\, n_e^0$ as constants, we obtain

$$\frac{\mathrm{d}\Delta}{\mathrm{d}t} = \left(S_i q_i - R_{i+1} q_{i+1}\right) n_e \tag{12.95}$$

Substituting q_i from Eq. (12.87) and writing $n_e = n_{e0} + n_i \Delta_i$, and integrating

$$\Delta_i = \frac{Z B_i \left(1 - D_i\right)}{\left(Z + B_i D_i\right)} \tag{12.96}$$

where

$$B_i = \left(S q_i^n - R_{i+1} q_{i+1}^n\right) / \left(S_i + R_{i+1}\right) \quad \text{and} \quad D_i = \exp\left[-\left(Z + B_i\right)\left(S_i + R_{i+1}\right) n_i \,\Delta t\right] \tag{12.97}$$

The relatively simple model integrating ion stage by ion stage within a stepped time step frame forms the foundation for simple, rapid calculations of the ionisation history of the plasma by repeated calculation. As the model is an exact solution of the diffusion equation with certain limited parameters, solutions obtained by successive calculation over the ion stages will have behaviour appropriate to that equation. Consequently, on successive time steps, the solution will be will be stable and convergent.

This treatment of the ionisation–recombination of the plasma by concentrating on the two dominant atom processes allows us to develop the history of ionisation in the plasma in two ways:

i) The more sophisticated method is to construct the set of ionisation rate equations as an implicit matrix with the populations as variables, which can be simply solved using the tri-diagonal matrix (Thomas) algorithm, thereby taking account the full development over a complete time step, but ignoring the change of rates due to the electron density.
ii) The alternative is to solve the ionisation rate equations step by step for each pair of states taking into account the change in electron density resulting from the ionisation of those states.

The former method allows us to take account of the history of the minor stages of ionisation, but at the expense of failing to include the role of the varying electron density in the implicit form of the calculation. On the other hand, the latter method explicitly follows each stage as the time-step develops. Each method has advantages and disadvantages, suited to the type of problem being studied, namely whether the detailed atomic state within particular ionisation stages is required. It follows from the general results concerning the diffusion equation that both methods are stable and convergent.

12.3.4 Ionisation–Recombination Rates

Many plasmas are optically thin to short wavelength radiation. In consequence, the only effective ionisation process is due to electron collisions. Apart from exceptional circumstances, which are easily handled, we may neglect any other ionisation process. Thus, we have

$$\mathcal{S} = S_2(T_e)n_i n_e \tag{12.98}$$

where S_2 is the two-body (ion/electron) ionisation coefficient.

The inverse process to collisional ionisation is a three-body collision process (ion +2 electrons) in which an ion and an electron recombine in the presence of a second electron which removes the excess energy and momentum, needed to satisfy the conservation laws. Thus, this process has a rate coefficient

$$\mathcal{R} = R_3(T_e)n_1 n_e^{\,2} \tag{12.99}$$

Since $S_2(T_e)$ and $R_3(T_e)$ are functions of the electron temperature alone and are clearly reciprocal processes, it follows that they must obey the principle of detailed balance.[4]

Electron recombination with an ion is the result of the two direct processes, namely radiative and three-body collisional recombination, although in non-hydrogenic ions, additional di-electronic recombination may also contribute but is usually weak. Radiative recombination is the spontaneous emission from a free (electron + ion) state to a bound level of lower ionisation state involving the emission of a photon whose energy exceeds the ionisation energy of the recombining ion. Consequently, radiative recombination favours transitions to lower states of the ion. On the other hand, three-body recombination, which is the complement to collisional ionisation, which involves energy transfer by collision. A classical picture involves elastic energy transfers between particles of similar mass, namely electrons. It therefore favours transitions with small energy gaps of the same order as the thermal energy, $\sim kT$. Such recombination therefore is to high-lying states, which cascade down through the excited states to the ground state by collisions (Hinnov and Hirschberg, 1962) or radiatively (Griem, 1964), a process known as a *recombination cascade*. These are two radically different processes having different scalings. At low density, direct radiative processes dominate with few emissions, the major contribution due to the direct transition to the ground state. On the other hand, three-body recombination is a random walk process as electrons gain and lose energy – a diffusive process in energy space. In recombination, the diffusive flux must move downwards in energy reflecting the population $q_1 \to 0$. This is a process well described by a Fokker–Planck model in energy (Hinnov and Hirschberg, 1962; Pitaevskii, 1962).

The relative strengths of radiative/collisional processes in recombination are determined by the parameter $\beta = \pi\alpha^3/n_e$ such that $\beta = 0$ corresponds to a collisional cascade and $\beta = \infty$ a radiative.

---

4 In a steady detailed balance, the forward and backward processes must balance if the drivers (n_i, n_e, T_e) satisfy the equilibrium condition, namely $n_e = n_i$ Saha(T_e), where Saha(T_e) is the Saha function,

$$\text{Saha}(T_e) = \frac{n_{i+1}}{n_i n_e} = 2\frac{u_{i+1}}{u_i}\left(\frac{2\pi m_e kT_e}{h^2}\right)^{3/2} \exp\left(-\chi/kT_e\right) \tag{12.100}$$

where u_{i+1} and u_i are the statistical weights (strictly the partition functions) of the ground states of ions $(i+1)$ and i, respectively, and χ_i the ionisation energy of ion stage i.

Accurate values for the recombination rate by this process are difficult to obtain and involve the solution of the collisional–radiative equations over a multiplicity of states, see Pert (1990). This is not consistent with our aim for a rapid, efficient calculation. Unfortunately, for real atoms/ions the energy levels, transition probabilities and cross sections are not simple expressions nor, in most cases, well known. There is however one conspicuous exception to this rule for which the necessary values are readily available, namely hydrogenic ions. As a result, the excited state populations are known, and expressions for the recombination rate can be derived via a Fokker–Planck analysis (Pert, 1990). Once the excited state population is known at the boundary between the upward and downward drift of the electrons in energy space, this is known as the *bottleneck* or the *collision limit* for a collision or radiation-dominated cascade, respectively. The recombination rate is then the energy diffusion flux through that level.

Furthermore, if the upper levels are dominant, most atoms have a quasi-hydrogenic set of Rydberg levels suggesting that the rates generated from the recombination rate are probably reasonably accurate, provided direct recombination to the ground state, namely direct radiative recombination, and the Saha inverse to ionisation from detailed balance are small. This conjecture is well supported by checking (Pert, 1990) against collisional-radiative modelling.

12.3.4.1 Low-Temperature Bottleneck ($\beta \to 0, Z^2 I_H \gg kT$)

At low temperatures, the cascade is almost entirely due to collisional de-excitation between closely spaced energy levels. The distribution function of an electron amongst the bound levels is given by the Fokker–Planck model (Pert, 1990). This yields an analytic form, which allows the net flux of electrons downwards at high energy near the ionisation limit and upwards at low energy near the ground state. The 'bottleneck' occurs at the turning point in the flux where the upwards changes to a downward one. In Pert (1990, Appendix 2), it is shown that a Boltzmann distribution is maintained down to the 'bottleneck' where the Boltzmann weighted population is a minimum (Pert, 1990) i.e.

$$\frac{\mathrm{d}}{\mathrm{d}n}\left[n^2 \exp\left(-Z^2 I_H/n^2 kT\right)\right] = 0 \tag{12.101}$$

where I_H is the ionisation energy of the hydrogen atom. Hence, the bottleneck is found at flux minimum where

$$\nu_{bottleneck} = Z^2 I_H / kT \tag{12.102}$$

and the corresponding collisional velocity space flux gives a recombination rate.

An electron in a state above the bottleneck $n > \nu_{bottleneck}$ will probably drift upwards towards the continuum and therefore become ionised. In contrast, one with $n < \nu_{bottleneck}$ goes to the ground state and is recombined. Thus, the recombination rate is just given by the flux across the bottleneck.

$$\begin{aligned} r &= (-j)/F_{\text{inf}}\, n_e \\ &= \frac{16\pi(2\pi)^{1/2}}{9\sqrt{3}} \frac{e^{10} Z^3 n_e}{m_e^{1/2} (kT)^{9/2}} \bar{g}_{III} \end{aligned} \tag{12.103}$$

where g_{III} is the free-free Gaunt factor which may be treated as a constant and whose asymptotic form is given in Pert (1990).

12.3.4.2 Radiative Cascade ($\beta \to \infty$)

The preceding analysis has neglected the role of radiation in recombination. A downward recombination cascade is limited in practise by radiative transitions from levels above a bottleneck state to below the bottleneck state. Radiative cascades are important if the bottleneck state is a high quantum state. The effective bottleneck is found at the state n where the downward radiative transition just balances the upward collisional excitational rate. This level is known as the *collision limit*. A simple approximation for the collision limit principal quantum state is given by Griem (1964)

$$n_{cl} \approx (27/256) BYn_e^{-1} g_{III}^{-1} \tag{12.104}$$

12.3.5 Summary

The cascade model of recombination was surveyed in detail by Pert (1990) and 'best' values for the rates in hydrogenic ions derived. General expressions for the rates and the state numbers of the collision limit and the bottleneck were given and should be adequate for most purposes; the work should be consulted for further details and values for the corresponding terms. The expressions given for the ionisation and recombination are messy, but reasonably easy to apply computationally. They are likely to be accurate within the accuracy of the hydrogenic approximation for the upper state. One further point concerns thermodynamic balance in that levels below the bottleneck/collision limit decay to the ground state and above to ionisation continuum. The effective ionisation energy is therefore the excitation energy of the bottleneck/collision limit. In practice, this normally makes little difference in calculations.

12.3.6 Energy Limit

The step length is chosen so that variations due to the plasma motion are relatively small during each interval. Thus, the electron density n_e is approximately constant during the integration of Eq. (12.86). However, as energy is required for ionisation, the electron temperature may vary significantly over a single time step and cause a significant change in the corresponding rates when the system is included in the total energy balance. When the time step is large as allowed by the ionisation–recombination routine.

To elucidate this, assume that at each time step of the ionisation calculation the system equilibrates and consider an iteration for the temperature deviation ΔT_e. The change in ionisation over the step

$$\Delta q_i \approx \left(\mathrm{d}q_i/\mathrm{d}T_e\right) \Delta T_e \tag{12.105}$$

and at the next iteration

$$\begin{aligned} \tfrac{3}{2} Z k T_e' &\approx -\sum_i \left(\tfrac{3}{2} k T_e + V_i\right) \Delta q_i \\ &= -\sum_i \left(\tfrac{3}{2} k T_e + V_i\right) \left(\mathrm{d}q_i/\mathrm{d}T_e\right) \Delta T_e \end{aligned} \tag{12.106}$$

where Z is the average charge.

If $\sum_i \left(\tfrac{3}{2} k T_e + V_i\right) \left(\mathrm{d}q_i/\mathrm{d}T_e\right) > \tfrac{3}{2} Z k$, the iteration grows without limit. In a similar manner, such iterations occur in the course of time-step integration and lead to unstable growth. To prevent this from taking place we introduced (Pert, 1978) an *energy limit* to limit the amount of energy transferred in a single time step.

Returning now to the full calculation, if the temperature rises $\Delta T_e'$ is calculated without the inclusion of the ionisation energy changes, the true incremental rise ΔT_e is given by

$$\frac{3}{2} Z \mathcal{k} T_e \approx -\sum_i \left(\frac{3}{2} \mathcal{k} T_e + V_i \right) \Delta q_i + \frac{3}{2} Z \mathcal{k} T_e' \tag{12.107}$$

Calculation of the fractional ionisation change with increments $\Delta T_e'$ and ΔT_e yields $\Delta q_i'$ and Δq_i, respectively, given by Eq. (12.105)

$$\Delta q_i \approx \Delta q_i' \Big/ \left\{ 1 + \sum_i \left(\frac{3}{2} \mathcal{k} T_e + V_i \right) \Delta q_i' \Big/ \frac{3}{2} Z \mathcal{k} T_e' \right\} \Delta T_e' \tag{12.108}$$

Since $\sum \left(\frac{3}{2} \mathcal{k} T_e \right)$ is the local heat exchange in ionisation, we may regard this equation as specifying a limiting heat exchange $\frac{3}{2} Z \, \Delta T_e'$, i.e. associated with the temperature change $\Delta T_e'$ which will prevent oscillation. We further note that this term is the first correction in a Newton–Raphson iteration.

In cases far from equilibrium, some form of physical energy limit will apply (although it is unlikely to be particularly important under these conditions) due to the rapid variation of the ionisation–recombination rates with temperature. This will limit the energy transfer to some value

$$\Delta E \sim \frac{3}{2} Z \mathcal{k} \left(\Delta T_e; + \alpha \, T_e \right) \tag{12.109}$$

Experience has suggested that

$$\alpha = 0.5 \sum_i \beta_i \left(q_i + q_{i-1} \right) \Big/ \sum_i \left(q_i + q_{i-1} \right) \tag{12.110}$$

where $\alpha \sim 1/2$ is a weighted parameter depending on the strength of the transition. A satisfactory empirical value is given in Pert (1978).

The algorithm is well-behaved throughout the run.

Supplementary Material

Appendix 12.A Solution of the Collisional-Radiative Equations

The collisional-radiative differential equations (12.2) may be directly integrated using an integrating factor. This is achieved by writing the equation as a first-order equation for the population of state i

$$\frac{dq_i}{dt} + P q_i = Q \quad \rightarrow \quad \mathrm{e}^{-\int P \mathrm{d}t} \frac{\mathrm{d}}{dt} \left[\int q_i(t) \, \mathrm{e}^{\int P \, \mathrm{d}t} \right] = Q \tag{12.A.1}$$

$$\rightarrow \quad \frac{\mathrm{d}}{\mathrm{d}t} \left[q_i(t) \, \mathrm{e}^{\int P \, \mathrm{d}t} \right] = Q \mathrm{e}^{\int P \, \mathrm{d}t} \quad \rightarrow \quad q_i = \left\{ \int_{t_0}^{t} Q \mathrm{e}^{\int P \mathrm{d}t} \mathrm{d}t + q_i^0 \right\} \mathrm{e}^{-\int_{t_0}^{t} P \, \mathrm{d}t} \tag{12.A.2}$$

where $P(t) = \sum_{j \neq i} X_{ji}$ and $Q(t) = \sum_{j \neq i} X_{ij} q_j$.

The limits on the integrals are found from the boundary conditions $q_i(0) = q_i^0$ and on the inner integral by noting that if $Q = 0$ there are no transitions into i from other states and that the population of i decays as $q_i(t) \rightarrow q_i^0 \exp -(P\,t)$; P is therefore the decay rate of state i. The term Q is the growth rate of level i due to transitions to i from all other states.

Appendix 12.B A Theorem on Determinants

Lemma 12.1 If a or its transpose is an irreducibly diagonally dominant real matrix of order n, i.e.

$$a_{ii} \geq \sum_{j \neq i} |a_{ij}| \qquad \text{or} \qquad a_{ii} \geq \sum_{j \neq i} |a_{ji}| \tag{12.B.1}$$

with equality in no more than $(n-1)$ cases, then the determinant of a

$$|a_{ij}| > 0 \tag{12.B.2}$$

Since a is a real matrix, its eigenvalues are real or occur in complex conjugate pairs. Furthermore, the eigenvalues of an irreducibly diagonally dominant matrix have positive real parts (Varga, 1962; Taussky, 1949), and their product, the determinant, is therefore positive. If equality occurs in all cases the matrix is singular and the determinant zero (Taussky, 1949).

Lemma 12.2 If b is a real matrix of order n and

$$b_{ii} > 0 : b_{1i} \geq 0 : b_{ij} \leq 0 \qquad 1 \leq j \leq n \tag{12.B.3}$$

and

$$b_{ii} \geq \sum_{\substack{j \neq i \\ j=2}} |b_{ij}| \qquad \text{or} \qquad b_{ii} \geq \sum_{\substack{j \neq i \\ j=2}} |b_{ji}| \qquad 2 \leq i \leq n \tag{12.B.4}$$

with equality in no more than $(n-2)$ cases, then the determinant of b is non-negative.

Furthermore, if equality occurs in all $(n-1)$ cases, then the determinant of b is non-negative.

The cofactor of b_{1i} is the determinant of an irreducibly diagonally dominant matrix and is therefore positive. The cofactor of b_{1j} is

$$(-1)^{j+1} \begin{vmatrix} b_{21} & b_{22} & \cdots & b_{2,(j-1)} & b_{2,(j+1)} & \cdots & b_{2n} \\ \vdots & & & & & & \end{vmatrix} = (-1)^{2j-2} \begin{vmatrix} -b_{j1} & b_{j2} & \cdots & b_{j_n} \\ -b_{21} & b_{22} & \cdots & b_{2n} \\ \vdots & & & \end{vmatrix} \tag{12.B.5}$$

and since $b_{ij} \leq 0 (i \neq j, 2 \leq i \leq n, 1 \leq j \leq n)$, this is a determinant of the same form as b, but order $(n-1)$. Since the lemma is clearly true for $n = 2$, it follows by induction that it holds for all orders n.

Theorem 12.3 The real matrix r is of order $(n-1)$ and is of the form

$$r_{ij} = s_i + t_{ij} \qquad s_i \geq 0 \tag{12.B.6}$$

and t_{ij} or its transpose is an irreducibly diagonally dominant M-matrix, i.e.

$$t_{ii} \geq \sum_{j \neq 1}^{(n-1)} |t_{ij}| \quad \text{or} \quad t_{ii} \geq \sum_{j \neq 1}^{(n-1)} |t_{ij}| \quad \text{and} \quad t_{ij} \leq 0, i \neq j \tag{12.B.7}$$

with equality in no more than $(n - 2)$ cases.

The determinant, which may be written as

$$\left|r_{ij}\right| = \begin{vmatrix} 1 & 0 & 0 & \cdots \\ -s_1 & r_{11} & r_{12} & \cdots \\ -s_2 & r_{21} & r_{22} & \cdots \end{vmatrix} = \begin{vmatrix} 1 & 1 & 1 & \cdots \\ -s_1 & t_{11} & t_{12} & \cdots \\ -s_2 & r_{21} & t_{22} & \cdots \end{vmatrix} \tag{12.B.8}$$

is therefore a determinant of order n of type b. Therefore, by Lemma 12.2

$$|r_{ij}| > 0 \tag{12.B.9}$$

and the matrix r is non-singular. The matrix r is only singular if there is equality in all $(n - 1)$ terms t_{ii}, and the terms s_i are all zero.

Appendix 12.C An Algorithm Using the Exact Solution

In principle, the set of differential equations (12.2) may be exactly integrated if the transformation of R to its Jordan normal form is known (Wilkinson, 1965). If R is not defective, this is readily accomplished by the projection of the vector, q, onto the biorthogonal set of eigenvectors of R. In practice, it is necessary to first remove the singularity of R by solving the set of $(n - 1)$ linear equations

$$\sum_{j=1}^{(n-1)} S_{ij} q_j^s = -R_{in} \tag{12.C.1}$$

where S is given by Eq. (12.15) to determine the equilibrium state vector, q_n^s. Then, if S is not defective, the deviation vector, Q, is given by

$$\sum_{k=1}^{(n-1)} (y_k, Q^0) x_k \exp[-\lambda_k (t - t_0)] \tag{12.C.2}$$

where Q^0 is the value of Q at time t_0, and y_k and x_k are left- and right-hand eigenvectors of S with eigenvalue λ_k.

This form can only be used if the matrix is not defective; i.e. no two eigenvectors are parallel (Wilkinson, 1965). In practice, numerical evaluation introduces errors so exact parallelism is unlikely to occur. However, when the matrix is nearly defective, attention must be paid to the round-off error terms. Consider the example:

$$S = \begin{pmatrix} a & 1 \\ 0 & b \end{pmatrix} \tag{12.C.3}$$

which has eigenvalues and eigenvectors

$$\lambda = a\colon \begin{pmatrix} 1 \\ 0 \end{pmatrix} : (1, 1/(a - b)) \tag{12.C.4a}$$

or

$$\lambda = b\text{: } \begin{pmatrix} 1/(b-a) \\ 1 \end{pmatrix} : (0, 1) \tag{12.C.4b}$$

respectively.

The solution $Q = \begin{pmatrix} x \\ y \end{pmatrix}$ is therefore

$$x = [x_0 + y_0/(a-b)]e^{at}t + y_0/(b-a)e^{bt} \tag{12.C.5a}$$

$$= x_0 - y_0/(b-a)[1 - e^{(b-a)t}] \tag{12.C.5b}$$

$$y = y_0 e^{bt} \tag{12.C.5c}$$

Due to evaluation errors, the denominator term $(a - b)$ is not exactly equal to the corresponding one $(b - a)$, nor is the difference of the exponential arguments. Thus, proceeding to the limit $b \to a$, as the matrix becomes defective,

$$x \to [x_0 + y_0 t]e^{at} \tag{12.C.6}$$

which is an exact solution. The numerical calculation will not achieve this result unless care is taken to ensure the limit by a reduction to the exact solution. This may be accomplished by a series of tests on the eigenvalues (for degeneracy) and subsequently on the eigenvectors (for parallelism). If the order of the non-linear divisors is greater than two, the appropriate higher-order analytic solution must be used (Wilkinson, 1965). The determination of the eigenvectors of S can be performed by a standard method. Since R is, in general, an asymmetric matrix with no clearly defined sparsity pattern, a general eigenvalue method must be used. The most efficient such method involves a reduction to Hessenberg form followed by an iterative evaluation of the eigenvalues using the LR or QR schemes (Wilkinson, 1965). The reduction to Hessenberg form is generally the most computationally expensive involving approximately $2(n - 1)^3$ multiplications for Householder's method (Wilkinson, 1965) per iteration step (Wilkinson, 1965). Thus, including the necessary preliminary solution of the equilibrium state, this method involves approximately $2(n - 1)^3$ multiplications.

In view of the complicated structure of the scheme outlined above and the iterative nature of the calculations, it is natural to consider the suitability of an iterative scheme for calculating eigenvalues, for which the transformation obtained in the previous time step may be used as a first approximation. Jacobi Eberlein (1962) method provides such a scheme. However, the general method involves about $8(n - l)^3$ multiplications persweep and is therefore considerably more expensive than the QR iterative method.

M.1 Quasi-Linear Equations

The partial differential equation

$$a(x, y, w)\frac{\partial w}{\partial x} + b(x, y, w)\frac{\partial w}{\partial y} + c(x, y, w) = 0 \tag{M.1.1}$$

is called quasi-linear provided all terms with the highest-order derivatives of the dependent variables appear linearly. That is the coefficients of such terms are functions of lower order derivatives of the dependent variables only.

M.1.i General Form of First-Order Partial Differential Equations

The general form of a first-order quasi-linear partial differential equation in two variables may be expressed as

$$F\left(x_1, x_2, \ldots, x_n, w\right)\left[\frac{\partial w}{\partial x_1}\frac{\partial w}{\partial x_2}, \ldots, \frac{\partial w}{\partial x_n}\right] = 0 \tag{M.1.2a}$$

where $w = w(x_1, x_2, \ldots, x_n)$ is the unknown function and $F(\cdots)$. The system of ordinary differential equations may often be cast into a characteristic form

$$\frac{\mathrm{d}x}{f(x,y,w)} = \frac{\mathrm{d}y}{g(x,y,w)} = \cdots = \frac{\mathrm{d}w}{h(x,y,w)} \tag{M.1.2b}$$

It is not difficult to show that Eq. (M.1.2c) may be cast into many equivalent forms, of which Eq. (1.12) is particularly useful, containing the eigensolutions of ideal flow and consequently the characteristic form. This is discussed in Section 1.3.1.

Such equations are encountered in various applications (continuum mechanics, gas dynamics, hydrodynamics, heat and mass transfer, wave theory, acoustics, multiphase flows, chemical engineering, etc.). If the functions f, g and h are independent of the unknown w, then Eq. (M.1.2) are called *quasi-linear*, most of the equations with which we shall be concerned are of this type.

M.1.ii Characteristic System – General Solution

Suppose that two independent particular solutions of this system have been found in the form $u_1(x,y,w) = C_1$, $u_2(x,y,w) = C_2$, where C_1 and C_2 are arbitrary constants, such particular solutions are known as *integrals of system* (Eq. (M.1.2)). Then, the general solution to Eq. (M.1.2) (first-order quasi-linear PDE) can be written as $\Phi(u_1, u_2) = 0$, where Φ is an arbitrary function of two variables. With Eq. (M.1.2c) solved for u_2, one often specifies the general solution in the form $u_2 = \Psi(u_1)$, where $\Psi(u)$ is an arbitrary function of one variable. If $h(x,y,w) \equiv 0$, then $w = C_2$ can be used as the second integral in Eq. (M.1.2c).

M.1.iii Linear Second-Order Partial Differential Equations

The general second-order partial differential equation differs significantly in form from the first. The system of ordinary differential equations in n variables (x_1, x_2, …, x_n) and with unknown $w\left(x_1, x_2, \ldots, x_n\right)$ takes the form (Courant and Hilbert, 1962)

$$F\left(x_1, x_2, \ldots, x_n, \frac{\partial w}{\partial x_1}, \frac{\partial w}{\partial x_2}, \frac{\partial w}{\partial x_n}\right) = 0 \tag{M.1.3}$$

which if linear takes the form

$$a(x,y)\frac{\partial^2 w}{\partial x^2} + 2b(x,y)\frac{\partial^2 w}{\partial x \partial y} + c(x,y)\frac{\partial^2 w}{\partial y^2} + d(x,y)\frac{\partial w}{\partial x} + e(x,y)\frac{\partial w}{\partial y} + g(x,y) = 0 \tag{M.1.4}$$

If g = 0 the equation is said to be a *homogeneous linear equation*. Such equations have a number of useful properties which may help in constructing solutions:

i) The particular solution $w = 0$.
ii) The *principle of linear superposition*: if $w_1(x, y)$ and $w_2(x, y)$ are solutions, then $w_1 w_2$ is also a solution.
iii) If the equation has a particular solution $\overline{w}(x, y; \mu)$ depending on a parameter μ, then the derivatives with respect to μ are also solutions.
iv) If the preceding particular solution is multiplied by a function of μ, say $\phi(\mu)$, and then integrated over some range $\mu_1 \rightarrow \mu_2$, then the resulting function $\int_{\mu_1}^{\mu_2} \overline{w}(x, y); mu)\phi(\mu)d\mu$ is also a solution.
v) If the coefficients $a, b, \ldots, f$ are all independent of x, then there is a solution $w = e^{\lambda x}u(y)$. Substituting this solution in the original differential equation yields $u(y)$ as a second-order linear differential equation. Differentiating the solution with respect to x yields a particular solution to the original equation.

The linear homogeneous equation can be transformed into a standard canonical form by a series of translations and rotations (in a manner similar to that the general conic is reduced to a standard form) (Courant and Hilbert, 1962). These are:

Equation type	Condition	Canonical form	Representative
Elliptic	$b^2 - ac < 0$	$\dfrac{\partial^2 w}{\partial x^2} + \dfrac{\partial^2 w}{\partial y^2} = 0$	Laplace's equation
Parabolic	$b^2 - ac = 0$	$\dfrac{\partial w}{\partial x} + \dfrac{\partial^2 w}{\partial y^2} = 0$	Diffusion equation
Hyperbolic	$b^2 - ac < 0$	$\dfrac{\partial^2 w}{\partial x^2} - \dfrac{\partial^2 w}{\partial y^2} = 0$	Wave equation

If the right-hand side of the canonical form is non-zero, the equation is *inhomogeneous*. For example Poisson's equation is the inhomogeneous form of Laplace's equation

$$\nabla^2 \phi = -\frac{1}{\epsilon_0}\rho \tag{M.1.5}$$

where ϕ is the electrostatic potential and ρ the electrostatic charge density.[5]

The nature of the solutions follows immediately from the canonical forms.

- **Laplace's equation** is characteristic of steady-state potential problems encountered in classical potential theory in electro-statics, magneto-statics and incompressible fluid flow in the absence of a source. The properties of the solutions to the equation established in the late nineteenth century are known as *harmonic functions*. They are subject to boundary conditions on the surface S enclosing the calculation space V, which are

5 Second-order linear partial differential equations are treated extensively by Courant and Hilbert (1962) and Morse and Feshbach (1953). The elliptic and parabolic equations are treated thoroughly by Friedman (2008).

easily established, namely (Pert, 2013) (for a more formal discussion see Courant and Hilbert, 1962). The different boundary conditions are:

i) *Dirichlet,* where the value of the potential on the surface S is specified.
ii) *Neumann*, where the component of the potential gradient normal to the surface S is specified.
iii) *Mixed*, where at every point on S either the value or the normal potential gradient *delete* is specified.

- **Diffusion equation** is typical of heat flow into the volume V through the surface S. Typically, the heat flux through the surface S, namely $\int_S -\kappa\nabla T \cdot \mathrm{dT}$, leads to an increase in the total heat content of the volume $\int_V cT dV$ or

$$-\int_S \kappa \nabla T \cdot \mathrm{dS} = \int_V c\, \partial\mathbf{T}/\partial t\, dV$$

Applying Gauss' theorem yields the inhomogeneous equation

$$\partial T/\partial t = -\chi\, \nabla^2 T \tag{M.1.6}$$

where $\chi = c/\kappa$ is the thermal diffusivity. The boundary conditions require Dirichlet, Neumann or mixed conditions, but this time the parameter W is specified at time t in addition to the initial state w_0 at time $t = 0$ throughout the volume.

- **Hyperbolic equations**: The canonical equation can be factorised as

$$\frac{\partial^2 w}{\partial x^2} - \frac{\partial^2 w}{\partial y^2} = \left(\frac{\partial}{\partial x} + \frac{\partial}{\partial y}\right)\left(\frac{\partial}{\partial x} - \frac{\partial}{\partial y}\right) w \tag{M.1.7}$$

namely two repeated first-order elliptic equations each generating a set of characteristics, both propagating along $x \pm y = 1$. The solutions of the second-order equation therefore are described by Riemann invariants on the appropriate characteristic. Since the characteristics span the entire domain V, the boundary conditions on the solution require the values on the boundary S giving the starting values of the invariants, see Section 1.4 for a full explanation.

M.1.iii.a Existence Theorem for Hyperbolic Equations

Since any hyperbolic equation may be reduced to the canonical form

$$\frac{\partial^2 w}{\partial x^2} - \frac{\partial^2 w}{\partial y^2} = 0 \tag{M.1.8}$$

we may finite difference the x and y derivatives on a grid of $x_{i,j} = ij\Delta x^2$ and $y_{ij} = ij\delta y$, where i and j integers to give

$$\frac{\{w_{i-1,j} + w_{i+1,j}\}}{\Delta x^2} - \frac{\{w_{i,j-1} + w_{i,j+1}\}}{\delta y^2} = 4\left\{\frac{1}{\Delta x^2} - \frac{1}{\delta y^2}\right\} w_{i,j} \tag{M.1.9}$$

It is easy to see that given an initial condition on a boundary surface $\mathcal{I}$ where both w and $\frac{\partial w}{\partial n}$ are specified on the initial line that the solution may be obtained as j increases step by step through the mesh by unity. Note that the values $w_{i,j+1}$ and $\{\partial w/\partial n\}_{i+1,j}$ at the initial line are required to start the solution. As the calculation proceed it is easy to see that the solution is in two independent parts w_{odd} with $i+j$ odd and w_{even} with $i+j$ even. The two sets are separated in y by δy. Since the wave speed is unity we conclude that the solution can only exist if the even and odd sets are separated, i.e. by δy.

M.1.iv Green's Function

The *Green's function* for an inhomogeneous equation is determined by the solution for a unit source at the point $\mathbf{r}'$: $K(\mathbf{r}|\mathbf{r}')$, namely the solution of

$$\mathcal{L}(\phi) = \delta(\mathbf{r} - \mathbf{r}') \tag{M.1.10}$$

where $\mathcal{L}$ is the homogeneous second-order differential operator.

For example for Poisson's equation, the Green's function is obtained from elementary electro-statics

$$K(\mathbf{r}|\mathbf{r}') = \frac{1}{4\pi\epsilon_0}\frac{1}{|\mathbf{r} - \mathbf{r}'|} \tag{M.1.11}$$

and a general solution to Poisson's equation in the full space of $\mathbf{r}'$ is

$$\phi = \int_V K(\mathbf{r}|\mathbf{r}')\rho(\mathbf{r}')\,\mathrm{d}\mathbf{r}' = \int_V \frac{1}{4\pi\epsilon_0}\frac{\rho(\mathbf{r}')}{|\mathbf{r} - \mathbf{r}'|}K(r|r')\,\mathrm{d}\mathbf{r}' \tag{M.1.12}$$

the familiar result from electro-statics.

The general solution to an inhomogeneous equation can be constructed as the sum of a particular integral, namely the Green's function integral, and complementary function, a solution to homogeneous equation (e.g. a harmonic function). Hence, it follows that same form of boundary conditions for the inhomogeneous equation are required as those for the homogeneous. In classical potential theory, the harmonic function is provided by charges lying outside the calculation space V. These may be either *image charges* or *surface sheets of charge (Dirichlet) or infinitesimal dipoles (Neumann)* lying just outside the boundary S of calculation. However, this topic lies outside the scope of this text and therefore we refer the interested reader to the discussions by Morse and Feshbach (1953) for a clear physical introduction or Courant and Hilbert (1953) for a more rigorous mathematical treatment.

M.1.v Separation of Variables

In many cases, the linear, second-order partial differential equations describing physical problems may be solved by the method of separation of variables based on a restricted form of Eq. (M.1.4),

$$a(x)\frac{\partial^2 w}{\partial x^2} + b(x)\frac{\partial w}{\partial x} + c(x)w + d(y)\frac{\partial^2 w}{\partial y^2} + e(y)\frac{\partial w}{\partial y} + f(y)w = g(x, y) \tag{M.1.4a}$$

Equations of this form occur widely in physics: examples include Helmholtz' equation (and its variants), wave equation, Laplace's equation and the heat equation, provided the boundary condition themselves are separable. This usually requires that the boundary conditions be defined on a surface defined by one of the co-ordinates, for example on a simply connected, orthogonal Cartesian mesh, the boundary conditions take the form of Dirichlet or Neumann boundary condition on the mesh surfaces $x = i = \text{const}$, etc.

M.1.v.a Homogeneous Equations

The solution of a partial differential equation of this type can be expressed as a linear combination of the eigenvectors of the homogeneous equation (left-hand side). Substituting the product $w(x, y)X(x)Y(y)$ yields the equations:

$$a(x)\frac{d^2X}{dx^2} + b(x)\frac{dX}{dx} + c(x)X = nX \tag{M.1.13}$$

$$d(y)\frac{d^2Y}{dY^2} + e(y)\frac{dY}{dy} + f(y)Y = -nY \tag{M.1.14}$$

for the eigenfunctions X, Y with eigenvalue n,

Let us examine this process in more detail with a simpler system. Consider an initial boundary value problem for the function $u(x, t)$ on $\mathcal{D}(x, t); x \in (0, t), t \geq 0$ in two variables, which may be written as[6]

$$(\mathcal{S}[u)(x, t)] = (\mathcal{T}[u)(x, t)] \tag{M.1.15}$$

where $\mathcal{S}$ is a differential operator with respect to x and $\mathcal{T}$ a differential operator with respect to t with boundary data

$$\mathcal{S}[u(0, t)] = \mathcal{S}[u(a, t)] = 0 \qquad \text{for} \qquad t \geq 0 \tag{M.1.16}$$

$$\mathcal{T}[u(x, 0)] = h(x) \qquad \text{for} \quad 0 \leq x \leq a \tag{M.1.17}$$

where $h(x)$ is a known function.

Inserting a trial solution $u(x, t) = f(x)g(t)$ and dividing through by $f(x)g(t)$ gives

$$\frac{\mathcal{S}f}{f} = \frac{\mathcal{T}g}{g} \tag{M.1.18}$$

Since the left-hand side depends only on t and the right only on x, both must equal a constant K i.e.

$$\mathcal{S}f = Kf \qquad\qquad \mathcal{T}g = Kg \tag{M.1.19}$$

which we recognise as the eigenvalue equations for the operators $\mathcal{S}$ and $\mathcal{T}$ subject to the appropriate boundary conditions. If $\mathcal{S}$ is a compact, self-adjoint operator on the space $\mathcal{L}^2[0, l]$ along with the relevant boundary conditions, then by the spectral theorem there exists a basis for $\mathcal{L}^2[0, l]$ consisting of eigenfunctions for $\mathcal{S}$. Let the spectrum of $\mathcal{T}$ be $\mathcal{E}$ and let f_λ be an eigenfunction with eigenvalue $\lambda \in \mathcal{E}$. Then for any function which at each time t is square-integrable with respect to x, we can write this function as a linear combination of the f_λ. In particular, we know the solution u can be written as

$$u(x, t) = \sum_{k \in \mathscr{E}} c_\lambda \; t_0 \, f_\lambda(x) \tag{M.1.20}$$

For some functions $c_\lambda(t)$ in the separation of variables, these functions are given by solutions to $\mathcal{T}g = Kg$, i.e. eigenfunctions of $\mathcal{T}$.

6 Note $\mathcal{S}$ and $\mathcal{T}$ must be in the form of Eq. (M.1.14): if necessary multiplication by an appropriate factor may be required.

M.1.v.b Inhomogeneous Equations

When the right-hand side of Eq. (M.1.4a) is not zero, we may proceed in a similar fashion to identify the eigensolutions of the homogeneous equations, and then using the spectral theorem the right-hand side is represented by a linear combination of those eigenfunctions.

It is clear that the separation of variables is applicable to a range of physics problems provided the boundary conditions are themselves separate. When appropriate this allows the dimensionality of the equations to be reduced, e.g. from a single two-dimensional system to two separate one dimensional ones. This may be a great saving in computer simulation in that one-dimensional problems are usually simpler to solve.

A good example is the Poisson's equation which may be solved in terms of its eigensolutions, namely Fourier expansions (Supplement M.6).

M.2 Stiff Equations

In mathematics, a stiff equation is a differential equation for which certain numerical methods for solving the equation are numerically unstable, unless the step size is taken to be extremely small (Gear, 1971; Press et al., 1986): Lambert (1991, ch. 6) gives a thorough discussion of stiff equations in a context appropriate to this work, to which reference may be made. It has proven difficult to formulate a precise general definition of stiffness, but the main idea is that the equation includes some terms that can lead to rapid variation in the solution.

When integrating a differential equation numerically, one would expect the requisite step size to be relatively small in a region where the solution curve displays much variation and to be relatively large where the solution curve straightens out to approach a line with a slope of nearly zero. For some problems this is not the case. In order for a numerical method to give a reliable solution to the differential system, sometimes the step size is required to be at an unacceptably small level in a region where the solution curve is very smooth. This phenomenon is known as stiffness. In some cases, there may be two different problems with the same solution, yet one is not stiff and the other is. The phenomenon cannot therefore be a property of the exact solution, since this is the same for both problems and must be a property of the differential system itself. Such systems are thus known as stiff systems.

Stiff sets of equations occur when one deals with more than one first-order differential equation in a set where the solution may contain several different functions of the independent variable which contain two or more very different scales of the independent variable on which the dependent solution variables depend. For example consider the simple second-order equation

$$\frac{d^2y}{dx^2} = f(x)y \tag{M.2.1}$$

which is equivalent to the two first-order equations

$$y = y_2 = \frac{dy_1}{dx} \qquad \text{and} \qquad y_1 = \frac{dy}{dx} = f(x)y \tag{M.2.2}$$

For example if we set $f(x) = const$ then the solution is subject to the boundary condition $x = 0, (\dot{y}) = \dot{y}_0$, when $y = y_0$ then the two solutions are a decaying and a growing term

$$y = A\exp(-\dot{y}_0 x) + B\exp(\dot{y}_0 x) \qquad \text{where} \qquad A + B = 0 \tag{M.2.3}$$

In many physical problems, we require the solution to decay to zero at infinity ($y(x) \to 0$ as $\to \infty$), namely the decaying exponential solution.

For large values of $f \approx 100$, y is the sum of two exponentials with large arguments. For example if we approximate f as a large constant then the solution contains both a growing and a decaying term. If, as is often the case, the physical solution is the decaying one, then the growing solution represents an error term, which is induced by small errors in the initial condition. This differentiation makes the solution difficult to calculate by standard numerical methods, which need a small time step ($\sim 1/\dot{y}_0$). Otherwise the growing (error) term will rapidly overtake the decaying one and the latter will be lost. This difficulty is circumvented if an analytic form is available, which may be used to form the basis of an approximate routine, thereby avoiding the stiffness problems.

Plasma physics is full of stiff equations. Perhaps the best known is the single-particle model where there are two characteristic times, widely different in scale. The first is the particle cyclotron period and the second is the laboratory time scale of any variation in the applied fields. It is axiomatic that these two are very different due to the large mass discrepancy between ions and electrons, cyclotron motion being controlled by the electron mass and the laboratory time scale, typically determined by the hydrodynamic time scale (i.e. the sound speed) associated with the ion (heavy) particle mass. In the single-particle model, the problem of treating the slow field-induced changes, simultaneously is resolved by averaging over the rapid motion and including the cyclotron terms as perturbations in the resultant drift (guiding centre motion) (Morozov and Solev'ev, 1966). Major issues arise from the coupled nature of several elements in a system, each with a widely differing time constant. This leads to a set of ordinary differential equations with markedly different eigenvalues (as in the collisional–radiative equations considered here). Numerical integration requires that the time development follows the slowest process. The integration of the full set of equations is therefore limited by the slowest terms and the overall integration is in consequence *stiff*.

A central problem arises when many of the individual equations are stiff. In that, the time step is difficult to determine such that any change over a step is small, and potentially leads to an arbitrarily large number of steps to integrate over predetermined time. This inhibits the requirement of a reasonably efficient scheme and usually prevents simple two-step schemes with reasonably long time steps.

Fortunately, the form of many of the equations is that of diffusion equations which have the property of converging uniformly to a steady state in the asymptotic limit. As a result, if the details of the approach to the steady state are not important we may essentially average over this phase by using an implicit scheme with a relatively long time step.

M.3 Weak Solutions

In mathematics, a weak solution (also called a generalised solution) to an ordinary or partial differential equation is a function for which the derivatives may not all exist but which is nonetheless deemed to satisfy the equation in some precisely defined sense. There are many different definitions of weak solution, appropriate for different classes of equations. One of the most important is based on the notion of distributions.

Avoiding the language of distributions, one starts with a differential equation and rewrites it in such a way that no derivatives of the solution of the equation show up (the new form is called the weak formulation, and the solutions to it are called weak solutions). Somewhat surprisingly, a differential equation may have solutions which are not differential; and the weak formulation allows one to find such solutions.

Weak solutions are required in many problems in physics and engineering where sufficiently smooth solutions do not exist to model the actual situation. In fluid mechanics, they provide an essential formal mathematics method for handling discontinuities and shocks (see Lax (1954) and also Pert (2013, Appendix A.10.4)). The only way to solve such problems is to generalise the solution of the differential equation as a weak solution. We shall meet such solutions in solving the hydrodynamics of compressible fluids and in formulating the equations of finite element solutions.

The general idea is that, when solving a differential equation in u, one can rewrite it using a *test function* φ, such that whatever derivatives in the solution u which show up in the equation are 'transferred' via integration by parts to the test function φ, resulting in an equation without derivatives of u. This new equation generalises the original equation to include solutions which are not necessarily differential.

The approach works in great generality. Indeed, consider a linear differential operator in an open set $\mathbf{W} \in \mathbf{R}^n$:

$$\mathcal{L} = \sum a_{\alpha_1, \alpha_2, \ldots, \alpha_n}(x)\, \partial^{\alpha_1} \partial^{\alpha_2} \ldots \partial^{\alpha_n} u(x) \tag{M.3.1}$$

where the multi-index $a_1, a_2, \ldots, a_n$ varies over some finite set in $\mathbf{N}^n$ and the coefficients $a_{\alpha_1, \alpha_2, \ldots, \alpha_n} \partial^{\alpha_1} \partial^{\alpha_2}, \ldots, \partial^{\alpha_n}$ are smooth enough functions of $x \in \mathbf{R}^n$.

The differential equation $\mathrm{P} = (\mathcal{L}(x, \partial) u(x)) = 0$ can, after being multiplied by a smooth test function φ with compact support in $\mathbf{W}$ and integrated by parts, be written as

$$\int_{\mathbf{W}} u(x) Q(x, \partial) \varphi(x) \mathrm{d}x = 0 \tag{M.3.2}$$

where the differential operator $Q(x, \partial)$ is given by the formula

$$Q(x, \partial)\varphi(x) = \sum (-1)^{\tilde{\alpha}} \partial^{\alpha_1} \partial^{\alpha_2} \ldots \partial^{\alpha_n} \left[a_{\alpha_1} a_{\alpha_2} \ldots a_{\alpha_n}(x) \varphi(x) \right] \tag{M.3.3}$$

The number

$$(-1)^{\tilde{\alpha}} = (-1)^{\alpha_1 + \alpha_2 + \cdots + \alpha_n} \tag{M.3.4}$$

shows up because one needs $\alpha_1 + \alpha_2 + \cdots + \alpha_n$ integration parts to transfer all the partial derivatives from u to φ in each term of the differential equation, and each integration by parts entails a multiplication by -1.

The differential operator $Q(x, \partial)$ is the formal adjoint of $P(x, \partial)$ (cf. adjoint of an operator).

In summary, if the original (strong) problem was to find $a\tilde{\alpha}$ times the differentiable function u defined on the open set $\mathbf{W}$ such that

$$P(x, \varphi)u(x) = 0 \tag{M.3.5}$$

for all $x \in \mathbf{W}$: when $P(x, \partial)\, u(x) = 0$ for all $x \in \mathbf{W}$.

The equation $P(x, \partial), u(x) = 0$ when $x \in W$ generates a so-called strong solution, and an integrable function u is said to be a weak solution if

$$\int_{\mathbf{W}} u(x)Q(x, \partial)(\varphi)\mathrm{d}x = 0 \tag{M.3.6}$$

for every smooth function φ with compact support in $\mathbf{W}$.

NB. The notion of a weak solution based on distributions is sometimes inadequate. In the case of hyperbolic systems, the notion of a weak solution based on distributions does not guarantee uniqueness, and it is necessary to supplement it with entropy conditions or some other selection criterion.

M.4 Operator Splitting

These methods are also known as *split-time-step* or *the method offractional steps* (Press et al., 1986).

Let us suppose that the time development of a complex system is described by a variable vector u whose time development is given by an operator $\mathcal{L}$ so that

$$\frac{\partial u}{\partial t} = \mathcal{L}\left(u, \frac{\partial \mathbf{u}}{\partial \mathbf{x}}, \frac{\partial^2 \mathbf{u}}{\partial \mathbf{x}^2}, \dots\right) \tag{M.4.1}$$

We suppose the operator $\mathcal{L}(u)$ may be subdivided into m-independent operators $\mathcal{L}_m(u)$:

$$\mathcal{L}\,u = \sum \mathcal{L}_1\,u + \mathcal{L}_2\,u + \cdots + \mathcal{L}_m\,u \tag{M.4.2}$$

where the operators $\mathcal{L}_i(u)$ are the components of $\mathcal{L}(u)$.

Suppose a set of numerical approximation schemes exist for each of the $\mathcal{L}_s$, which satisfy the necessary conditions of consistency, convergence and stability, and other physical constraints such positivity, conservation, etc. within a known range of time-step Dt. The order of the whole is that of the lowest integration scheme. Then, we may update each component independently of the others in turn

$$\begin{aligned} u^{n+1} &= \mathcal{U}'_1(u^n, Dt) \\ u^{n+1} &= \mathcal{U}'_2(u^n, Dt) \\ &\vdots \\ u^{n+1} &= \mathcal{U}'_m(u^n, Dt) \end{aligned} \tag{M.4.2:1}$$

where u^n is the value of u after n time steps and $\mathcal{U}'_i(u, Dt)$ the result of operating on the state u with the finite difference from the operator $\mathcal{L}_i$ for the time step Dt.

Suppose that a successive evaluation of each well-behaved operator $\mathcal{L}_i$ for a time Dt will correspond to one evaluation of the full-time development operator $\mathcal{L}$ for a single-step Dt; and that the overall result will be well behaved. This very simple method of splitting is widely used to time integrate the complex set of processes shown in Figure I. We further assume that the operators $\mathcal{L}_i$ are linearly independent of the time step Dt.

It is easy to show that provided the individual operators are independent and stable this overall scheme is stable. Consider a set of operators $\mathcal{U}_i$ operating on a set of elements u_j, each with error ϵ_j. Stability of each operator requires that amplification factor G_{ij} of the element j due to the ith operation satisfy the condition $|G_{ij}| \leq 1$. The further condition that the operators be independent requires that ϵ_i does not depend on $\epsilon_j (i \neq j)$ at every stage (thereby avoiding non-linear instability). The error in the element j is increased as a result of operation i by the amplification factor G_i so that the overall amplification factor due to m operators over a single-step Dt:

$$G'_j = \prod_{i=1,m} G_{ij} \leq 1 \tag{M.4.3}$$

This leaves the order in which the individual integrations are performed open and undecided at this stage. It is left to the experience and knowledge of the code designer to select an appropriate route. Although, this is usually a fairly easy decision as many of the process commute

$$\mathcal{L}_i \, \mathcal{L}_j = 0 \qquad \text{if} \quad i \neq j$$

and their ordering is irrelevant. Each term is evaluated using the data from the previous time step similar to the Jacobi method for matrix solution by iteration.

We may however seek to obtain an improved value by always using the current best value, u_i^{n+1} at the ith operator. Hence, become Eq. (M.4.2:1)

$$\begin{aligned} u^{n+1} &= \mathscr{U}'_1(u^n, Dt) \\ u^{n+1} &= \mathscr{U}'_2(u^n, Dt) \\ &\vdots \\ u^{n+1} &= \mathscr{U}'_m(u^n, Dt) \end{aligned} \tag{M.4.3:2}$$

At this point, we must note a further problem if the operators $\mathcal{L}_i$ do not commute. In fluid mechanics, most of the peripheral operators take the form of $\mathcal{L} = \mathbf{I} + \mathcal{L}'\Delta t$ where $\mathbf{I}$ is the identity operator. If the commutator $\mathcal{L}_i - \mathcal{L}_j \neq 0$ then the order of the terms associated with $\mathcal{L}_i$ and $\mathcal{L}_j$ is reversed depending on the order in which i and j are accessed leading to a difference in the sign of the terms i and j of order $O(\Delta t^2)$.

M.4.i Split-Time-Step

We may remove much of the subjectivity by treating each split step in an equivalent way by using a split-time-step Dt/m, and simultaneously further refining this solution by successively changing the operator U_i to include the full set (Yanenko, 1971). Thus, let $\mathcal{U}_i$ be result when the full operator $\mathcal{L}$ in which the element only $\mathcal{L}_i$ is treated implicitly, then

$$\begin{aligned} u^{(n+1/m)} &= \mathcal{U}_1(u^n, Dt/m) \\ u^{(n+2/m)} &= \mathcal{U}_2(u^{n+1/m}, Dt/m) \\ &\vdots \\ u^{(n+i/m)} &= \mathcal{U}_i(u^{n+(-1)1/m}, Dt/m) \\ u^{n+1} &= \mathcal{U}_m(u^{n+(m-1)/m}, Dt/m) \end{aligned} \tag{M.4.4}$$

The time step in each case is taken to be Dt/m to allow for the fact that each individual iteration is repeated m times.

This method is closely related to the ADI method[7] used particularly for elliptic and parabolic methods – indeed for $m = 2$ the two methods are identical. The latter approach is discussed in Section 10.1, which deals with diffusion in two or more dimensions (Section 10.2.2.1).

The stability of the final algorithm is easily established as the operators commute. Since each operation is stable, each individual amplification matrix (namely over one sub-step) satisfies $|g_i| \leq 1$. Further, since each operator commutes the total amplification factor per complete time step (namely over m sub-steps) $G_i = mg(i)/m \leq 1$. Hence, the operators are linear in Dt. The amplification matrix of the full operator $\mathcal{L}$ satisfies $|G| \leq 1$. Hence, the full operation $\mathcal{L}$ is stable.

A simple example (Yanenko (1971, §2,6), Richtmyer and Morton (1967, §8.9)) will make this approach clear.

Consider the three-dimensional diffusion equation

$$\frac{\partial X}{\partial t} + D\nabla^2 X = 0 \tag{M.4.5}$$

which is differenced in three steps ($S = 1, 2, 3$) in x, y and z, each sequentially using the one-dimensional unconditionally stable Crank–Nicholson or centred-difference scheme in each direction as foundation. Thus, $Dt/m = \frac{1}{3}Dt$ is the modified time-step and $mD = 3D$ the modified constant $D' = \frac{Dt}{m}\left\{D_1 + D_2 + D_3\right\} = D$

$$\frac{D'_s \Delta t'}{\delta x_s^2}\left\{\left[X_{m+1}^{n+s} - 2X_m^{n+s} + X_{m-1}^{n+s}\right]\left[X_{m+1}^{n} - 2X_m^{n} + X_{m-1}^{n}\right]\right\} = 0 \tag{M.4.6}$$

Since this involves a series of independent Crank–Nicholson equations each of which is stable, due to the individual growth factors $|G_S| \leq 1$, the overall integration growth factor $|G| = |\prod G_s| \leq 1$, and the algorithm is therefore stable.

M.5 Statistics Primer

For those not familiar with basic statistical nomenclature and results, we have included a short primer of statistical concepts and terms used in Monte-Carlo methods. This is not intended as an introduction to the formal theory of statistics or probability, but rather as an introduction to the key concepts as used to interpret physics experimental and therefore simulation results (Levy and Preidel, 1944, see for example)

M.5.i Basic Stochastic Nomenclature and Results

Random event is an event which has a chance of happening.

Probability is a measure of that chance.

Strictly probability is a Boolean or logical variable, it takes two values TRUE or FALSE. If TRUE the event happens and if FALSE it does not.

---

7 The two methods were developed independently by Peaceman and Rachford (1955) and Douglas (1955) in the United States and Yanenko (1971) in the USSR at the same time. There is little to choose between them either for accuracy or for efficiency.

More generally, we broaden the definition to a variable in the range $0 \leq P \leq 1$ which measures the likelihood of the particular event taking place: higher values indicate greater chance and smaller less. Unit probability events certainly take place. Zero probability events never occur. We shall use probability in this sense.

If $P(A)$ is the probability of event A, then

$$P(A+B+\cdots) \text{ is the probability of either } A \text{ or } B$$
$$P(AB) \text{ is the probability of } A \text{ and } B$$
$$P(A|B) \text{ is the probability that } A \text{ occurs when } B \text{ occurs} \tag{M.5.1}$$

the latter is known as the conditional probability of A given B.

The basic rules of probability theory[8] are

$$P(A+B+\cdots) \leq P(A)+P(B) \qquad P(AB) = P(A|B)P(B) \tag{M.5.2}$$

If only one of A or B can occur they are *exclusive* and

$$P(A+B+\cdots) = P(A)+P(B)+\cdots \tag{M.5.3}$$

If at least one of A, B must occur,

$$P(A+B+\cdots) = 1 \tag{M.5.4}$$

an *exhaustive* set of events.

The conditional probabilities obey Bayes' theorem

$$P(A|B) = \frac{P(A)P(B|A)}{P(B)} \tag{M.5.5}$$

If the chance of A occurring is independent of B, A and B are *independent*, then

$$P(A|B) = P(A) \tag{M.5.6}$$

Random Variables

Consider a set of exhaustive and exclusive events, each characterised by a number η. The number η is called a *random variable*. Associated with it is its *cumulative distributive function* $F(y)$ defined as the probability that the event has a value η not exceeding y

$$F(y) = P(\eta \leq y) \tag{M.5.7}$$

Since the event takes place the probability of it occurring is $0 \leq P \leq 1$.

---

8 These are not the formal set of axioms introduced by Kolmogorov which are

i) Probability is a non-negative real number

$$P(E) \in \mathbb{R}, P(E) \geq 0 \qquad E\forall F$$

where E is the event space.

ii) One elementary event has a probability 1-unit measure

$$P(\omega) = 1$$

iii) **Additivity**: Any countable sequence of disjoint sets, or mutually exclusive events $E_1, E_2, \ldots$ satisfies

$$P\left(\bigcup_{i=1}^{\infty}\right) = \sum_{i=1}^{\infty} P(E_i)$$

The *expectation or mean value* of a function g(y) is

$$\langle \eta \rangle =: \mathbf{E}(y) = \int g(y)\, \mathrm{d}F(y) \tag{M.5.8}$$

the integral being taken over all values of y.

i) If η is continuous, then $F(y)$ has a derivative and

$$f(y) = \mathrm{d}F(y)/\mathrm{d}y \qquad \langle g \rangle = \int g(y) f(y)\, \mathrm{d}y \tag{M.5.9}$$

ii) If η is discrete so that $F(y)$ is a step-function with steps f_i at points y_i,

$$\langle g \rangle = \sum_i g(y_i) f_i / \sum_i g(y_i) \tag{M.5.10}$$

Clearly, $F(-\infty) = 0$, $F(\infty) = 1$ and we require that $F(y)$ is a non-decreasing function of y. The terms $f(y)$ and f_i are called the *frequency functions* of the random variable η. $f(y)$ is also called the *probability density function*; f_i or $f(y)\mathrm{d}y$ give the probability of a value at y_i or $y - y + \mathrm{d}y$, respectively.

We may generalise to consider a pair of exhaustive and exclusive events each characterised by a pair of numbers η and ζ for which $F(y, z)$ is the

$$F(y, z) = P(\zeta \le y, \eta \le z) \tag{M.5.11}$$

From this set, we may construct a new set, each event being the aggregate of all events in the full set which have a prescribed value η. The new set has a distribution function $G(y)$. Similarly, for the random variable ζ we have a distribution $H(z)$,

$$G(y) = P(\eta \le y) \qquad H(z) = P(\zeta \le z) \tag{M.5.12}$$

If it happens that

$$F(y, z) = G(y)\, H(z) \tag{M.5.13}$$

the variables η and ζ are *independent*.

M.5.ii Moments

We define

$$\langle \eta^r \rangle = \int y^r\, \mathrm{d}F(y) = \int y^r f(y) \mathrm{d}y \tag{M.5.14}$$

when $\int G(y)dy = 1$ as the rth moment of η and $\mu_r = \langle \eta \rangle^r$ as the rth central moment. Hence, we have:

- The first moment $\langle \mu \rangle = \overline{\mu}$ is the *mean* or *expectation value* of μ.
- The second moment $\langle \mu^2 \rangle = \sigma^2$ is the *coefficient of variation* or *variance*. Its square root $\sigma = \sqrt{\langle \sigma^2 \rangle}$, approximately equal to the root mean squared (RMS) deviation, measures the width of the distribution of the values about the mean.

Algebraic sums and functions of random variables

Given a set of M random variables x_i with mean $\overline{x}_i$ and a second independent set of N random variables y_j with mean $\overline{y}_j$. The deviations from their means are $\delta_{xi} = x_i - \overline{x}$ and

$\delta_{y_j} = y_j - \bar{y}$ respectively, then clearly the variances

$$\sigma_x^2 = \frac{1}{M}\sum_{i=1}^{M}\delta_{xi}^{\;2} \qquad \sigma_y^2 = \frac{1}{N}\sum_{j=1}^{N}\delta_{yj}^{\;2} \tag{M.5.15}$$

More generally, if $F(x)$ and $G(y)$ are functions of x and y it follows for an arbitrary function $F(x)$ that

$$\sigma_{F(x)}^2 = \left(\frac{\partial F}{\partial X}\right)^2 \sigma_X{}^2 \tag{M.5.16}$$

and similarly for a function $G(y)$.

M.5.iii Covariance, Correlation and Regression

These are three closely related terms identifying the relationship of one set of variables with another. Given two real-valued random variables η and ζ with finite second moments (variances).

- *Covariance* (cov) is the expected value (mean) of their individual deviations from their joint mean. It is a measure of the width of the distribution of the values about the mean

$$\begin{aligned}\mathrm{cov}(\eta,\zeta) =: \sigma(\eta,\zeta) &= < (\eta - <\eta>) >< (\zeta - <\zeta>) > \\ &= < \eta\zeta - \eta <\zeta> - \zeta <\eta> + <\eta><\zeta>> \\ &= < \eta\zeta - <\eta><\zeta>>\end{aligned}$$

  or changing the notation to that commonly used in statistics where $\mathbf{E}[X]$ is the mean or expected value of the set X

$$\begin{aligned}\sigma(\eta,\zeta) &= \mathbf{E}[(\eta - \mathbf{E}[\eta])]\mathbf{E}[(\zeta - \mathbf{E}[\zeta])] \\ &= \mathbf{E}[\eta\,\zeta] - \eta\mathbf{E}[\zeta] - \zeta\mathbf{E}[\eta] + \mathbf{E}[\eta]\mathbf{E}[\zeta] \\ &= \mathbf{E}[\eta\zeta]L1 - \mathbf{E}[\eta]\mathbf{E}[\zeta]\end{aligned}$$

  If positive the sign of the covariance tells us whether one variable is increasing with the other and vice versa if otherwise. The covariance therefore gives an indication of the extent to which the variation of one set of variables Y follows the other X.
- *Correlation* is closely related to the covariation between two sets of values X and Y. The correlation is the measure of this co-variation. It is the dimensionless normalised form of covariance

$$\mathrm{corr}(X,Y) =: \rho(\eta,\zeta) = \frac{\mathrm{cov}(X,Y)}{\sqrt{\mathrm{var}\,X}\sqrt{\mathrm{var}\,Y}} \tag{M.5.17}$$

  It follows from the Cauchy–Schwarz inequality that the correlation coefficient satisfies the condition[9]

$$-1 \le \mathrm{corr}(X,Y) \le 1 \tag{M.5.18}$$

9 Defining an inner product on the set of random variables using the expectation of their product

$$\rho(X,Y) = \mathbf{E}(XY)$$

the Cauchy–Schwarz inequality

$$\mathbf{E}(XY)^2 \le \mathbf{E}(X^2)\mathbf{E}(Y^2)$$

- **Regression**: Suppose in an experiment we measure a set of dependent values y_i at each independent point x_i, each value of y_i consists of a random variation and a systematic error. Suppose the systematic value is linearly dependent on the control variable x_i, i.e.

$$y_i = y_i' + \beta x_i \tag{M.5.19}$$

Can we use the random nature of y' to find β and hence eliminate the systematic error from the random reading? The best estimate of the deviation may be obtained by the least-square deviation of the values y_i from the line

$$y_i = mx_i - c \tag{M.5.20}$$

by varying m and c. The deviation of the value y_i from the line is which we minimise by varying m and c,

$$\begin{aligned}\frac{\mathrm{d}\delta^2}{\mathrm{d}m} &= 2\sum x_i(y_i - mx_i - c)\\ &= 2\sum x_i y_i - m\sum x_i^2 - c\sum x_i = 2n\left\{\sigma_{x,y} - m\sigma_x^2 - c\bar{x}\right\} = 0\end{aligned} \tag{M.5.21}$$

$$\begin{aligned}\frac{\mathrm{d}\delta^2}{\mathrm{d}c} &= 2\sum (y_i - mx_i - c)\\ &= 2\left\{\sum y_i - m\sum x_i^2 - c\sum x_i\right\} = 2n\left\{\bar{y} - m\bar{x} - c\right\} = 0\end{aligned} \tag{M.5.22}$$

Substituting for c from Eq. (M.5.22) in (M.5.20) we obtain

$$y - \bar{y} = m(x - \bar{x})$$

so that the line passes through the centroid of the points $(\bar{x}, \bar{y})$. The gradient follows from substituting Eq. (M.5.22) in Eq. (M.5.21)

$$m = \sigma_{x,y}/\sigma_x^2 \tag{M.5.23}$$

and Eq. (M.5.20) may be written as

$$\frac{y - \bar{y}}{\sigma_y} = r\frac{x - \bar{x}}{\sigma_x} \tag{M.5.24}$$

where

$$r = \sigma_{x,y}/\sigma_x \sigma_y \tag{M.5.25}$$

is the *coefficient of correlation*. This line, where y is assumed to be accurate, is known as the *regression of x on y*, and vice versa if x is accurate. This equation may be interpreted in several ways:

1) As the best-fit line when y_i contains a random element.
2) Interchanging y and x to consider the case when x_i contains a random element in the line through the centroid is

$$\frac{(x - \bar{x})}{\sigma_x} = r'\frac{(y - \bar{y})}{\sigma_y} \tag{M.5.26}$$

where $r' = \sigma_{y,x}/\sigma_y \sigma_x 1 = /r$ to obtain the *regression line of y on x*.
3) Identifying the two regression lines with gradients r and r' passing through the centroid. Noting $r = \sigma x, y\sigma_y/\sigma_x$ and $r' = \sigma_{y,x}\sigma_x/\sigma_y$.

If the two lines coincide, $r = r' = 1$ then x and y are totally correlated and one is fully dependent on the other.

M.5.iv Elementary Statistical Results

Consider a set of N randomly distributed values $x_1, x_2, \ldots, x_N$.

- Firstly the set of values x_i may be formed from a well-defined distribution which follows a mathematical distribution function $\phi(x)$ with well-defined true value equal to its mean, and distribution determined by the mean $\mathbf{E}(x)$ and the variance σ^2, which are easily calculated by the appropriate integrations.
- Secondly the set x_i are a set of measurements generated in an experiment. In this case, the distribution is not well defined and the spread of the values is defined by the RMS value. Consider a set of N randomly distributed values $x_1, x_2, \ldots, x_n$. for a quantity x. The 'true' value $\tilde{x}$ is unknown, so the question becomes what is the best approximation to the true value? The total squared deviation with respect to the assumed 'true value' summed over all values gives a measure of error $\sum \tilde{\delta}_i^2 = (x_i - \tilde{x})^2$. Since $\tilde{x}$ is unknown we may seek the value of the unknown $\tilde{\delta}$ such that the net deviation is minimised $\sum(x_i - \tilde{x}) = 0$. Therefore the best approximation to the true value is
$$\tilde{x} = \frac{1}{N}\sum x_i = \overline{x} \tag{M.5.27}$$
i.e. the mean.

What is the relationship between the 'variance' σ^2 of the well-defined distribution and the mean squared deviation of the actual values? Clearly, $\sum \delta_i = 0$ and therefore

$$0 = \frac{1}{N^2}\sum_i \delta_i \sum_j \delta_j = \frac{1}{N^2}\left\{\sum_i \sum_{j \neq i} \delta_i\, \delta_j + \sum_i {\delta_i}^2\right\} \tag{M.5.28}$$

so we obtain the useful result that $\sum_i \sum_{j \neq i} \delta_i\, \delta_j = -N\sigma^2$. Hence, the mean squared deviation

$$\overline{\delta}^2 = \frac{1}{N}\left\{\sum_i \sum_{j \neq i} \delta_i\, \delta_j = \sum_i \sum_j \delta_i\, \delta_j - \sum_i {\delta_i}^2\right\} = \frac{N-1}{N}\sigma^2 \tag{M.5.29}$$

A result which is familiar to experimental physicists assigning errors to a set of measured results. The factor $N/(N-1)$ is due to the constraint introduced on their sum by a condition that deviations are taken with respect to the measured best (mean) value not to the (unknown) true one.

M.5.v Variance of a Sum of Samples

Suppose that all the values in a sample X are multiplied by a constant a, then is obvious that

$$\mathrm{var}(aX) = a^2 \mathrm{var}(X) \tag{M.5.30}$$

Suppose that the values of the sample X are added to a second sample Y of equal number, then it is straightforward to show

$$\mathrm{var}(aX + bY) = a^2 \mathrm{var}(X) + b^2 \mathrm{var}(Y) + 2\,a\,b\,\mathrm{cov}(X, Y) \tag{M.5.31a}$$

$$\mathrm{var}(aX - bY) = a^2 \mathrm{var}(X) + b^2 \mathrm{var}(Y) - 2\,a\,b\,\mathrm{cov}(X, Y) \tag{M.5.31b}$$

Generalising to consider multiple sets we have

$$\text{var}\sum_{i=1}^{N}(X_i) = \sum_{i=1}^{N}\text{var}(X_i) \tag{M.5.32}$$

If the sets are independent $\text{Cov}(X_iX_j) = 0$ when $i \neq j$, this reduces to Bienaymé's formula

$$\text{var}\sum_{i=1}^{N}(X_i) = \sum_{i=1}^{N}\text{var}(X_i) + \sum_{i=1}^{n}\sum_{\substack{j=1\\ j\neq i}}^{n}\text{cov}(X_iX_j) \tag{M.5.33}$$

M.5.vi Variance of the Mean

We may consider the sample set of values obtained in an experiment to be one from a series of N repeated measurements each of n values. Then the best value is obtained from the overall mean of the full nN total set ($N \geq 1$)

$$\overline{x} = \frac{1}{nN}\sum_{i=1}^{nN}x_i = \frac{1}{N}\sum_{j=1}^{N}\frac{1}{n}\sum_{i=1}^{n}x'_{ij} = \frac{1}{N}\sum_{i=1}^{N}\overline{x}_i \tag{M.5.34}$$

Since the individual distributions x'_{ij} are identically distributed, they each have the same variance $\sigma = \sigma_i$. Therefore, from the Bienaymé's formula (M.5.33)

$$\text{var}(\overline{x}) = \frac{1}{N}\sigma^2 \tag{M.5.35}$$

where σ^2 is the variance of the mean obtained in the experiment.

Hence, the best value of the set of readings is $\overline{x} \pm \sqrt{\overline{\delta^2}/(N-1)}$. This value is normally taken as the error of a set of results.

M.5.vii Weighted Averaging

Let us suppose we have a set of measurements of a quantity in separate independent experiments each given as a best value and an error range $x_i \pm \epsilon_i$. Some measurements will be more reliable than others if their error is small. However, this does mean that we should disregard the others. Do we have enough information to weigh the values according to their accuracy to obtain an improvement in 'best value' and overall error?

If we presume that following the convention that the stated error is the square root of the variance of the experimental set of values, we may consider that the error of different members of each value is due to the different numbers of samples taken from a large overall set, i.e. the number of set i is proportional to σ_i^2 from Eq. (M.5.35). We may therefore assign a weight w_i to the set i where $w_i\sigma_i^2 = \text{const}$.

Alternatively, let us weight the set i by w_i, then the weighted mean of a large set M

$$\overline{x} = \sum_i w_ix_i \Big/ \sum_i w_i \tag{M.5.36a}$$

and the weighted variance of the mean

$$\sigma_M^2 = \lim_{M\to\infty}\frac{1}{M}\overline{x}^2 = \lim_{M\to\infty}\frac{1}{M}\sum_{\text{sets}}\left(\sum w_ix_i\right)^2 \Big/ \left(\sum w_i\right)^2 = \sum_i w_i^2\sigma_i^2 \Big/ \left(\sum w_i\right)^2 \tag{M.5.36b}$$

since each result is statistically independent $\lim_{M\to\infty} w_iw_jx_ix_j = 0$.

To minimise the overall variance and therefore error

$$\frac{\partial \sigma_M^2}{\partial w_i} = \frac{2w_i\sigma_i^2\left(\sum_i w_i\right)^2 - 2\sum_i w_i \sum_i w_i^2\sigma_i^2}{\left(\sum_i w_i^4\right)} \tag{M.5.37}$$

Therefore there is a turning point if

$$w_i\sigma_i^2 \sum_i w_i - \sum_i w_i^2\sigma_i^2 = 0$$

i.e.

$$w_i\sigma_i^2 = \text{const} \tag{M.5.38}$$

as before.

M.6 Numerical Solution of Poisson's Equation

Poisson's equation is an inhomogeneous elliptic equation satisfied by the potential $\phi(\mathbf{r})$ of the spatial distribution of charge $\rho(\mathbf{r})$, namely

$$\nabla^2\phi = -\frac{\rho(\mathbf{r})}{\epsilon_0} \tag{M.6.1}$$

The solution of Poisson's equation plays an important role in finding the solution to plasma physics problems by particle-in-cell methods. The nature of the method adopted to find the solution is very dependent on the boundary conditions, namely specified or periodic.

M.6.i One Dimension

In one dimension the solution is relatively straightforward and several simple methods are available. Given a set of values of the potential function $\phi(x)$ defined at a set of N discrete points $x_n = n\Delta x$, find the solution of the differential equation

$$\frac{d^2\phi}{dx^2} = -\frac{\rho}{\epsilon_0} \tag{M.6.2}$$

subject to either prescribed or periodic boundary conditions, namely:

- **Dirichlet boundary condition**: Prescribed potential boundary conditions, for example $\phi = 0$ at each boundary $x = 0$ or $N\Delta x$.
- **Neumann boundary condition**: Prescribed field boundary condition, e.g. $E = -d\phi/dx = 0$ at each boundary $x = 0$ or $N\Delta x$.
- **Mixed boundary conditions** or a combination of the two.
- **Periodic boundary conditions**: $\phi(x) = \phi(x + N\Delta x)$.

To solve

a) Use Gauss theorem to obtain the electric field strength at each cell boundary $E_{n+1/2}$

$$E_{n+1/2} - E_{n-1/2} \approx \left.\frac{dE}{dx}\right|_n \Delta x = \frac{\rho_n}{\epsilon_0}\Delta x \tag{M.6.3}$$

where $E = -\mathrm{d}\phi/\mathrm{d}x$ and ρ the net charge difference per unit area. By finite differencing

$$E_n \approx \frac{1}{2}\left(E_{n+1/2} + E_{n-1/2}\right) \tag{M.6.4}$$

Finally, integrating the potential across the plasma

$$\phi_N = \frac{1}{2}E_{1/2} + \sum_{m=1}^{N-1} E_{m+1/2} + \frac{1}{2}E_{N-1/2} = \sum_{m=1}^{N} E_m \tag{M.6.5}$$

Some obvious conditions follow from this solution:

- If the potential across the plasma is specified as a boundary condition, the fields at the boundary must be specified to match the total net charge $Q = \epsilon(E_{N+1/2} - E_{1/2})$.
- If the mesh is periodic $E_{1/2} = E_{N-1/2}$ and from Eq. (M.6.3) the total charge:

$$Q = \epsilon_0 \sum_{n=1}^{N} \rho_n \,\Delta x = \epsilon_0 \sum_{n=1}^{N} \left(E_{n+1/2} - E_{n-1/2}\right) = \epsilon_0 \left(E_{N+1/2} - E_{3/2}\right) = 0 \tag{M.6.6}$$

- If the mesh is periodic and infinite then $\phi_M = \cdots = \phi_N = \phi_0$, which can only be true if $\phi_N = 0$, i.e. the potential across one period is zero.

b) To obtain the potential directly finite difference to obtain

$$\phi_{n+1} - 2\phi_n + \phi_{n-1} = -\epsilon_0 \rho_n \Delta x^2 \tag{M.6.7}$$

which is tri-diagonal and can be solved by the Thomas algorithm, Appendix 10.A.

c) Take the Fourier transform of ρ_n using a fast Fourier transform (Cooley and Tukey, 1965): the number of elements N should be a power of 2 for efficiency. Dirichlet boundary conditions are easily fitted using the ***sin*** transform and Neumann using the ***cos*** transform.

$$\tilde{\rho}_k = \Delta x \begin{cases} \sum_{n=1}^{N} \rho_n \sin(kn) & \text{Dirichlet} \\ \sum_{n=1}^{N} \rho_n \cos(kn) & \text{Neumann} \\ \sum_{n=1}^{N} \rho_n \exp(\imath\, kn) & \text{Mixed} \end{cases} \tag{M.6.8}$$

and hence taking the inverse transform

$$\tilde{\phi}(k\Delta x) = \begin{cases} \sum_{n=1}^{N} \rho_n \sin(2\pi\, kn/N) & \text{Dirichlet} \\ \sum_{n=1}^{N} \rho_n \cos(2\pi\, kn/N) & \text{Neumann} \\ \sum_{n=1}^{N} \rho_n \exp(-\imath\, 2\pi\, kn/N) & \text{Mixed} \end{cases} \tag{M.6.9}$$

$$\tilde{\phi}_k = -\frac{1}{\left\{2\epsilon_0\left[1 - \cos(k)\right]\right\}}\tilde{\rho}_k \approx \frac{1}{\epsilon_0}\frac{1}{k^2}\tilde{\rho}_k \tag{M.6.10}$$

Taking the inverse Fourier transform

$$\phi_n = \frac{1}{N}\sum_{k=0}^{N-1} \tilde{\phi}_k \exp(\imath\, 2\pi\, k\, n/N) \tag{M.6.11}$$

M.6.ii Two Dimensions

In most cases, the configuration uses a simple orthogonal, singly connected grid representing orthogonal co-ordinates with equal spacing: Cartesian (x, y), cylindrical (r, z), cylindrical polar (r, θ), spherical polar (r, θ) and spherical (θ, ϕ). In each case, if the boundary conditions are simply defined on a bounding surface, as in the problem described earlier, it is easy to establish that the solution is separable, i.e. $\phi(x, y) = \phi_x(x)\,\phi_y(y)$, etc. which makes the solution much easier to find by splitting the separate operators in (x, y), and finding ϕ_x and ϕ_y independently.

As a prototype let us consider a Cartesian system. The centred difference form of Poisson's equation is the five-point scheme:

$$\phi_{i+1,j} + \phi_{i,j+1} - 4\phi_{i,j} + \phi_{i,j-1} + \phi_{i-1,j} = \rho_{i,j} \tag{M.6.12}$$

This matrix equation may be solved by standard methods for treating sparse matrix equations discussed earlier. More efficient methods involve the use of fast Fourier transform routines. Taking the Fourier transform in j,

$$\tilde{\phi}_{i,k} = \sum_{j=1}^{N} \phi_{i,j} \exp(2\pi\iota jk/N) \tag{M.6.13}$$

so that

$$\phi_{i,j} = \frac{1}{N}\sum_{k=1}^{N} \tilde{\phi}_{i,k} \exp(2\pi\iota jk/N) \tag{M.6.14}$$

and substituting in the finite-difference equation

$$\left\{\tilde{\phi}_{i+1,k} - (4 - 2\cos(2\pi k/N))\,\tilde{\phi}_{i,k} + \tilde{\phi}_{i-1,k}\right\} = \tilde{\rho}_{i,k} \tag{M.6.15}$$

which give a set of tri-diagonal equations for $\tilde{\phi}_{i,k}$.

We may proceed in a number of different ways to solve these: discrete Fourier transforms, Thomas solution or cyclic reduction. Hockney (1970) investigated the relative efficiency of these three methods for the case of periodic boundary conditions in the $y(j)$ direction and $\phi = 0$ (Dirichlet) boundary conditions in $x(i)$. He found the best solution was Fourier transform in y and a limited (two or three steps) cyclic reduction in x – a method called Fourier analysis cyclic reduction (DECRY), which has been widely adopted in 2D particle codes: see Hockney and Eastwood (1988) for a full description of the method.

Poisson's equation

$$\nabla^2\phi = -\frac{\rho(r)}{\epsilon_0} \tag{M.6.16}$$

plays an important role in the solution of Maxwell's equations in PIC modelling. To that end, methods for its evaluation on a simply connected orthogonal grid with Dirichlet or Neumann boundary conditions are required. Under these conditions, the solution is separable $\phi(\mathbf{r}) = \phi_x(x)\,\phi_y(y)\,\phi_z(z)$; each of which is subject to a one-dimensional equation of the form

$$\frac{\partial^2\phi_x}{\partial x^2} = -\frac{\rho_x(x)}{\epsilon_0} \tag{M.6.17}$$

It is therefore appropriate to consider the efficient evaluation of finite-difference forms of Eq. (M.6.17), from which the multidimensional solution may be built.

M.6.ii.a Discrete Fourier Transform

Consider the discrete function V_i with N elements defined on a discrete set of points $x_i = i\Delta x$ and periodic with a period over N equal intervals: $V_{i+N} = V_i$. The discrete Fourier transform of the set V_i is $\tilde{V}_k$ and its inverse V_i' are defined by

$$\tilde{V}_k = \sum_{i=0}^{N-1} V_i \exp(-\imath 2\pi\, i k/N) \tag{M.6.18a}$$

$$V_i' = \frac{1}{N}\sum_{k=0}^{N-1} \tilde{V}_k \exp(\imath 2\pi\, i k/N) \tag{M.6.18b}$$

To verify substitute $\tilde{V}_k$ from (M.6.18a) in (M.6.18b)

$$V_i' = \frac{1}{N}\sum_{k=0}^{N-1}\sum_{i=0}^{N-1} \exp\left[\imath 2\pi(i-i')k\right] V_i = \frac{1}{N}\sum_{i'=0}^{N} V_i' N \delta_{i,i'} = V_i \tag{M.6.19}$$

Since the sum of a geometric series with initial value $a_0 = 1$ and multiplier $r = \exp(\imath 2\pi/N)$ is $\sum_0^{N-1} a_0(1-r^N)/(1-r)$, the following result is easily established:

$$\sum_{k=0}^{N-1} \exp\left(\imath 2\pi(i-i')k/N\right) = \begin{cases} N & \text{if } \; i = i' \\ 0 & \text{otherwise} \end{cases} \tag{M.6.20}$$

$$= N\delta_{i,i'} \tag{M.6.21}$$

where $\delta_{i,i'}$ is the Kronecker delta.

The sets V_i and $\tilde{V}_k$ therefore form a consistent conjugate pair, with each being the transform of the other. They form the discrete analogue of the familiar Fourier transform for continuous functions.

A more complete discussion of the discrete Fourier transform is given by Stoer and Bulirsch (1991) together with a description of the fast Fourier transform routines which efficiently calculate the transform. Programme listings in FORTRAN for the transforms using these techniques will be found in Birdsall and Langdon (1991) and Press et al. (1992).

M.6.ii.b Cyclic Reduction

The finite-difference approximation to a general second-order separable partial differential equation, which is finite differenced in one dimension y, takes the form of the set of tri-diagonal equations for each element, namely ϕ_j

$$l_j\boldsymbol{\phi}_{j-1} + \mathcal{A}_j\boldsymbol{\phi}_j + m_j\boldsymbol{\phi}_{j+1} = \rho_j \tag{M.6.22}$$

The general solution to these equations using cyclic reduction has been given by Swarztrauber (1974) but is complex, and not appropriate to our purpose; namely the solution of Poisson's equation. In this case, $l_j = m_j = 1$, greatly simplifying finding the solution. For convenience, we assume a reduced charge density and drop the dielectric constant in Poisson's equation.

The finite-differenced set of tri-diagonal equations simplify to the form

$$\boldsymbol{\phi}_{j-1} + \mathcal{A}\boldsymbol{\phi}_j + \boldsymbol{\phi}_{j+1} = \rho_j \tag{M.6.23}$$

where $\boldsymbol{\phi}_j$ and ρ_k are column vectors and $\mathcal{A}$ a five-point finite-difference operator matrix.

The algorithm proceeds by the reduction of the three consecutive equations $(j-1), j$ and $(j+1)$ to eliminate ϕ_{i-1} and ϕ_{i+1}, thereby reducing the number of unknowns from 5 to 3, repetitively, until only 1 is left.

The three neighbouring equations:

$$\left.\begin{aligned} \boldsymbol{\phi}_{j-2} + \mathcal{A}\boldsymbol{\phi}_{j-1} + \boldsymbol{\phi}_j \qquad\qquad &= \boldsymbol{\rho}_{j-1} \\ \boldsymbol{\phi}_{j-1} + \mathcal{A}\boldsymbol{\phi}_j + \boldsymbol{\phi}_{j+1} \qquad &= \boldsymbol{\rho}_j \\ \boldsymbol{\phi}_j + \mathcal{A}\boldsymbol{\phi}_{j+1} + \boldsymbol{\phi}_{j+2} &= \boldsymbol{\rho}_{j+1} \end{aligned}\right\} \tag{M.6.24}$$

Multiplying the middle even line equation by $-\mathcal{A}$ and adding eliminates $\boldsymbol{\phi}_{j-1}$ and ϕ_{j+1} to give

$$\boldsymbol{\phi}_{j-2} + \left(2\mathcal{I} - \mathcal{A}^2\right)\boldsymbol{\rho}_j + \boldsymbol{\phi}_{j+2} = \boldsymbol{\rho}_j^{(1)} \tag{M.6.25}$$

where

$$\mathcal{A}^{(1)} = 2\mathcal{I} - \mathcal{A}^{(0)2} \tag{M.6.26}$$

and

$$\boldsymbol{\rho}_j^{(1)} = \boldsymbol{\rho}_{j-2} - \mathcal{A}^{(0)}\boldsymbol{\rho}_j + \boldsymbol{\rho}_{j+2} \qquad\qquad 0 < j < N \tag{M.6.27}$$

Equation (M.6.27) is again a tri-diagonal matrix equation, which may be successively reduced to a single equation plus the boundary values after m cycles[10]

$$\mathcal{A}^{(m)}\boldsymbol{\phi}_{N/2} = \boldsymbol{\rho}_{N/2}^{(m)} - \boldsymbol{\phi}_0 - \boldsymbol{\phi}_N \tag{M.6.28}$$

and

$$\mathcal{A}^{(m)} = -\prod_{l=1}^{2^m}\left(\mathcal{A}^{(0)} - \beta_l\mathcal{I}\right) \tag{M.6.29}$$

In the case of Poisson's equation ($l_j = m_j = 1$) in Eq. (M.6.22) the coefficient β_l has the form (Buzbee, Golub and Nielson, 1970)

$$\beta_l = 2\cos\left[\frac{2(l-1)\pi}{2^{(l+1)}}\right] \tag{M.6.30}$$

Alternatively, Eq. (M.6.26) may be solved directly by the Thomas algorithm at any stage.

Finally, knowing $\boldsymbol{\phi}_{N/2}$ we may use Eq. (M.6.26) and the boundary values to solve for $\boldsymbol{\phi}_{N/4}$ and $\boldsymbol{\phi}_{3/N/4}$ successively and similarly for the complete set of values $\boldsymbol{\phi}_j$.

What level of reduction is required to leave only one equation? Let N be the number of unknowns plus two boundary values ϕ_0 and ϕ_{N+1} (Figure M.6.1). The number of equations is therefore $N + 1$. Since each set of three equations is reduced by two at each stage, we may work either forwards or more simply backwards from the desired result of a single equation subtracting two equations in each interval to obtain successively T_n new equations in stage n and a total of $S_n = S_{n-1} + T_n = \sum_{m=0}^{n} S_m$. Since the internal equations are reduced from 3 to 1 at each stage, i.e. the total loss of equations in each step is $T_n = 2^n$ forming a geometric series with multiplier 2 and initial value 1 (as shown in Table M.6.1).

The method clearly works best if the number of equations $N + 1 = 2^n$ is a power of 2, and therefore the number of unknowns $N = 2^n - 1$ plus the two known boundary values, e.g. for a Dirichlet problem $\phi_0 = \phi_N = 0$, are the complete set of $(2^n + 1)$ variables.

10 Note, it is only in this final equation that the boundary values ϕ_0 and ϕ_N are used.

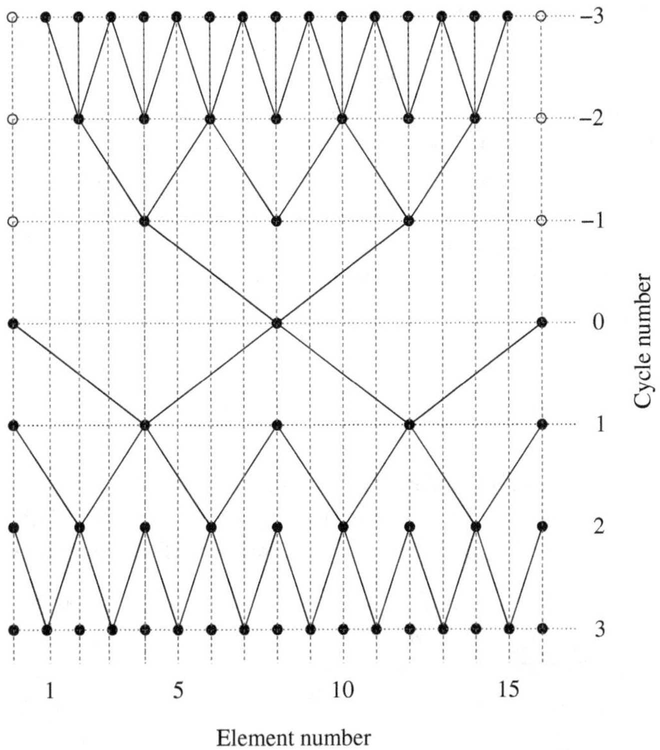

Figure M.6.1 The progressive reduction of the number of equations and unknowns is shown graphically for a set of 15 (=$2^4 - 1$) unknowns plus two boundary values. The active elements are shown by closed circles and inactive boundary values by open. The number of variables at successive cycles reduces to 1 before being restored to generate the solution. The number of unknowns N is set to 16 to ensure the reduction to a single equation.

Table M.6.1 Number of unknowns required per cycle.

n	0	1	2	3	4	5	6	⋯	n
T_n	0	2	4	8	16	32	64		2^n
S_n	1	3	7	15	31	63	127		$2^{(n+1)} - 1$

There remains one further problem, namely the calculation of $\rho_i^{(l)}$ at the lth cycle which is unstable due to the generalisation of Eq. (M.6.27). To avoid this, the value is calculated using the routine originally given by Bunemann (1969) (see also Hockney and Eastwood, 1988): namely define the vectors $\mathbf{p}_j$ and $\mathbf{q}_j$ which obey the recurrence relation

$$\mathbf{p}_j^{(l)} = \mathbf{p}_j^{(l-1)} + (\mathcal{A}^{(n-1)})^{-1}\left(\mathbf{q}_j^{(l-1)} - \mathbf{p}_{j-2^{(l-1)}}^{(l-1)} - \mathbf{p}_{j+2^{(l-1)}}^{(l-1)}\right) \tag{M.6.31}$$

$$\mathbf{q}_j^{(l)} = \mathbf{q}_j^{(l-1)} + \mathbf{q}_{j-2^{(l-1)}}^{(l-1)} + \mathbf{q}_{j+2^{(l-1)}}^{(l-1)} - 2\mathbf{p}_j^{(l)} \tag{M.6.32}$$

$$\mathbf{p}_j^{(0)} = 0 \qquad \text{and} \qquad \mathbf{q}_j^{(0)} = \rho_j \tag{M.6.33}$$

M.6.ii.c Fourier Analysis Cyclic Reduction (FACR) Method

It is not necessary to complete the full reduction to obtain a solution. The reduction may be terminated after ℓ steps leaving a tri-diagonal set of equations of the form of Eq. (M.6.27), which may be solved in a conventional manner, e.g. using the Thomas algorithm. It has been shown by Hockney (1970) in one dimension that it is most efficient to limit the number of cyclic reduction steps to 2 or 3 and complete the solution by a Fourier decomposition.

However, since the Fourier method offers additional attractions in terms of filtering and spectral analysis it is more generally used to solve Poisson's equations, particularly as the Poisson solver is a relatively insignificant part of the programme compared to the particle mover.

M.7 Compressible Gas Potential Flow

The steady flow of gas after passing through a shock wave is in general rotational. However, if the shock is uniform across its surface, the flow remains irrotational if it was so initially. This useful result follows from the following argument. Since we may write for inviscid flow

$$\frac{1}{2}\nabla \mathrm{v}^2 - \underline{\mathrm{v}} \wedge (\nabla \wedge \underline{\mathrm{v}}) = -\frac{1}{\rho}\nabla p \tag{M.7.1}$$

it follows that

$$\nabla(h + \frac{1}{2}\mathrm{v}^2) - \underline{\mathrm{v}} \wedge (\nabla \wedge \underline{\mathrm{v}}) = T\nabla s \tag{M.7.2}$$

However, since Bernoulli's equation holds both outside and across the shock

$$h + \frac{1}{2}\mathrm{v}^2 = const \tag{M.7.3}$$

and

$$\underline{\mathrm{v}} \wedge (\nabla \wedge \underline{\mathrm{v}}) = -T\nabla s \tag{M.7.4}$$

In general, the entropy change across the shock varies as the intensity of the shock varies. However, if the shock has a constant strength, the flow behind the shock will have a uniform entropy and the overall flow will be isentropic. Hence, $\nabla s = 0$ and therefore $\nabla \wedge \underline{\mathrm{v}} = 0$, or $\underline{\mathrm{v}}$ is parallel to $\nabla \wedge \underline{\mathrm{v}}$. The latter is obviously incorrect at the shock since the transverse velocity component is continuous. Hence, we conclude that the vorticity $\underline{w} = \nabla \wedge \underline{\mathrm{v}} = 0$ and the flow remains irrotational.

Alternatively, if the shock is weak, the entropy jump across the shock is of order $(\Delta p)^3 \to 0$. Since the entropy is uniform downstream of the shock, it will remain so upstream, at least to terms of the second order in the pressure jump Δp.

In these cases, Euler's equation takes the simpler form

$$\frac{1}{2}\nabla \mathrm{v}^2 = -\frac{1}{\rho}\nabla p = -\frac{c^2}{\rho}\nabla \rho \tag{M.7.5}$$

in terms of the sound speed c. Since the equation of continuity requires $\nabla \cdot (\rho \underline{\mathrm{v}}) = 0$, we obtain

$$c^2 \nabla \cdot \underline{\mathrm{v}} - \underline{\mathrm{v}}(\underline{\mathrm{v}} \cdot \nabla)\underline{\mathrm{v}} = 0 \tag{M.7.6}$$

Introducing the velocity potential $\underline{v} = \nabla\phi$ we find

$$\begin{aligned}(c^2 - \phi_x^2)\phi_{xx} + (c^2 - \phi_y^2)\phi_{yy} + (c^2 - \phi_y^2)\phi_{yy} \\ -2(\phi_x\phi_y\phi_{xy} + \phi_y\phi_z\phi_{yz} + \phi_z\phi_x\phi_{zx}) = 0\end{aligned} \tag{M.7.7}$$

M.7.i Perturbation Flow

If the effect of the body is weak, introducing only a small perturbation to the incoming uniform flow, then we may consider the changes in the flow by linearised perturbation methods. Thus, writing the total velocity as the sum of the incoming flow $\underline{v}_\infty$ and the perturbation, which we express as a potential ϕ, the total velocity $\underline{v} = \underline{v}_\infty + \nabla\phi$. The overall potential is therefore $x\underline{v}_\infty + \phi$, where the incoming flow is taken along the x-direction.

For Mach numbers not close to unity the perturbation potential flow equation can then be reduced to Landau and Lifshitz (1959)

$$(1 - M_\infty^2)\frac{\partial^2\phi}{\partial x^2} + \frac{\partial^2\phi}{\partial y^2} + \frac{\partial^2\phi}{\partial z^2} = 0 \tag{M.7.8}$$

for a flow incoming along the x-direction with Mach number M_∞.

If $M_\infty > 1$, the flow is supersonic, and if $(1 - M_\infty^2) < 0$, the equation is hyperbolic. It therefore admits wave-type solutions (weak shock waves), which leave the surface of the body. We shall not consider these solutions further here.

For sub-sonic flow, $M_\infty < 1$, $(1 - M_\infty^2) > 0$ the equation is elliptic and may be reduced to Laplace's equation by a transformation of the x co-ordinate, i.e. writing x' for x in Eq. (M.7.8), where

$$x' = x/\sqrt{(1 - M_\infty^2)} \tag{M.7.9}$$

M.7.ii The General Solution

Any arbitrary vector field (e.g. velocity) can be written quite generally as the sum of a constant vector, a scalar and a vector potential (Helmholtz' theorem)

$$\underline{v} = \underline{v}_\infty + \nabla\phi + \nabla\wedge\underline{A} \tag{M.7.10}$$

where the scalar and vector fields at the point $\underline{r}$ can be expressed in terms of contributions from the volume V external to the surface S.

Using the analogies with electromagnetism for the scalar (electro-statics) and vector (magneto-statics) potentials, we may write

$$\phi(r) = \int_S \frac{1}{|\underline{r} - \underline{r}'|}\sigma\,\mathrm{d}s' + \int_V \frac{1}{|\underline{r} - \underline{r}'|}\rho\,\mathrm{d}V' \tag{M.7.11}$$

and

$$\underline{A}(r) = \int_S \frac{1}{|\underline{r} - \underline{r}'|}\underline{\omega}\,\mathrm{d}s' + \int_V \frac{1}{|\underline{r} - \underline{r}'|}\underline{\zeta}\,\mathrm{d}V' \tag{M.7.12}$$

where $\underline{r}'$ is the position vector of $\underline{\mathrm{d}s}'$ or $\mathrm{d}V'$, ρ and $\underline{\zeta}$ the volume sources in V of fluid and vorticity, respectively, and σ and $\underline{\omega}$ the surface sources on S.

For an incompressible, irrotational flow, we have $\nabla \cdot \underline{v} = \nabla \wedge \underline{v} = 0$ and hence the volume source terms are zero.

$$\rho = \frac{1}{4\pi}\nabla \cdot \underline{v} = 0 \qquad \text{and} \qquad \underline{\zeta} = \frac{1}{4\pi}\nabla \wedge \underline{v} = 0 \tag{M.7.13}$$

Taking the appropriate gradients of the above expression, the velocity can be written in the general form as

$$\underline{v} = \underline{v}_{\infty} + \int_S \sigma \frac{\underline{r} - \underline{r}'}{|\underline{r} - \underline{r}'|^3}\,\mathrm{d}s' + \int_S \underline{\omega} \wedge \frac{\underline{r} - \underline{r}'}{|\underline{r} - \underline{r}'|^3}\,\mathrm{d}s' \tag{M.7.14}$$

The first integral represents the contribution from surface sources of strength σ and the second a vortex sheet of strength ω.

By applying the general uniqueness theorem for Laplace's equation, any solution which satisfies the boundary conditions for the incoming flow and over the surface S will be the correct one. The solution can be represented by a distribution of simple sources (monopoles), doublets (dipoles) or both over the boundary, distributed in such a way to satisfy the boundary conditions. In this case, we require that the source distribution be such that the velocity component normal to the surface is zero.

M.7.ii.a Potential Solution

Assuming a simple source distribution of strength $\sigma(j)$ per unit area at the point j on the surface S (Figure M.7.1), this leads to the integral equation for the velocity at any point i on the surface

$$2\pi\sigma(i)\underline{n}_i - \int_S \sigma(j) \cdot \nabla(1/r_{ij})\,\mathrm{d}s_j = 0 \tag{M.7.15}$$

where $\underline{n}_i$ is the unit surface normal at i and r_{ij} is the distance between i and j.

This equation is easily understood from electro-statics, where $2\pi\sigma(i)$ is the field, normal to the surface, at i due to its own charge density, and $\sigma(j) \cdot \nabla(1/r_{ij})\,\mathrm{d}s_j$ the field due to the element $\mathrm{d}s_j$ at i. A similar equation can be found for doublets.

The complete integral equation (M.7.15) is reduced to a set of matrix equations by dividing the surface into N quadrilateral panels. A constant strength is assumed over each. The integral in Eq. (M.7.15) is evaluated by a sum over all the panels j not including i plus the term for i, so that

$$\underline{v}_i = \sum_j \underline{V}_{ij}\sigma_j \tag{M.7.16}$$

where

$$\underline{V}_{ij} = \begin{cases} \int_{S_j} \nabla\left(\frac{1}{r_{ij}}\right)\mathrm{d}s_j & \text{if } i \neq j \\ 2\pi\underline{n}_i & \text{otherwise} \end{cases} \tag{M.7.17}$$

are determined solely by the geometry of the system.

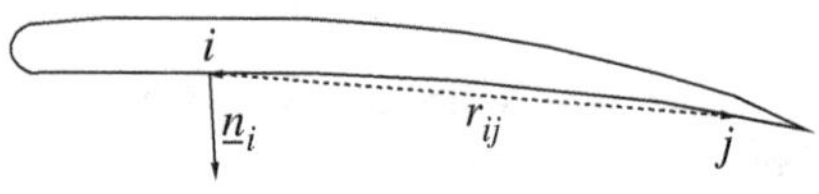

Figure M.7.1 Diagram illustrating the relationship of the points i and j on the surface of the body.

The condition of zero normal velocity is imposed at the control point taken as the centre of each panel i, by requiring that the normal component of the velocity due to the sources balance that from the incoming flow, namely

$$\sum_j \underline{V}_{ij} \cdot \underline{n}_i \sigma_j = \sum_Q A_{ij}\sigma_j = -\underline{v}_\infty \cdot \underline{n}_i \tag{M.7.18}$$

where $A_{ij} = \underline{n}_i \cdot \underline{V}_{ij}$, leading to a set of N linear simultaneous equations for the N unknown source strengths.

M.7.ii.b Vorticity Solution

We may also generate an alternative solution using the vortex sheet distribution, $\underline{\omega}_i$, at each panel. This is particularly useful for dealing with circulating flows around a body, rather than with an external incoming flow, i.e. a closed rather than open flow. Although the calculation proceeds in a manner similar to that for potential flow it is complicated by the vector nature of the vortex sheet.

However, in two dimensions, the vector potential has a single scalar component, perpendicular to the plane of calculation. This is in fact the stream function. This scalar nature markedly simplifies the calculation as the strength of the vortex sheet is also a vector perpendicular to the plane, i.e. a scalar. In this case, we may also write the velocity at panel i in terms of the vortex strength at panel j

$$\underline{v}_i = \sum_j \underline{V}'_{ij}\omega_j \tag{M.7.19}$$

where $\underline{V}'_{ii} = 0$ and $|\underline{V}'_{ij}| = |\underline{V}_{ij}|$ but is rotated by $\pi/2$. Thus, the boundary condition of zero velocity normal to the surface at each control point is

$$\sum_j \underline{V}'_{ij} \cdot \underline{n}_i \omega_j = \sum_j B_{ij}\omega_j = -\underline{v}_\infty \cdot \underline{n}_i \tag{M.7.20}$$

where $B_{ij} = \underline{t}_i \underline{V}_{ij}$ and $\underline{t}_i$ is the unit tangent vector at panel i.

M.7.ii.c Applications in Aerodynamics

Panel methods are very widely used in aeronautical design. The basic properties of an aerofoil in two dimensions can be calculated by the standard Smith–Hess method (Hess and Smith, 1967) outlined earlier. The flow is represented by a series of sources, taken as constant across each panel (line element) in the simplest case, plus a constant strength vortex sheet across the entire wing surface $\omega_i = \omega$. The velocity at each control point is taken as the centre of the panel, where the normal velocity is zero. This gives N equations although we have $N + 1$ unknowns – the set σ_i and ω.

The equations are completed by a Kutta (Joukowski) condition that the flow be well behaved at the trailing edge, which can be expressed in a number of ways. Since the pressure at the trailing edge must be the same on both the top and bottom surfaces, Bernoulli's equation requires that the tangential velocities are also equal. The condition we use is therefore that the tangential velocity in the upper and lower surface trailing edge panels be equal. In order that the control points be equivalent and near the edge, the trailing edge panels should be small and of equal length.

An alternative, but entirely equivalent, formulation uses the fact that the stream function is constant on the surface. Thus, using Eq. (M.7.12) and remembering that $\underline{\omega}$ is a vector perpendicular to the plane, we can write the stream function as a sum of contributions from the panels, to give a set of N simultaneous equations in $N+1$ unknowns, since the value of the stream constant is unknown. Again the equations are completed by including the Kutta condition, which takes the simple form

$$\omega_1 + \omega_N = 0 \tag{M.7.21}$$

at the trailing edge.

These methods are much simpler and faster than general fluid methods, which solve over the entire domain. Panel methods are essentially one-dimensional, solving only over the domain of the aircraft surface. However, the complete three-dimensional flow field is easily calculated from the potential once the source distribution is known. The methods involve the solution of a dense set of simultaneous equations, in comparison to the simpler sparse systems of the direct methods. Nonetheless the gain in speed is clear.

These general methods are readily extended into three dimensions, where they form the basis of a range of successful and widely implemented modelling tools. They find wide applications not just to aircraft design, but also for cars and ships, for example in the design of America's Cup yachts.

This technique has been extended by including higher order sources, which allow larger and therefore fewer panels to be required, with consequent improvement in computer speed and accuracy. Despite their restricted domain of applicability, to low Mach number flows, where compressibility is not important, they are widely used for solving complex systems, e.g. a complete aircraft in the ideal flow limit with no boundary layer (Hess and Smith, 1967). In particular, they are very appropriate for investigating the flow around an aircraft of complex profile, for example with a complex external payload. In general, panel methods are well suited to complex systems where the surface is easily divided into quadrilaterals, but the full domain may not be easily discretised.

M.8 Viscous Incompressible Flow

M.8.i Introduction

Incompressible flow brings new problems. These stem from the fact that the sound speed is infinite. As a result, pressure waves are transmitted instantaneously from one point in the fluid to another. The character of the differential equations governing the motion changes from hyperbolic to elliptic, reflecting this difference. Thus, although the flow may be time-varying, it passes through a series of stationary states. In consequence, we can no longer use the time stepping approach in the same way as with compressible flows.

The essential difference between compressible and incompressible flows is the nature of the equation of state. The condition that the density of a fluid element is constant requires that density is independent of pressure and temperature. As a result, the flow pattern is determined solely by the requirement to conform to the geometry of the system, i.e. by the streamlines and the boundaries. The pressure gradients are then simply the consequences of the dynamics of the velocity profile, since the absolute value of the pressure is determined instantaneously throughout the fluid by a boundary condition on the streamlines.

This behaviour is exemplified by the solution of inviscid, incompressible flows in two dimensions using conformal mapping with (complex) analytic functions, where the solutions are obtained from Laplace's equation (elliptic). The panel method described earlier is essentially an example of this approach.

We must distinguish two classes of problem using incompressible fluid.

a) **Flows with fixed boundaries**: These are typically flows in a duct or as earlier around a wing in a closed environment. These problems are normally steady state.
b) **Flows with free boundaries**: These are flows typically involving an interface or a surface, such the impact of a drop on a surface or the release of a dam into a restricted channel. These problems are usually time-varying.

The former are generally easier to solve as the latter have the complication of identifying the surface. All involve the solution of a set of simultaneous equations, implicit in any finite difference or finite element solution of an elliptic problem. For this purpose the regular grids established in finite-difference methods are particularly convenient matching to the efficient preconditioned methods for handling sparse diagonal matrices exemplified by ICCG (Kershaw, 1978).

M.8.i.a Vorticity–Stream Function Method

In two dimensions a straightforward method uses the vorticity and stream functions as the principal variables. Thus, for incompressible flow the equations

$$\nabla \cdot \underline{v} = 0 \tag{M.8.1}$$

and

$$\frac{\partial \underline{v}}{\partial t} + (\underline{v} \cdot \nabla)\underline{v} = -\frac{1}{\rho}\nabla p + \nu\nabla^2 \underline{v} \tag{M.8.2}$$

are simplified by the use of the vorticity $\underline{\zeta} = \nabla \wedge \underline{v}$ and the stream function ψ.

Taking the curl of the Navier–Stokes equation (M.8.2)

$$\frac{\partial \underline{\zeta}}{\partial t} - \nabla \wedge (\underline{v} \wedge \underline{\zeta}) = \frac{\partial \underline{\zeta}}{\partial t} + (\underline{v} \cdot \nabla)\underline{\zeta} - (\underline{\zeta} \cdot \nabla)\underline{v} = \nu\nabla^2 \underline{\zeta} \tag{M.8.3}$$

This equation takes a simpler form in two dimensions. Since the vorticity has only a single component (normal to the flow plane) and can therefore be treated as a scalar quantity, which can be expressed in a conservation law form

$$\frac{\partial \zeta}{\partial t} + \nabla \cdot (\zeta \underline{v}) = \nu\nabla^2 \zeta \tag{M.8.4}$$

The stream function satisfies Poisson's equation with the vorticity as source

$$\nabla^2 \psi = \zeta \tag{M.8.5}$$

and the velocity is given by the stream function

$$v_x = \frac{\partial \psi}{\partial y} \quad : \quad v_y = -\frac{\partial \psi}{\partial x} \tag{M.8.6}$$

The overall routine then takes a time-stepping development:

- **Step 1**: Given the velocity field from the previous time-step, calculate the vorticity field in the interior of the mesh at the next time step, using some appropriate form of the conservative advection and Laplacian operators. This may be either explicit or implicit. If the former a Courant–Friedrichs–Lewy time step limit must be employed.
- **Step 2**: Solve Poisson's equation for the stream function using the new values of the vorticity on the interior of the mesh.
- **Step 3**: Include the boundary conditions on the stream function, and hence the vorticity values at the edge of the mesh.
- **Step 4**: Calculate the new values of the velocity from the stream function.

M.8.ii Marker-in-Cell

The introduction of free boundaries on which, for example a constant pressure, is prescribed introduces additional problems. In particular, it becomes necessary to identify the space which contains fluid separate from that which is void, both lying within the computational mesh. One of the first successful methods of achieving this used a set of marker particles, Lagrangian fluid elements, which play no role in the calculation but simply define the location of the fluid, and hence filled, partially filled and void cells (Harlow and Welch, 1965; Amsden and Harlow, 1970).

The calculation falls into two distinct elements: the dynamics of the fluid and the identification of the surface through the marker particle positions.

M.8.ii.a Dynamics

We consider two-dimensional incompressible flow with the equations written in a slightly unconventional manner:

$$\frac{\partial u}{\partial t} + \frac{\partial u^2}{\partial x} + \frac{\partial uv}{\partial y} = -\frac{\partial p}{\partial x} + g_x + \nu \frac{\partial}{\partial x}\left(\frac{\partial u}{\partial x} - \frac{\partial v}{\partial y}\right) \tag{M.8.7}$$

$$\frac{\partial v}{\partial t} + \frac{\partial uv}{\partial x} + \frac{\partial v^2}{\partial y} = -\frac{\partial p}{\partial y} + g_y - \nu \frac{\partial}{\partial y}\left(\frac{\partial u}{\partial x} - \frac{\partial v}{\partial y}\right) \tag{M.8.8}$$

where p is the pressure and ν the kinematic viscosity. The equations are completed by the equation of continuity

$$\frac{\partial u}{\partial x} + \frac{\partial v}{\partial y} = 0 \tag{M.8.9}$$

The finite-difference representation is performed on a staggered grid. The two velocity components are calculated at different points on the cell faces, namely $u_{(i+1/2),j}$ and $v_{i,(j+1/2)}$ and the pressure $p_{i,j}$ at the cell centre. The cell centre velocity squared terms are written in ZIP form, namely

$$(u_{i,j})^2 = u_{i-1/2,j}u_{i+1/2,j} \tag{M.8.10}$$

which improves momentum conservation at a rigid wall. The difference form for the time development from step n to $n+1$ takes the form

$$\frac{\tilde{u}^{n+1}_{i+1/2,j} - u^n_{i+1/2,j}}{Dt} = \frac{u^n_{i+1/2,j}u^n_{i-1/2,j} - u^n_{i+3/2,j}u^n_{i+1/2,j}}{\Delta x} + \frac{u^n_{i+1/2,j-1/2}v^n_{i+1/2,j-1/2} - u^n_{i+1/2,j+1/2}v^n_{i+1/2,j+1/2}}{\delta y} + \frac{\phi_{i,j} - \phi_{i+1,j}}{\Delta x} + g_x + \nu\left[\frac{1}{\Delta x^2}\left(u^n_{i+1/2,j+1} + u^n_{i+3/2,j-1} - 2u^n_{i+1/2,j}\right) + \frac{1}{\Delta x\delta y}\left(v^n_{i+1,j+1/2} - v^n_{i+1,j-1/2} - v^n_{i,j+1/2} + v^n_{i,j-1/2}\right)\right] \quad \text{(M.8.11)}$$

$$\frac{\tilde{v}^{n+1}_{j,j+1/2} - v^n_{i,j+1/2}}{Dt} = \frac{u^n_{i-1/2,j+1/2}v^n_{i-1/2,j+1/2} - u^n_{i+1/2,j+1/2}v^n_{i+1/2,j+1/2}}{\Delta x} + \frac{v^n_{i,j+1/2}v^n_{i,j-1/2} - v^n_{i,j+3/2}v^n_{i,j+1/2}}{\delta y} + \frac{\phi_{i,j} - \phi_{i,j+1}}{\delta y} + g_y + \nu\left[+\frac{1}{\Delta x\delta y}\left(v^n_{i+1/2,j+1} - v^n_{i-1/2,j+1} - v^n_{i+1/2,j} + v^n_{i-1/2,j}\right) \frac{1}{\delta y^2}\left(v^n_{i+1,j+1/2} + v^n_{i-1/2,j+3/2} - 2u^n_{i,j+1/2}\right)\right] \quad \text{(M.8.12)}$$

and

$$D^{n+1}_{i,j} = \frac{u^{n+1}_{i+1/2,j} - u^{n+1}_{i-1/2,j}}{\Delta x} + \frac{v^{n+1}_{i,j+1/2} - v^{n+1}_{i,j-1/2}}{\delta y} \quad \text{(M.8.13)}$$

M.8.ii.b Boundary Conditions

In these equations, we have replaced the true pressure p by an arbitrary scalar field ϕ. The velocities at the conclusion of this sweep through the mesh are provisional. Boundary conditions are most easily applied by specifying a set of virtual cells beyond the physical boundaries of the problem, which are used to prescribe the boundary conditions.

- The boundary conditions applied at the internal and free surface depend on the nature of the boundary and if a cell is full, void or mixed. For a full cell, the velocity is assigned to prevent flow through the wall and either allow free-slip or no-slip along the wall. Thus, for a wall parallel to the y plane at $i - 1/2$, we have

$$\left.\begin{aligned} u_{i-1/2,j} &= 0 \\ v_{i-1/2,j+1/2} &= \pm v_{i,j+1/2} \end{aligned}\right\} \quad \text{(M.8.14)}$$

  the plus being for a free slip-wall and the minus for a no-slip.
- For an inflow or prescribed outflow, the boundary velocity is set equal to prescribed value.
- For a continuative outflow boundary, the velocity in the virtual cell is set equal to its neighbour.
- At the boundary between a full cell and an empty one, the tangential stress on the face is zero so that for a boundary $j + 1/2$ we have the velocity in the cell $i, j + 1$

$$u_{i+1/2,j+1} = u_{i+1/2,j} - \frac{\delta y}{\Delta x}(v_{i+1,j+1/2} - v_{i,j+1/2}) \quad \text{(M.8.15)}$$

  and the normal stress is the applied pressure plus a viscous term

$$p_{i,j} = p_{i,j}(applied) + \frac{2\nu}{\Delta x}(v_{i,j+1/2} - v_{i,j-1/2}) \quad \text{(M.8.16)}$$

M.8.ii.c Pressure Correction

The divergence of the velocity calculated with these velocities is not zero. We therefore correct the velocities to bring them into agreement with Eq. (M.8.9) by adjusting the scalar 'pressure' function, ϕ. Thus, we adjust the velocity by adding the gradient of a new function ψ, thus

$$\begin{aligned} u^{n+1}_{i+1/2,j} &= \tilde{u}^{n+1}_{i+1/2,j} - \frac{1}{\Delta x}(\psi_{i+1,j} - \psi_{i,j}) \\ v^{n+1}_{i,j+1/2} &= \tilde{v}^{n+1}_{i,j+1/2} - \frac{1}{\delta y}(\psi_{i,j+1} - \psi_{i,j}) \end{aligned} \tag{M.8.17}$$

so that

$$D^{n+1}_{i,j} = \tilde{D}^{n+1}_{i,j} - \frac{1}{\Delta x^2}\left(\psi_{i+1,j} - 2\psi_{i,j} + \psi_{i-1,j}\right) - \frac{1}{\delta y^2}\left(\psi_{i,j+1} - 2\psi_{i,j} + \psi_{i,j-1}\right) \tag{M.8.18}$$

The boundary conditions on ψ are:

- At a rigid wall

$$\psi_{i-1,j} = \psi_{i,j}$$

- At a free-surface cell

$$\psi_{i,j} = 0$$

- At a continuative outflow wall, the outside cell has

$$\psi_{i-1,j} = 0$$

Since the equation for ψ is sparse diagonal, it is readily solved using an appropriate routine as discussed earlier.

The actual pressure in the fluid is easily obtained by adjusting the arbitrary pressure function ϕ with the correction ψ to match the value at the free boundary to the applied pressure.

M.8.ii.d Particle Movement

The particle positions are retained parameters and updated each time step. They are moved using the mesh velocities calculated in the dynamics. Since the particles are positioned centrally within the cells, an area weighting scheme is used to derive an improved value of the particle velocity dependent on its position.

The particle positions are used to identify the position of the free boundary and identify full, surface and void cells. A free-surface cell is identified as one containing particles, but having at least one empty cell as neighbour.

M.8.ii.e Volume of Fluid Methods

A serious handicap of the MAC method is the computational expense of maintaining the large number of marker particles necessary to identify the filled cells and the surface. To that end, an alternative method of tracking the interface has been developed using a parameter F which measures the fractional volume of fluid. Thus $F = 1$ corresponds to a full cell and $F = 0$ to an empty one. The parameter F will be convected with the fluid so that its Lagrangian derivative is zero, i.e.

$$\frac{\partial F}{\partial t} + u\frac{\partial F}{\partial x} + v\frac{\partial F}{\partial y} = 0 \tag{M.8.19}$$

This transfer is carried out with the same advection operator as that used for the dynamics. Thus, the total flux of fluid and void crossing the cell face from the donor to the acceptor cell is $V = uDt$. The sign of u determines the acceptor (A) and donor (D) cells. The actual flux must be limited so that the flux of both fluid or void does not exceed that available. Thus, the flux of F per time step is taken as

$$\delta F = MIN\{F_{AD}|V| + CF, F_D\Delta x\} \tag{M.8.20}$$

where

$$CF = MAX\{(1.0 - F_{AD})|V| - (1.0 - F_D)\Delta x, 0.0\} \tag{M.8.21}$$

The double subscript allows that flux to be determined by either the donor or acceptor cell. If $F_{AD} = F_D$ the method is the normal donor cell.

The first limit (M.8.20) is taken to ensure that the quantity of F leaving the donor cell does not exceed that contained by it. The second limit (M.8.21) ensures that there is sufficient void in the donor, otherwise additional fluid is convected.

Whether the donor or the acceptor cell is used depends on the orientation of the surface. The acceptor is used to determine F_{AD} when the surface is convected mostly normal to itself, otherwise the donor cell is used. However, if the acceptor cell is empty, or if the cell upstream of the donor cell is empty, then the acceptor cell value is used. This ensures that a donor cell must fill before any fluid can enter an empty cell.

To identify the surface we need to identify its position and slope within the cell. The position and slope are calculated based on the orientation. Thus, if the surface is represented as $Y(x_i)$ we have the position relative the bottom edge of the cells $j-1$

$$Y(x_i) = Y_i = (F_{i,j-1} + F_{i,j} + F_{i,j+1})\delta y \tag{M.8.22}$$

and slope

$$(\mathrm{d}Y/\mathrm{d}x)_i = 0.5(Y_{i+1} - Y_{i-1})/\Delta x \tag{M.8.23}$$

Similarly, for $X(y_j)$

$$X(y_j) = X_j = (F_{i-1,j} + F_{i,j} + F_{i+1,j})\Delta x \tag{M.8.24}$$

and

$$(\mathrm{d}X/\mathrm{d}y)_j = 0.5(X_{j+1} - X_{i-1})/\delta y \tag{M.8.25}$$

If $|\mathrm{d}Y/\mathrm{d}x|$ is less than $|\mathrm{d}X/\mathrm{d}y|$, then the surface is more nearly parallel to the x-axis. The derivative with the smallest magnitude gives the best approximation to the slope.

If $\mathrm{d}X/\mathrm{d}y$ is negative, fluid lies below the surface and cell $(i, j-1)$ is used as the interpolation neighbour for (i, j). On the other hand, if $|\mathrm{d}X/\mathrm{d}y|$ is positive $(i, j+1)$ would be used as the fluid lies above the surface.

The method may be made more flexible by the use of non-uniform mesh as described by Hirt and Nichols (1981).

M.9 Equation of State

When writing fluid dynamic equations, it is often assumed that fluid state variables such as pressure, volume, temperature, etc. are known from point to point at all times. In fact of course this is not the case in simulations as the equation of state parameters need to be implicitly evaluated within the overall motion of the system. A basic axiom of thermodynamics is that the thermodynamic state of a system in thermal equilibrium is uniquely specified by just two of the state variables. The remainder are specified by the equation of state, which therefore takes the form $Z = F(X, Y)$, where X, Y, Z are all state variables and F the equation of state function. In a fluid problem, the natural state functions are (ρ, T) and the required equation of state is of the form (ρ, T). The form of the equation of state is given by one of the characteristic functions:

- Heat function $\mathrm{d}U = T\mathrm{d}S - P\mathrm{d}V$
- Enthalpy $\mathrm{d}H = T\mathrm{d}S + V\mathrm{d}P$
- Helmholtz free energy $\mathrm{d}F = -S\mathrm{d}T + P\mathrm{d}V$
- Gibb's free energy $\mathrm{d}G = -T\mathrm{d}S + V\mathrm{d}P$

In general, aside from a simple gas where the equation of state is $PV = R_g T$, the equation of state is not a simple analytic expression.

In the fluid set of dynamical equations the natural set are the thermo-dynamic functions for which quasi-analytic expression can be found, which makes the Helmholtz free energy $F(P, \rho)$ the best choice to characterise the equation of state. The thermodynamic state variables are given by the derivatives of the Helmholtz free energy:

$$P = \rho^2 \frac{\partial F}{\partial \rho} \qquad S = -\frac{\partial F}{\partial T} \tag{M.9.1}$$

$$\epsilon = -T^2 \frac{\partial}{\partial T}\left(\frac{F}{T}\right) \quad c_V = -T\frac{\partial^2 F}{\partial T^2} \tag{M.9.2}$$

However, the free energy is not a simple analytic function of ρ and T. In order to simplify the problem it may be conveniently assumed that (Zel'dovich and Raizer, 1966) the free energy may be split into three non-interacting terms

$$F(\rho, T) = E_c(\rho) + F_n(\rho, T) + F_e(\rho, T) \tag{M.9.3}$$

where $E_c(\rho)$ is the zero temperature isotherm which includes all atomic and electronic interactions at zero temperature. $F_n(\rho, T)$ accounts for the thermal motion, excitation and ionisation of atoms and ions and $F_e(\rho, T)$ accounts for the thermal motion of the electrons. The thermal motion of atoms and ion component also takes account of the phase properties of the material. In most cases of laser–plasma interaction, the detailed phase changes from and within solid targets (including elastic–plastic deformation) do not appear to be important, except in studies of laser machining or welding where appropriate phase changes must be included. We shall assume that either a set of tabular values for an equation of state is available or the ideal gas law is sufficiently accurate.

The splitting of the equation of state into atomic and electronic components is extremely convenient as electrons and heavy particles may not be in thermal equilibrium. The assumption is made, therefore, that the contribution to the total pressure of each may be calculated according to separate temperatures for atoms/ions and electrons. Free electron populations are described satisfactorily by Maxwell–Boltzmann statistics. The number of free electrons is determined by the ionisation balance at the appropriate temperature which is determined by either coronal or Saha equations depending on the density. The pressure, etc. can then be determined by the perfect gas laws: $p_e = n_e \kappa T_e$, $\epsilon_e = 3/2 Z\kappa T_e$ and $n_e = Z\rho/m_i$ where Z is the mean ion charge and m_ith heavy particle mass (see Thompson and Lauson, 1971).

As the electron density increases, two effects become important: the energy levels are broadened and the depression of the ionisation edge becomes large. Finally, the electrons become degenerate. In this situation when the Fermi energy $E_F \gtrsim \kappa T_e$, the distribution is determined by the Thomas–Fermi model (Zel'dovich and Raizer, 1966). It is difficult to obtain a simple match of ionisation energies at the intersection of the ideal gas and Fermi-Dirac models. This difficulty is relaxed if only free electrons need to be included, for example, in a hydrogen target for then only a smooth transition into the degenerate Fermi–Dirac statistics is met. Alternatively, if the density is not too high the energy levels are well defined and may be estimated using an average atom model.

Application of the equation of state routines within a hydro model is accomplished in one of the two ways:

- A set of tabulated values is available and may be interpolated.
- **Analytic models**: These are approximations to the functional relations of the governing equations of the electrons using either Thomas–Fermi if near degenerate [XEOS] (More et al., 1988) or ionisation balance otherwise [ANEOS] (Thompson and Lauson, 1971). The solid equation of state in both cases is based on the Grüneisen pressure law and Dulong–Petit law.

References

Amsden, A A and F H Harlow. 1970. The SMAC method for calculating incompressible fluid flows. Technical Report LA-4370. Los Alamos National Laboratory.

Bates, D R, A E Kingston and R W P McWhirter. 1962. "Recombination between electrons and atomic ions: 1 optically thin plasmas." *Proc R Soc A* 267:297–312.

Birdsall, C K and A B Langdon. 1991. *Plasma Physics Via Computer Simulation*. Bristol, U.K.: Adam Hilger.

Bunemann, O. 1969. A compact non-iterative Poisson solver. Technical Report SUPRI 294. Stanford Univ. Inst for Plasma Research.

Buzbee, B L, G H Golub and C W Nielson. 1970. "On direct methods for solving Poisson's equations." *SIAM J Numer Anal* 7:627–656.

Collatz, L. 1960. *The Numerical Treatment of Differential Equations*. Berlin: Springer-Verlag.

Cooley, J W and J W Tukey. 1965. "An algorithm for the machine calculation of complex Fourier series." *Math Comput* 19:297–301.

Courant, R and D Hilbert. 1953. *Methods of Mathematical Physics: Vol* I. New York: Interscience.

Courant, R and D Hilbert. 1962. *Methods of Mathematical Physics: Vol II, Partial Differential Equations.* New York: Interscience.

Dahlquist, G G. 1963. "A special stability problem for linear multistep methods." *BIT* 3:27–43.

Douglas, Jr, J. 1955. "On the numerical integration of $\frac{\partial^2 u}{\partial x^2} + \frac{\partial^2 u}{\partial y^2} = \frac{\partial u}{\partial t}$ by implicit means." *J Soc Ind Appl Mathe* 3:42–65.

Eberlein, P J. 1962. "A Jacobi-like method for the automatic computation of eigenvalues and eigenvectors of an arbitrary matrix." *SIAM J Ind Appl Math* 10:74–88.

Friedman, A. 2008. *Partial Differential Equations of the Parabolic Type.* New York: Dover Books.

Gear, C W. 1971. *Numerical Initial Value Problems in Ordinary Differential Equations.* Englewood Cliffs, NJ.: Prentice-Hall.

Griem, H R. 1964. *Plasma Spectroscopy.* New York: McGraw-Hill.

Harlow, F H and J E Welch. 1965. "Numerical calculation of time-dependent viscous incompressible flow." *Phys Fluids* 8:2182–2189.

Hess, J L and A M O Smith. 1967. "Calculation of potential flow about arbitrary bodies." *Prog Aeronaut Sci* 8:1–138.

Hinnov, E and J G Hirschberg. 1962. "Electron-ion recombination in dense plasma." *Phys Rev* 125:795–801.

Hirt, C W and B D Nichols. 1981. "Volume of fluid (VOF) method for the dynamics of free boundaries." *J Comput Phys* 39:201–225.

Hockney, R W. 1970. The potential calculation with some applications. In *Plasma Physics*, ed. B Alder S Fernbach and M Rotenberg. Number 9 *in "Methods in Computational Physics"* New York: Academic Press pp. 136–211.

Hockney, R W and J W Eastwood. 1988. *Computer Simulation Using Particles.* Abingdon, U.K.: Taylor and Francis.

Kershaw, D S. 1978. "The incomplete Cholesky-conjugate gradient method for the iterative solution of systems of linear equations." *J Comput Phys* 26:43–65.

Lambert, J D. 1991. *Numerical Methods for Ordinary Differential Systems.* Chichester, Sussex, U.K.: Wiley.

Landau, L D and E F Lifshitz. 1959. *Fluid Mechanics.* London: Pergamon.

Lax, P D. 1954. "Weak solutions of nonlinear hyberbolic equations and their numerical computation." *Commun Pure Appl Math* 7:159–193.

Levy, H and E E Preidel. 1944. *Elementary Statistics.* London: Nelson. A very basic (old) text introducing statistics.

McWhirter, R W P and A G Hearn. 1963. "A calculation of the instantaneous population densities of the excited levels of hydrogen-like ions in a plasma." *Proc Phys Soc* 82:641–655.

More, R M, K H Warren, D A Young and G B Zimmerman. 1988. "A new qotodianequation of state (QEOS) for hot dense matter." *Phys Fluids* 31:3059–3078.

Morozov, A I and L S Solev'ev. 1966. Motion of charged particles in electromagnetic fields. In *Reviews of Plasma Physics*, ed. M A Leontovich. Vol. 2 New York: Consultants bureau pp. 201–297.

Morse, P M and H Feshbach. 1953. *Methods of Theoretical Physics.* New York: McGraw-Hill.

Peaceman, D W and H H Rachford, Jr. 1955. "The numerical solution of parabolic and elliptic differential equations." *J Soc Ind Appl Math* 3:28–41.

Pert, G J. 1978. "Algorithms for the calculation of ionization in rapidly changing plasmas." *J Comput Phys* 27:241–255.

Pert, G J. 1981. "Efficient integration of the time dependent collisional-radiative equations." *J Comput Phys* 39:251–271.

Pert, G J. 1990. "Models of collisional-radiative recombination." *J Phys B: At Mol Opt Phys* 23:619–650.

Pert, G J. 2013. *Introduction to Fluid Mechanics.* London: Wiley.

Pitaevskii, L P. 1962. "Electron recombination in a monatomic gas." *Sov Phys-JETP* 15:919–921.

Press, W H, B P Flannery, S A Teukolsky and W T Vetterling. 1986. *Numerical Recipes - The Art of Scientific Computing.* Cambridge, U.K.: Cambridge University Press.

Press, W H, B P Flannery, S A Teukolsky and W T Vetterling. 1992. *Numerical Recipes - The Art of Scientific Computing.* Second ed. Cambridge, U.K.: Cambridge University Press.

Richtmyer, R D and K W Morton. 1967. *Difference Methods for Initial Value Problems.* New York: Interscience.

Stoer, J and R Bulirsch. 1991. *Introduction to Numerical Analysis.* Second ed. New York: Springer-Verlag.

Swarztrauber, P N. 1974. "A direct method for the discrete solution of separable elliptic equations." *SIAM Numer Anal* 11:1136–1150.

Taussky, O. 1949. "A recurring theorem on determinants." *Am Math Mon* 56:672–676.

Thompson, S L and H S Lauson. 1971. Improvements in the Chart D radiation-hydrodynamic code III Revised analytic equations of state. Technical Report SC-RR-71 0714. Sandia Laboratories.

Varga, R S. 1962. *Matrix Iterative Analysis.* Englewood Cliffs, NJ: Prentice-Hall.

Wilkinson, J H. 1965. *The Algebraic Eigenvalue Problem.* Oxford: Oxford University Press.

Yanenko, N N. 1971. *The Method of Fractional Steps: The Solution of Problems of Mathematical Physics in Several Variables.* Berlin: Springer-Verlag.

Zel'dovich, Ya B and Yu P Raizer. 1966. *Physics of Shock Waves and High Temperature Hydrodynamic Phenomena.* Seventh ed. New York: Academic Press.

Index

Note: *Italicized* page numbers refer to figures.

d

e

j

k

l

n

o

p

t

u

v

Printed in the USA
CPSIA information can be obtained
at www.ICGtesting.com
CBHW080916190824
13226CB00004B/120